Systems Optimization Methodology

Part I

Systems Optimization Methodology

Part I

Translated from Russian
by Mr Y M Donets

V V Kolbin
Leningrad University

World Scientific
Singapore • New Jersey • London • Hong Kong

Published by

World Scientific Publishing Co. Pte. Ltd.

P O Box 128, Farrer Road, Singapore 912805

USA office: Suite 1B, 1060 Main Street, River Edge, NJ 07661

UK office: 57 Shelton Street, Covent Garden, London WC2H 9HE

British Library Cataloguing-in-Publication Data
A catalogue record for this book is available from the British Library.

SYSTEMS OPTIMIZATION METHODOLOGY — Part 1

ISBN 981-02-1589-4

This book is printed on acid-free paper.

Printed in Singapore by Uto-Print

I dedicate this book to Dr. Galina Kolbina,
my wife and friend.

PREFACE

Teleological problems are wide in scope and perhaps go back to pre-Socratic times when science and myth were still one. Dealing with system optimization necessarily implies the problem of goal-oriented behavior. The term "goal-oriented" behavior has at least two meanings: "the behavior directed towards a goal" and "the behavior directed by a goal". The statement that an object is moving toward a goal or is directed by a goal contains no teleology, for we have merely an answer to the question: where is the object moving? In the second case, when the object (or a system) is directed by the goal, the situation is completely different. The goal now becomes an external cause of movement of the object. Teleology serves as a basis for the functional approach to describing reality in such a way that all phenomena, systems, and processes are penetrated by mutual utility: one thing arises for the sake of the other. Teleological problems came to the fore when cybernetics, general systems theory, system approach and other scientific and quasi-scientific disciplines whose names contain the word "system" evolved.

This monograph defines the notion of a "system" by reference to those systems which exhibit the goal-oriented behavior and utilize the notion of decision making and controls. Such systems allow for phenomenological description and fix the nature of causal transformations of input effects into output quantities. The study of consequences of the fact that the systems possess some properties constitutes the content of systems optimization methodology which goes beyond the scope of descriptive classification of systems.

Chapter 1 deals with philosophical problems of systems methodology. An attempt is made to systematize and analyze the problems of scientific methodology as applied to systems modeling methodology which is viewed as the most general concept utilized in modern science.

Chapter 2 focuses on problems of qualitative analysis in natural and social sciences. Attention is drawn to problems of measurement theory and quantitative analysis of systems.

Approaches and methods of systems analysis and synthesis form the central portion of the book. Much study is given to the methods of systems decomposition, an integration using both the discrete and the continuous description of objects, processes, and phenomena. Examples of complex goal-oriented systems are also provided.

The remaining part of the book is largely centered around methodology of multiobjective systems optimization.

The book is provided with references to relevant lines of investigation.

I would like to thank V. I. Zubov, member of the Russian Academy of Science, whose interest in this book stimulated my writings. In particular, I wish to express my indebtedness to N. L. Kirichenko, typist, Y. M. Donets, translator, the editorial and production staff of World Scientific Publishing, and all those who have contributed to the preparation of this book.

St. Petersburg, November 1996
V. V. Kolbin

CONTENTS

Preface vii

Chapter 1. Philosophical Problems of the Methodology for Systems Modeling 1

§1. Philosophical Problems of the Methodology of Scientific Cognition 2
§2. Philosophical Problems of Modeling 6
§3. Theoretical Aspects of Modeling Systems in the Process of Scientific Cognition 11
§4. Systems Modeling 19
§5. Mathematical Logic as a Means of Cognition 28

Chapter 2. Problems of Quantitative Analysis in Natural and Social Sciences 34

§6. Problems of Systems Quantitative Analysis 35
§7. Problems of Systems Description 42
§8. Definitions of Systems 48

Chapter 3. Dantzig-Wulf Decomposition 58

§9. Dantzig-Wulf Decomposition Method 58
§10. Dual Approach in Block Programming 67

§11. Transportation Problem Solution by the Decomposition Method 75
§12. Decomposition for Problems with a Block-staircase Structure 80
§13. Solution of the Interval Programming Problem 85
§14. Extension of the Dantzig-Wulf Decomposition Principle to Nonlinear Programming Problems 92

Chapter 4. Parametric Decomposition 101

§15. Kornai-Liptack Method 101
§16. Solution Technique for Block-Separable Nonlinear Problems 105
§17. On Parametric Decomposition of the Resources Allocation Problem 116
§18. One Method of Parametric Decomposition for Linear and Separable Programming Problems 123

Chapter 5. Decomposition Based on Separation of Variables 127

§19. Constraint Relaxation Method for the Convex Programming Problem 127
§20. The Ritter Method for the Block Problem with the Tying Variables and Constraints 130
§21. The Rosen Division Method for Linear Programming Problems 136
§22. The Rosen Division Method for Nonlinear Programming 142
§23. Benders Method for a Special Mathematical Programming Problem 152

Chapter 6. Decomposition Based on Optimization Technique 166

§24. Application of the Componentwise Descent Method for Solving the Problems of Mathematical Programming and Optimal Management 166
§25. Conditional Gradient Method and Decomposition of Problems of Mathematical Programming and Optimal Control 170
§26. Utilization of a Penalty Constant in Decomposition of the Mathematical Programming Problem 175
§27. Decomposition Based on Simplex Method Modifications 180

Chapter 7. Decomposition and Aggregation 194

§28. Aggregation Method for Solving a System of Linear Equations 194
§29. Aggregation Method for the Block Problem of Linear Programming 201
§30. Decomposition and Aggregation Based on Perturbations Method 209
§31. Decomposition Method Based on Aggregation of Variables from Different Blocks 226

Chapter 8. Application of Solution Techniques for Large Dimension Problems to Grain Farming Optimization 239

§32. Grain Chain 239
§33. Grain Farming Optimization Problem 245
§34. Solution Technique 271
§35. Application of the Algorithm to the Grain Farming Optimization Problem 299

Chapter 9. Major Problems of Multiobjective Optimization 304

§36. Formulation of the Multiobjective Optimization Problem 304
§37. The Study of the Maximal Efficiency Principle 319
§38. The Study of the Parametric Principle of Maximal Efficiency in Multiobjective Optimization 326

Chapter 10. The Study of Improvability and Priority Issues in Multiobjective Optimization Problems 331

§39. The Study of the Problem of Feasible Solution Improvability 331
§40. The Study of the Priority Problem in Multiobjective Optimization 344

Chapter 11. Problems of Multiobjective Optimization under Information Deficiency 356

§41. Problems of Decision Multiobjective Optimization under Uncertainty 356

§42. Multiobjective Optimization Problems for Dynamic Control Systems 367

Chapter 12. Methodology of Vector Optimization 379

§43. Optimization Methodology for Hierarchical Sequence of Quality Criteria 379
§44. Optimization of Hierarchical Sequence of Quality Criteria 384
§45. Finding the Set of Unimprovable Points 399
§46. Determination of the Solution Based on a Particular Tradeoff 412

Conclusion 433

References 435

Chapter 1

PHILOSOPHICAL PROBLEMS OF THE METHODOLOGY FOR SYSTEMS MODELING

Every year the social importance of science is growing and its influence on all aspects of life is becoming stronger. At the same time, science attracts the attention of a wide circle of readers. Modern science represents an adequately developed system intended for mass dissemination of knowledge necessary for society. Human thought always aspires to attain knowledge, but under the conditions of scientific and technological revolution a demand arose for a special study of the methods of scientific cognition. Now it becomes necessary to acquire philosophical understanding of relations between various scientific methods on the basis of the theory of cognition. The problems of the methodology of scientific cognition have found their reflection in a number of publications, and many top scientists contributed to the development of various problems of the methodology of science and studied the main principles of the philosophical problems of scientific cognition. However, as we see it, the problems of the methodology of science should be systematic and analyzed on the basis of dialectical materialistic interpretation of the main concepts of the modern methodology of scientific cognition. From our standpoint, it might be of interest to conduct such studies on the methodology of systems modeling as the most general concept used by modern science.

§1. PHILOSOPHICAL PROBLEMS OF THE METHODOLOGY OF SCIENTIFIC COGNITION

Investigation into the fundamentals of science, its gnosiological problems and development of the methodology constitutes the subject of analysis in many branches of knowledge and, primarily, philosophy. It is precisely philosophy that develops materialistic dialectics, investigates the theory of knowledge and logic as well as the methodological problems of social, natural and technical sciences. These directions of investigation are closely related to the aspect of studying the science as such, which might be naturally called methodological, and it constitutes the subject of the methodology of science in a broad sense. Our work will deal with the philosophical problems of the methodology of scientific cognition proceeding from the principles and laws, categories and rules of the dialectical and historical materialism. In this case, the philosophical problems of science are studied with respect to the three main directions of methodological problems: ontological, logical-gnosiological and sociological.

The term "methodology" should be understood as a theory of method. Practice shows that every field of activity has developed its own specific methods differing from those applied in other fields. It is possible to talk about various levels of the methodology of scientific cognition in relation to the difference in the levels of generalization, the character of theoretical substantiation and a particular field of application of one or another of the methods of scientific cognition. Here the numerous techniques, regulations and operations peculiar to a narrow field of scientific research and bearing sometimes empirical character form the lowest level of the methodology of science coinciding with methodicalness.

The next level of methodology features not only the knowledge where and how to apply proper methods, but also the apprehension of their fundamentals. Here the scientific level rests on the knowledge of appropriate regularities and a theory of the given sphere of research. In this case the theory acquires the methodological function, which permits not only to explain data and predict new phenomena and facts, but to serve as a means of discovery of new features and intricate regularities.

At the next level the methodology is characterized by the development of the sufficiently general methods for scientific research, the application of which is not limited merely by one science but is based upon the existence of the regularities common to various fields of knowledge and activities.

Investigation of such regularities results in the creation of appropriate theories that underlie more general interdisciplinary methods. At this level the methodology of scientific cognition does not yet bear philosophical character, though, even here, the philosophical questions arise primarily with respect to the gnosiological analysis of one or another of the fundamental concepts, estimation of the gnosiological value and bounds of application of proper methods as well as elucidation of interdisciplinary relations.

The most general methods used in all sciences without exception constitute, in various forms and modifications, the subject of philosophical methodology. This methodology is naturally not the only science involved in the investigation of general methods of scientific cognition. The general scientific methods constitute the object of scientific investigation, substantiation and development in many fields of knowledge, though the matter does not rest here. In all instances the general scientific methods explicitly or implicitly employ particular philosophical principles, categories and laws as their theoretical bases. It is in the sphere of philosophy that the general scientific principles applied and developed in a number of scientific disciplines acquire their substantiation in the conceptual and theoretical, ontological and gnosiological aspects, which constitutes a philosophical level of methodological investigations.

Almost every philosophical doctrine implicitly formulates or suggests one or another philosophical principle defined by the gnosiological theory. An essential disadvantage of most of the philosophical doctrines resides in the erroneous belief in the existence of a unique method or technique suitable for all sciences and for all stages of scientific cognition. As a result, some theories eschewed the empirical methods for the theoretical, while the others preferred the inductive approaches to the deductive. It was not unusual to choose the methods of synthesis to those of analysis, description over logical constructions, systematization and generalization of the accumulated data. The methodologists of idealistic orientation, even Hegel, did not understand that the nature of cognition depended upon the specific features of the object of research, hence their philosophical constructions could not form a methodological foundation of concrete natural and social sciences.

Exponents of the rationalistic approach of the idealistic methodology disparaged the importance of an experiment and practice in scientific cognition. Representatives of the idealists of the empirical school of thought declared the materialistic conception of an experiment and practice to be "metaphysics" and, interpreting an experiment and practice as "sensory experience", they concluded that there is no relationship between the empirical (synthetic) knowledge and the theoretical (analytical). The methodology based on the principles of materialism and dialectics rests on the criterion of practice, when tackling scientific problems, and proceeds from the conception of organic relationship between philosophy and concrete sciences. The materialistic dialectical methodology features the concept of existence of many methodological techniques and means for scientific cognition, explanation and transformation of reality. Such methods, means and techniques are interrelated due to the interrelation of all phenomena of the objective world but, at the same time, they differ from one another since qualitatively different objects are used for research.

The general philosophical methodology, in the broad sense, comprises the processes, forms and methods of scientific cognition with an orientation towards gnosiological processes rather than ontological aspects.

Cognition is organically associated with practice and activity directed towards transformation of the world. The transformation of nature and society exerted influence on the development of intelligence and knowledge of man. From the moment of division of mental and physical labor it became possible to talk about the first step towards the rise of science as a specific sphere of activity focused on cognition and employment of the laws of the objective world. As the productive forces of society developed, scientific activity was transformed into a branch of mass production, propagation and application of scientific knowledge. Under modern conditions, science deals with the processes of cognition inherent in man as well as with the organizational forms, methods and techniques employed in scientific research. In this case the methodology of cognition aims at tackling the gnosiological problems oriented towards a general character of the scientific cognition as a whole, revealing its dialectics.

The methodology of scientific cognition distinguishes two levels: empirical and theoretical. The empirical level includes the numerous methods and forms of scientific cognition directly related to practical activity in the sphere of science. The theoretical level features the forms and methods securing the development of a scientific theory as logically organized knowledge about nature and society. Discrimination between these two levels reflects the internal dialectical relation existing in the objective world between general and particular, phenomenon and essence.

The current development of science involves, as one of its peculiarities, the knowledge of the main rules for constructing the artificial language used in it. Even at the empirical level it becomes necessary to use a special language, and here the language is treated in the methodology of science in unity with thinking and comes out as a form of its material embodiment. In the course of development of science the need for an adequate and precise language is satisfied by devising special terminology and scientific nomenclature, and these processes are inherent in all sciences. The tendency to develop a precise language, as it has been the case with a number of special languages, leads to the creation of special formalized languages featuring great adequacy.

Cognition generally starts with observation and experiment. Here we concern ourselves not only and even not so much with sensory perception but with the incorporation of the object of research in a total practical activity of the cognizing subject. In the given context the main and initial forms of cognition, i.e. observation and experiment, represent the essential elements of experience. From the dialectical materialistic standpoint experience is understood as a process of interaction of a subject and object of research, the latter being independent of the consciousness of the former, though it is actively influenced.

Observation will be treated here as an initial and elementary process occurring at the empirical level of scientific cognition. As a form of sensory knowledge, observation is simultaneously a purposeful and organized perception of an object (phenomenon) of the surrounding world. From a scientific aspect observation is called upon to give some actual data providing support for or against the advanced hypothesis and to serve as a basis for theoretical interpretation. Besides, scientific observation features a specific goal. Observation is not, however, attended with a transformative effect on the object of research, which is inherent in experiment. Today observation not infrequently involves transcending biological limitation of the senses. The range of sense perceptions is substantially extended due to the use of numerous devices and instruments. At the same time, we cannot say that the use of technical devices has eliminated the dependence of the subject upon the investigated object or phenomenon. Part of the disadvantages intrinsic in observation is obviated in the experimental form of the cognition process.

Notwithstanding that an experiment, along with observation, should be related to the empirical level of the cognition process, from the gnosiological standpoint there is no substantial difference between them. An experiment is one of the forms of practice directed towards transformation of the external world, whereas observation is somewhat characterized by contemplation. In an experiment, material practice is implemented by the action of some material objects on the others. Furthermore, an experiment displays an organic unity of practical action and theoretical work of intelligence. According to the goal and character of the strategy of an experiment it is possible to discriminate the experiments realized through the use of the trial-and-error method, the experiments based on a closed algorithm, the experiments conducted with the use of the "black box" method, etc. Among the numerous types and forms of experiments the measurement experiment seems to hold a prominent place.

In comparatively rare cases the acquisition of quantitative data on phenomena does not encounter substantial difficulties and can be easily accomplished by comparing and correlating some perceived objects with the others. The "measurement" should imply a process of comparison, by means of the physical experiment, of a given magnitude with a certain value taken as a standard of comparison. In this instance, we are faced with the problem of incorporating the experimental means of cognition and various measuring devices in the gnosiological object of research. The problem whether these devices refer to the subjective or objective side of an experiment (a gnosiological subject or object) should be tackled particularly with relation to the analysis of the structure of an experiment and character of the researcher's activity.

§2. PHILOSOPHICAL PROBLEMS OF MODELING

The first stage of the empirical level of cognition features direct investigation of the object or phenomenon, when in the course of observation the senses of the subject interact with the object. In general, investigation loses its direct character at the following stage. Thus, at the second stage of the empirical level of cognition the data derived from an experiment are used as initial information, whereas employment of the systematized initial data is inherent in the third stage. In both cases we find the substitutes of the object at hand: experimental data and their grouping. All substitutes of the object of research are the models of this object or its parts. Modeling is taken as a process of research carried out by means of such substitutes. In a broad sense, a model represents a natural or artificial, material or ideal substitute of the object at hand and has common features with it. The results of research on these features on a model have an objective value and can come out as the data on features of the investigated object itself. In this case, the model performs a double function: it is, simultaneously, the object of research, since it replaces a real object, and an experimental means, since it is used as a means (device) for cognizing the object. The experiment concerned with the employment of the model is based on the similarity theory. This theory contains formulations of the conditions, under which physical systems are thought to be similar, and the fields of occurrences, to which the results from particular experiments on the model can be transferred. In general, the similarity theory is applicable to the systems studied by the sciences dealing with inorganic nature. The spread of the similarity theory to biology, social and economic sciences depends on the level of knowledge in these fields.

In the case of mathematical modeling the model-object relation is based on such generalization of the similarity theory that disregards the qualitative heterogeneity of the model and object. Generalization of this kind assumes the form of an abstract theory of the systems isomorphism. The concept of isomorphism and the more general concept of homomorphism may be regarded as the formalized types of analogy. As a relation of similarity, the analogy features difference in the correlated elements and identity of relations: the laws of relation between the elements of two systems.

With respect to modeling cybernetics holds a special place. In cybernetics a researcher concentrates his attention on the general laws of functioning of the self-organized control systems, no matter whether these are the technical devices created by man, living organisms or human associations. Such functional approach to control systems follows from the peculiarities of cybernetics as a science and the level of its abstractions. From the gnosiological standpoint the cybernetic models may fall, as in other areas of research, into two groups: material and ideal. The importance of cybernetic modeling does not lie exclusively in realization of the functional approach and prediction of the behavior of a modeled object or phenomenon on the basis of examination of the model. Investigation into behavior of a cybernetic model is valuable in the aspect of relation of functions to the structure of the model and penetration into the structure of the modeled object. We may admit a certain relation between a structure and function (behavior) in a specified range of possibilities. With the similarity of the behavior of the model and object under given conditions, knowing the structure of the model it is possible to draw conclusions of a certain degree of reliability. Cybernetic modeling is not only an important instrument of empirical research allowing to transcend some limited possibilities of common experiments, but serves, in a sense, as a universal means. Cybernetic modeling within its possibilities determined by a general character of the laws of control penetrates into various spheres of reality and represents a specific form of experimental practice not only in technical science, but in biological and social sciences as well.

Observation, experimentation and examination result in scientific facts which, having been accumulated, can adequately serve as a foundation of theoretical knowledge. In scientific researches the concept "fact" has several meanings. This term is used in the meaning of a "phenomenon", "event" or "part of reality". It may also signify particular empirical statements or sentences. Sometimes the term is employed as a synonym for the words "correct" and "truth", although it is more extensively used in the first two meanings, which reflects, in general, the difference between philosophical trends on this problem.

According to the conception of the logical empiricism facts, firstly, signify everything given to a subject in sense perception (hence there is no essential discrepancy between facts as a total combination of this and facts as a total combination of the signs defining things); secondly, facts are isolated from each other ("atomistic"), and relation is introduced into them by a subject. This is inherent in the subjective idealistic concept of facts.

According to the materialistic concept facts represent the events existing independently of consciousness (sensation) and established in an experiment, observation. The latter is very important, because here the concept of facts comprises not only the objective events or phenomena irrespective of a subject, but also their objective relation to a subject as well as conditions and means to observe and fix them.

At the same time, the problem arises as to how to describe, treat, systematize and generalize facts, because the methodology of scientific cognition calls for fixation of separate facts and investigation of the "whole totality of the facts pertaining to the problem in hand without a single exclusion..."*)

Each research aims at developing the theory explaining facts and predicting new events and phenomena. Scientific facts feature the necessity to be checked not in one experiment but in a number of experiments independent of each other. It may be stated that systematization of facts forms the first step to their generalization and, consequently, to the development of a scientific theory.

Systematization of scientific facts involves their grouping, formation of classes on the basis of isolation of essential and common signs and characteristics. This process is based on the previously accumulated theoretical knowledge, without which, for example, it is impossible to estimate essential signs and characteristics. There is no need for a special discussion on the importance of making a right choice of signs for the systematization (classification) of facts, phenomena and events. The experience of scientific cognition accumulated by the moment allows the consideration of the major methods of generalization of scientific facts, including those obtained from investigation and experimentation with models.

As far as dialectical materialism is concerned, scientific cognition involves a complicated process of interaction of the empirical and theoretical methods involved in practice and experiment and based on the object-instrument activities of man and society. Due to this, it is possible to explain the fact that none of the general methods, not even in combination with other methods, can explain the essence of scientific cognition and solve the problem of transition from the empirical to the theoretical level. Here the general methods will be regarded as necessary but not sufficient for employment in the many-sided, hierarchically divided and dialectically integrated process of scientific cognition. For example, the relation between induction and deduction will be considered as a partial case of the dialectical unity of opposites. In so doing, it is possible to treat three aspects of relations between induction and deduction: gnosiological, methodological and logical. The first aspect manifests itself in relation to the statements present in the inductive and deductive reasoning upon the reality of the surrounding world in connection with the object of cognition. The problem of substantiation of transition from empirical knowledge to theoretical, from concrete

*V. I. Lenin, *Complete Works*, 5th edn, vol. 30, p. 351.

facts to general regularities, from phenomena to essence reflects the methodological aspect of relations. Analysis of induction as a probabilistic conclusion pertains to the logical aspect of relations, when relationships of induction and deduction are treated as two forms of logical reasoning: the first based on the concept of logical necessity (logical following) in the deductive logic and the second based on the concept of logical probability in the inductive or probabilistic logic.

At the same time, it might be well to point out the unity of the induction and deduction as the methods for attaining the probable and logically necessary knowledge, and this unity has an objective ontological foundation. Necessity and regularity do not exist in nature apart from an infinite diversity of events, and here necessity is realized as a dominant tendency manifesting itself in a large number of chances. The inductive methods aim at searching for regularity, necessity and penetration into the essence attainable in a theory, where the necessary, general and essential relations find their reflection and cannot be realized otherwise than by means of the deductive methods. The thesis of unity of induction and deduction is one of the fundamentals of the methodology of scientific cognition.

Among the many methods of induction and deduction, we are mostly interested in the method of model extrapolation based on a relation between a model and modeled object (system). Such a relation is that of similarity accessible to strict formalization. Within the given method, the relation of similarity is revealed and carefully formulated in the form of definite rules of geometrical similarity, criteria of physical and mathematical similarity or, in general, relations of homo- or isomorphism. Here we find not the similarity of accidentally selected characteristics, but that of laws, relations, structures, functions. Modeling represents experimental investigations based on the model that replaces a real object (phenomenon) and is in objective relation to the object. Thus understood, modeling comprises the following three interrelated problems: (a) construction of the model satisfying the criteria of similarity, (b) experimental investigation into the model, observation of its behavior in the circumstances of an experiment, (c) extrapolation of the data derived from experimental investigation into the model as compared with a real object of research. In such a situation, the question whether it is relevant to transfer the data from the model to the object is not idle from the methodological standpoint. The theory of similarity formulates the conditions of relevancy of the conclusions consisting in transferring characteristics of one object to the other. In general, the theory of similarity may be generalized for such cases where the occurrences in hand refer to various modes of the motion of the matter. These cases are, naturally, limited to the similarity of the mathematical formalism that describes the same behavior of the systems of different physical nature, although having a similar structure, performing similar functions and differing merely in the physical nature of their elements. The obligatory condition allowing transfer of data from the model to a real object is the availability of a certain degree of similarity fixed in the criteria of geometrical, physical and mathematical similarity.

The model extrapolation may refer to the empirical level of cognition, although development of the model and its examination require an appropriate level of a theory. However, modeling comes out at this stage as a specific form of a scientific experiment. According to this interpretation modeling demonstrates both the unity of induction and deduction and the unity of empirical and theoretical methods of scientific cognition.

§3. THEORETICAL ASPECTS OF MODELING SYSTEMS IN THE PROCESS OF SCIENTIFIC COGNITION

The movement of scientific cognition from a basic empirical level to theoretical constructions and scientific discovery results from a dialectical "jump". The process is rather complex and comprises such interrelated, controversial and opposite elements, necessarily complementing each other, as those of abstract thinking and clear images, induction and logical reasoning, the arising problem and its tentative solution produced by an analogy or experience, etc. Research into a structure of composition of such a complicated dialectical phenomenon represents one of the difficult problems of many sciences, including the dialectical logic, methodology of scientific cognition and psychology of scientific work.

Here the problem is not brought forth to investigate this complicated process, although we are interested in the unquestionable circumstance that on the way from the empirical to theoretical stage we find a hypothesis, its suggestion, formulation and development, substantiation and demonstration. The necessity to develop a hypothesis is conditioned in the process of scientific cognition by the fact that laws do not follow immediately from separate data of observation, because the essence does not coincide with phenomena. Before developing a theory, we observe the rise of various more or less plausible variants, more or less apt ideas coming out as assumptions for future theory. This period of preparation of the theory, ranging from the first guesses to practical tests of the consequences following from a conjecture, is that of formation, statement and development of a hypothesis. In such an event, the hypothesis represents one of the possible answers to the problem encountered, e.g. that of investigation of the system. Discrimination of the problem requires a certain amount of knowledge that allows the isolation of essential features of the system at hand, the development of an adequate model and finding necessary idealizations. To do this, a researcher is expected to possess a developed ability for abstract thinking and a certain amount of knowledge. He is also supposed to be able to notice the source of the problem and formulate it so that its solution might result in tackling many other problems. A total collection of such abilities is a sign of scientific talent. In our case, the hypothesis involves a conceivable explanation in the form of statements as to the law of a phenomenon, its cause and structure or some other essential relation. Such understanding of the hypothesis lays emphasis on its probabilistic character, since the problem, whether it is true or erroneous, has not been solved and the hypothesis constitutes not reliable but probable knowledge (gnosiological probability). In scientific usage it is possible to discern the following two meanings of the concept "hypothesis": (a) in a narrow sense, as designation of a certain sentence containing an assumption of a law-governed order or other essential connections and relations, (b) in a broad sense, as a system of sentences, some of which are the initial premises of probabilistic character, whereas others constitute deductive development of these premises. The hypothesis in the broad sense constitutes one of the versions of the theory when they are discriminated by the truth of initial premises rather than by their logical structure. In theory these premises are thought to be true, while in hypothesis (in the broad sense) they have probable (likely) character, and here the logical structure of the hypothesis in the broad sense and theory may be the same. The process of cognition as a dialectical unity of practical and theoretical activities consists in finding whether the hypothesis is true or false. A sentence with the same content and pertaining to the same object area comes out as a hypothesis or an element of a theory depending on the degree of its verification in an experiment, social and historical practice. It is impossible to establish a strict line of demarcation between a hypothesis and a theory. In the final analysis, only confirmation by practice transforms a hypothesis into a true theory and makes a probable piece of knowledge reliable, and vice versa refutation by practice, experiment discards a hypothesis as a false conjecture. Therefore, verification of a hypothesis is not only possible, but necessary, and it is realized in a complicated process of cognition characterized by dialectics of relative and absolute truths. Where verification is accomplished a hypothesis becomes a theory or its element.

The methodology of scientific cognition employs scientific practice and gnosiological principles of the theory of reflection to formulate the conditions which should be satisfied by any assumption in order to become a scientific hypothesis. An important condition for testing scientific hypotheses is constituted by the relation of a hypothesis to facts. A scientific theory should not contradict the well-known and verified facts. A hypothesis must not be discarded or reformulated if among the well-known facts only one disagrees with it. To tell the truth, the history of science knows the examples when confirmation of a hypothesis required the revision of facts. The scientific value of a hypothesis is determined by the extent to which it can explain a total collection of the well-known facts and predict the appearance of new, previously unknown data.

No less worthy of notice is the condition delineating the relation of a hypothesis to the familiar laws of science, in other words, to the existing scientific theories. Each new hypothesis interpreting the phenomena and laws of a given sphere of scientific activity should not come into contradiction with the other theories the truth of which has already been proved for the same sphere. In those cases when a new hypothesis contradicts the old one it is possible to state that their relations are regulated by the principle of agreement, according to which, with the emergence of a new and more general theory, the old one changes into a particular case of a new theory.

Agreement between the scientific hypotheses and principles of the scientific (dialectical materialistic) world view constitutes a condition of their validity.

Finally, an extremely important condition of a scientific character of the proposed hypothesis is its availability for experimental or, in general, practical tests. It is possible to distinguish a fundamental as well as technically and historically realizable test of the validity of a scientific hypothesis. In those cases when a hypothesis is formulated without violating the laws of nature its fundamental test becomes possible. When society has at its disposal technical means for verification of a scientific hypothesis we can say that it is technically verifiable.

The condition of verifiability of a hypothesis is closely connected with the possibility of deductive development of a hypothesis. The criterion of simplicity might well serve, to a certain extent, as an additional condition for reliability of the theory advanced (not to be confused with the Makh's principle of the economy of thought). The principle of simplicity should not be taken as gnosiological, giving the validity criterion for a hypothesis (theory). An experiment or practice serves as the gnosiological criterion for validity of a hypothesis (theory), whereas thought may be economical if it agrees with the objective laws of the world. In a sense, the principle of simplicity represents a methodological requirement reflecting a unity of the world. Thus, the process of making the probabilistic knowledge reliable is associated in the final analysis with various practical actions, e.g. an object – instrument experimental activity where we find a solid foundation for the validity (falsity) of a hypothesis (theory). Transformation of a hypothesis into a valid theory takes place within the dialectics of relative and absolute truths.

As far as gnosiology is concerned, the difference between a hypothesis and theory lies in the fact that a hypothesis represents probabilistic knowledge, whereas a theory constitutes reliable knowledge which is valid as far as it is confirmed by practice. A theory, as a qualitatively specific form of scientific knowledge, implies true knowledge which exists as a particular system of the logically interrelated sentences reflecting the essential, law-governed, general and necessary internal relations of one or another of the object area. A theory explains a diversity of the available data and can predict the emergence of new, still unknown facts forecasting the law-governed behavior of a system in the future or elements of the given system which have not been revealed by the moment of establishment of a system. Thus a theory performs two important functions: explanation and prediction.

A theory, as a system of knowledge, just as any other system, is determined by a composition (a total combination of the elements determining its conceptual content) and structure, or in other words, by a totality of links and relations between its elements. The specific character of a theory shows itself in the fact that the concepts (terms) and sentences (statements) forming its composition are arranged in a logical order, which reflects a particular symmetry of the composition.... The concepts and statements are logically related to each other so that by using the laws and rules of logic some sentences may be derived from other particular sentences. Many logical relations between concepts and statements of a theory form its logical structure which by and large is deductive. What is more, the theory features a gnosiological structure which constitutes a totality of relations between various modes, means and techniques of theoretical reflection of the reality. The gnosiological structure of theory features, for example, a relation between its factors such as conceptual content, ideas, language, logical means, mathematical apparatus and models, which bear witness to a complicated character of reflecting the reality of the surrounding world at the theoretical level of cognition.

One important gnosiological problem is that of relating the concepts of various types and level of abstraction to real objects (ontological correlatives).

There is an intimate relation between the problem of validity of concepts (constructs) and that of existence of the objects including appropriate concepts. In this respect the model of the object plays an important role.

The model has already been defined above as a substitute for the object which reproduces the latter in the process of cognition to obtain data on the object reproduced and reflected. As indicated above, there is a class of material models complying with this definition which constitute the means and object of a specialized form of an experiment — a model experiment. There is also another class of models which performs definite functions of cognition as mental images rather than its material instruments as ideal constructions created by a scientist in his mind. In so doing, a scientist performs mental manipulations and transformations with them, i.e. mental "experiments". Such mental models are closely related to construction and development of a theory, though not being theories themselves.

Many reasons determine the necessity and expediency of employment of the mental models in the theory of cognition, and complexity of the real world (object of research) is the main reason. Various techniques of abstraction are realized for the purpose of purging a complex object of unessential, accidental and secondary factors and selecting its essential connections and relations in a refined form. Abstractions of simplification and idealization may exemplify the processes. The mental models perform, simultaneously, the functions of simplification, idealization, representation and substitution of the real object under study. The models become necessary intermediate links between the totality of statements of a theory and the reality approximately reflected in appropriate models. The method of mental modeling is a science — a wide ranging method that plays an essential part in constructing, developing and interpreting a theory.

Application of the mental models in the theory of cognition fostered development of the axiomatic method as one of the forms of the deductive method. The axiomatic method presupposes division of the deductive system into the following two groups of statements: the group of axioms and the group of statements derived with the help of the rules of logic.

The deductive-axiomatic methods of formalization obtained the most complete development in mathematical sciences. Each mathematical theory is a deductively constructed system, which allows us, within its limits, to draw deductive conclusions. Successful application of the mathematical apparatus in a specific case requires establishment of the agreement between the concepts of the employed mathematical theory on the one hand and the objects, regularities and functions studied by a given science on the other.

The way of mathematization of sciences, however simple or complex models are involved, has its distinctive features. The initial stage of mathematization of any scientific theory is called quantitative characteristics and relations. Here it is necessary to select among the phenomena in hand particular classes of the qualitatively homogeneous elements with characteristics differing only in quantity and devise appropriate methods for measuring quantitative differences between the elements of a class. The further development of mathematization is usually supported by finding stable law-governed relations between the elements of one or different classes and searching for the suitable forms of functional dependency capable of serving as idealized expressions of the relations of this kind. Next it serves as a foundation for developing (or selecting) an appropriate mathematical theory by means of which a mathematical scheme of the phenomena in hand is devised; having in itself various characteristics. Thus in some instances it allows us to construct an abstract description of the phenomena at hand in mathematical language and, otherwise to devise their mathematical models.

Mathematical methods allow primarily reflection of the quantitative aspect of the phenomena studied by a particular science. However, it might be erroneous to reduce application of mathematical methods merely to quantitative descriptions and calculations. Modern mathematics has the theoretical means to allow representation and generalization in its language of many qualitative characteristics of real objects since it deals with their structure not only in a specific way, but in general with the abstract structure having law-governed order. Application of mathematics may prove to be very fruitful for an appropriate science because it leads to an accurate quantitative description of the phenomena and allows us to work out clear cut concepts and form conclusions that cannot be obtained otherwise. When a given science finds proper mathematical means for expression and interrelation of its concepts it becomes possible to discuss its higher logical and theoretical levels and more effective prediction possibilities. The modern level of development of science features a steady increase in its mathematization. At present, "machine mathematics" plays a special part in the spread of mathematical methods. Application of the information–computational technology gave rise to new methods of mathematical development of the problems that did not admit earlier even to expression in mathematical concepts.

Now as mathematics is being used with increasing frequency as an instrument of scientific discoveries, mathematical methods are growing in heuristic importance in modern science. Strong heuristic emphasis is placed on the introduction of new mathematical concepts with subsequent generalization to endow them with physical meaning. This method quite often comes out as what might be called a mathematical hypothesis. The gist of the method (methodology) lies in the fact that the mathematical formalism reflecting distinctive characteristics of one field of events is used as a hypothetical mathematical scheme to express the laws of the other new previously unknown field of events. The familiar mathematical scheme can be transformed to satisfy the conditions of a new object area, and its symbols then acquire another interpretation. In modern science the heuristic importance of mathematical methods is so great that they often become the main instruments of theoretical investigation rather than an auxiliary means.

Each scientific theory performs two main functions: explanation of the existing and prediction of the future. If the essence is taken as a total diversity of the substantial features defining, conditioning and determining a specified phenomenon, process, system of connections and relations, then explanation as revelation of the essence amounts to a comprehensive analysis of these connections and relations and, on this basis, to mental reproduction, synthesis of the objects explained. The sphere of the essence, as a totality of general and necessary relations, represents the whole system of causal, law-governed, structural-functional and other relations.

Among the relations constituting the content of the essence, the most important are as follows:

— causal relation (that of an active cause to the effect brought about by it);
— a law, as a stable relation between different aspects of reality bearing a universal and necessary character;
— a structural-functional relation, where the structure of a system determines its functions and characteristics, and functions constitute a condition intrinsically indispensable to the existence and objectively expedient behavior of the system;
— the genetical relations characterizing the process of development of phenomena, their emergence and transformations, as well as internal contradictions.

A particular type of explanation is formed in compliance with specific features of the relation at hand.

The use of models for the purpose of explanation differs from the common deductive and inductive schemes of explanation in that a specific part is played by the model in substantiating the transference of the theoretical explanation obtained for one class of phenomena to a new sphere of phenomena. The basis for that kind of transference is formed by the relations of similarity and analogy established by means of the models between the phenomena already explained and a new sphere of phenomena the explanation of which constitutes the object of research. The model explanation refers, by its logical character, to nondeductive reasoning, differing in its gnosiological nature both from a theory and hypothesis.

Another essential function of a scientific theory is the prediction inseparable from explanation. Prediction and explanation rest upon the same theory, regularities and essential relations and employ the same logical "mechanism". They differ in the fact that in the process of explanation a theory is found for previously unknown facts, from which we logically derive the sentences coinciding in content with the description of these facts, whereas prediction consists in deriving logically from a given theory the results describing the facts and events which are still unknown but can and must occur in the future and the facts already existing, though unknown, which should be found in the future. Prediction in science is one of the most prominent and significant manifestations of the creative activity of scientific theoretical thought. The ability of the theoretical thought to be considerably ahead of the empirical level of knowledge is revealed here. Dialectics of the empirical and theoretical comes out as one of the most general regularities reflecting the objective world in scientific concepts, laws of science and theories.

§4. SYSTEMS MODELING

Modeling, as a concept, can be employed in a broad, generally cognitive sense of the word as well as in a narrow, special sense. As noted above, modeling in the broad sense reflects a universal aspect of the cognitive process, in this case scientific theories, categories and concepts constitute models. Modeling in the narrow sense of the word represents a specific method of cognition when the object of research (one system) is substituted by another (reproduced in a model). In general, the range of likeness, similarity of the model to its object extends from complete (absolute) up to zero. In the case of absolute similarity we deal not with modeling but with identical systems. Such multiplication occurs, say, in engineering, series production, manufacturing process and other fields. With (zero) similarity lacking, we have to deal with different systems not repeating one another. It is the latest trend to distinguish between complete, incomplete, approximate and mathematical similarity. Complete similarity prevails provided that basic parameters of the original system and those of the model coincide. The existing differences generally apply to minor, qualitative characteristics which can be ignored. From the cognition standpoint the model of that kind, if it can be called such, is of no interest. The model can reflect only some features (parameters) of the original system, and here this signifies incomplete similarity. In this case other parameters of the original and the model may not coincide, nevertheless it does not interfere with the coincidence of the final result of the original system functioning and the final theoretical conclusion derived from construction of the model.

Similarity, when simplification of the model is sufficiently great, as against the original, but partial theoretical inferences dictated by the model correspond to equally partial characteristics of the original, is called approximate similarity. In such an event, a set of models (a family) is most commonly developed, each of which reflects particular parameters of the original and all together the models yield a more precise knowledge of the original. Similarity of a structural nature, wherein the models reflect characteristics of the original which can be expressed quantitatively, is called mathematical (cybernetic, formally logical) similarity. We may say that structural similarity is that of the model and the original in statics, internal organization, forms and relations of elements (components), whereas functional similarity is that of the model and the system in dynamics, activity, in a result. We recognise system-communicational similarity when interaction of the original with the medium is reflected in external manifestations of the model.

Mathematical (cybernetic, formally logical) modeling enables one to investigate social and economical, biological, psychological and other systems.

Since the model represents a simplified reproduction of the whole original, it alone represents something integrated or whole. A scientific model generally features simplicity and symmetry, convenience, an ordered arrangement of elements corresponding to some extent to the order and structure of the original element. Modeling constitutes a continuous process of successive development of a series (a family) of models changing each other, which provides an ever increasing approximation of the model to the original to be modeled. Such an order of developing models constitutes a concrete manifestation of cognition evolution from the relative to the absolute truth.

In the overwhelming majority of cases the modeling process starts with describing a system. The initial description can be generally nonconcretized and not always fundamental. Further, the initial description is transformed into a logical, semiformal, symbolic description enabling one to gain a deeper insight into the system, its structure and relations of elements. In the process of modeling, definition of restrictions comes next: the external are time and conditions, the internal are labor, material and financial resources. The following stage features definition of the means for accomplishing objectives and efficiency criteria of the system functioning with respect to specified objectives. We reveal and define then the factors of external environment which produce an effect on objectives, means, conditions and criteria. At the final stage we establish relations among the objectives, conditions and external restriction (characteristics), form the functional solution efficiency relative to the system, in other words, by using the functional dependence criteria we set a degree of attaining the objective with respect to the factors exerting an influence on it. Modeling can be regarded as a process of consecutive development of functional, informational and morphological submodels. Needless to say that the greatest difficulties encountered in modeling systems and system problems are constituted by integration of these aspects of the model and its submodels.

In the process of developing a functional submodel we employ the analogs of available experience and form the functional equations describing particular properties of the system. Here the set of formalized descriptions is gradually extended due to the efforts of researchers. Mathematical relations serve as a basis for formalizing the objectives, improving the criteria and incorporating the conditions. In the final analysis, the initial description, vague in the beginning of the modeling process, gradually acquires a strict, mathematically accomplished form. The model of activities of the original finds its reflection in the functional model which is essential to the development of the informational submodel. The latter takes into account the degree of uncertainty inherent in the system or the system problem as well as the level of our knowledge about them, the lack of information on the original and the methods of obtaining additional data. The importance of the informational model resides in determining how and which uncertainty factors exert influence on the efficiency of the solution.

The next stage of the modeling process involves development of the morphological submodel which constitutes description of the internal structure of the system or the system problem, isolation of subsystems or subproblems and determination of interconnection (relation) between them. Such a stage of modeling presupposes isolation of independent (autonomous) functions in the functional submodel, subsystems or subproblems; estimation of uncertainty of the subproblems revealed; development of the functional morphological model, isolation of elementary problems in each subproblem; systematization of models. The process of developing the morphological model culminates in representation of the structure, distribution of functions among the elements, classification of the relations and description of the dynamics of functioning of the elements and the whole system. As a result, we obtain the improved morphological model which corresponds to the functional and informational submodels as comprehensive as the level of knowledge and the mathematical and technical means make it possible. The process of modeling can be treated as that comprising some other stages, which takes into account other approaches to the objectives and value of modeling.

A researcher needs a model when a real system, for some reasons cannot be studied immediately. It can happen when the system is not available for direct investigation because it is bulky, remote and due to extremely high or low temperatures, pressures or other characteristics inherent in the system which are dangerous to man's health and life. Sometimes the immediate investigation involves a violation of the system functions or its destruction. The researcher may also encounter difficulties in isolating the functions, properties and characteristics of interest, if at all. In such cases the whole system is investigated via modeling. It should be noted that modeling enables one to study non-existent systems which permits the estimation of the consequences of decisions to create such systems including virtual embodiment of the original system.

In the process of modeling and at the stage of investigation, two types of model functioning processes can be distinguished, which are the simulation and regime types. At first the states and dynamics of the original system are simulated and new properties and characteristics are tested on the model under simulation conditions. As far as the regime (operation) type is concerned, the conditions under which a decision is made and comparison of available with possible resources is accomplished, are tested in the model. The researcher has the opportunity of comparing decision alternatives with real opportunities and prescribed standards; and this enables him to select the decisions most suitable under appropriate conditions. Modeling is of particular value for social-economic systems when the opportunity arises to estimate the consequences of planned decisions before their realization in practice, which enables one to foresee negative moments and eliminate possible errors in planning and management. As a result, modeling involves the possibility of calculating rational prices, standards, profitability, efficiency norms and other social-economic factors and characteristics. Basically, modeling allows optimal behaviors to be found in complex systems such as the social-economic one.

The limited opportunities of the human cognition and the relative character of knowledge in each historical period determine the existence of the same whole (original system) at different stages of its study. The models reproducing the original system undergo transformations, their development therewith goes in the direction of ever increasing adequate reproduction of the whole system. The degree of correspondence between the model and the original and the nature of modeling itself are determined by a general state of scientific knowledge of the given whole under study, the level of experimental technique, the degree of special and general erudition of the researcher and some other factors. The background of modeling allows the models to be classified in a particular way. Here the experience in modeling shows that the basis for classification and, as a result, the classes of models can be different. As noted above, from the substantial standpoint models can be material, comprising real elements (components), and ideal, comprising various mental (logical, mathematical, and, in the general case, sign) forms. The material models are most commonly homomorphic and reflect only some of the aspects of the original system and considerably simplify this system. The ideal models are of a more or less prominent isomorphic nature, which enables one to discuss the uniqueness of the relation between the model and the original system. With science developing and the knowledge of systems becoming keener, the ideal, sign, logical, mathematical modeling is growing in importance, which reflects one of the features of modern science.

We can distinguish models according to the direction of time, isolating the models of the present state of the system, its possible state and the desired state of the original system. The models of the first class are developed on the strength of investigating the system background and its present state. Taken together these factors enable us to reveal the tendencies, the principal directions of the system evolution and the results of the model elaboration movement in time. The possible state of the models of the system are based on investigating the present state, revealing the tendencies of its movement from the present to the future and determining the result to which these tendencies may lead in a particular period of time. Interest is created in the peculiarity of this class of the models which resides in the fact that the greater the period of time till the modeled result is obtained the more probable is the model, since the greater the period the more deviations from the tendencies are revealed at present and the more are the variants of the system evolution. The models of the desired state of the original system are designed on the basis of the specified objective to be accomplished in the imaginable future. Such models are also classed with the program (normative, ideal, and others) models taking into account of the available potentialities of the system and those which the system must possess proceeding from the purpose in hand. The class of problem models apparently relates different classes of temporal models. A necessity for the problem models exists when we have to make drastic changes in the state of the original system or the ordinary processes of its functioning. The models of problem situations are constructed when a contradiction arises between the present and the specified states of the system. These models formulate the problems to be solved so that the contradiction might be obviated. The directions and means for obviating the above contradiction are contained in the decision models. With respect to the model classification, we may employ the following terminology: the models of the system's present state will be identified as cognitive, heuristic; the models of the future state as predictive; the models of the desired, specified state as normative; the problem situation models as heuro-pragmatic, and the decision model as pragmatic, since, being created on the strength of cognition of the original system, they come out as the means for transforming this system in the direction desired by man. As the above statements indicate, the models feature heuristic, predictive and pragmatic functions.

Models are distinguished by the implementation procedure and the language in which the model is formulated. Among them we isolate the models written in the "mathematical language" as well as in the natural language. Under modern conditions the latter models give way to the formalized: linguistic and symbolic, theoretical-multiple and abstract-algebraic, topological, logical-mathematical, theoretical-informational, etc. The model comes out as the original system simulation expressed in signs. Among the formalized sign models, a special place is occupied by the mathematical models. The isomorphism of some quantitative characteristics of systems irrespective of their material substrate nature enables us to employ, for investigating the self-control systems, the modern mathematical apparatus, specifically, the theory of sets, the theory of categorizing the probability theory, mathematical logic and many other theories.

The mathematical models of social systems differ considerably from those of physical, technical, chemical and other systems. The content of mathematics as a science may be characterized as a system of formal, sign models of the objective world. Hence the applications of mathematics consist in describing different phenomena in the formal language as well as utilizing formal mathematical means to devise the best ("optimal") interferences with the spontaneously proceeding phenomena, plan the expedient actions (behavior) and predict the course of phenomena. The objections to specific applications of mathematics in general cover the practical application of mathematics for modeling the phenomena and systems of the real world. The main objection of that kind applies to the important problem of susceptibility of some branches of knowledge to formalization, which, on the one hand, might be sufficiently consistent to be beyond the scope of an appropriate system of designations and yield deductively new scientific facts, and, on the other, it should be sufficiently comprehensive to cover all basic facts accumulated in a given science in the course of its mathematical development.

It is claimed that mathematics, created for the needs of sciences dealing with inorganic nature, cannot ("it will never be able") cover with its descriptions (models) such complex phenomena as those treated by biological or social sciences. This claim is correct only in that mathematics has been devised as an instrument for modeling (describing) truly physical phenomena and systems, and its available apparatus is not adapted to modeling phenomena and systems of another, more complex nature. The attempts to be confined, as applied to biology, to the apparatus of classical mathematics oriented to physical problems are as little justified as the mechanical attempts to reduce biological processes to physical. This holds, being expressed more sharply, for psychology, economics, sociology and other fields of science. The way out of the above situation is to be found in developing new divisions of mathematics more suited to modeling the phenomena and systems irreducible to truly physical phenomena. The theory of games should be considered the first mathematical theory of nonphysical origin of that kind. At the same time, internal processes are going on in biological and social sciences to facilitate mathematization of these sciences. A number of authors supported the need for a certain "maturity" of science to apply fruitfully ideas and appropriate mathematical apparatus to its subject.

The second objection, which proceeds from K. Hedel's well-known result on "nonformalizibility of arithmetic", is of a "more scientific nature". In this case it is said that even if "such simple" science as arithmetic is not amenable to complete formalization in the sense that we can formulate in it unprovable and incontrovertible statements, and each proof of its consistency must be inevitably inconsistent, then the prospects of formalizing other, more complex and profound sciences are quite hopeless. With respect to the above statements it should be noted that the concept of simplicity of arithmetic is much exaggerated, as arithmetic admits, within its framework, extremely complicated and subtle constructions. Simultaneously if some profound (say, economic or social) theory exhibits its nonformalizibility via the construction of a conceptual and formally unprovable statement, then only this would bear testimony to such a high level of theory formalization that we could not imagine.

The overwhelming majority of trends in modern mathematics have physical origins. The mathematical apparatus devised, roughly speaking, describes various phenomena, associated with the shifting of physical bodies in physical space, in a number of cases at a fairly high level. The same circumstance, i.e. the physical origin of the traditional mathematical apparatus, restricts its application in the fields of knowledge wherein the phenomena under study generally do not obey physical laws as some others do. In order that nonphysical sciences could be mathematized in full, we should devise for them their own special, sufficiently independent mathematical apparatus. Naturally, it cannot and must not be completely new or cut off from the traditional mathematics and its methods. Moreover, the divisions of mathematics, appearing to mathematize new branches of knowledge, should employ many achievements of classical division much to the extent that simple physical phenomena are essential for the existence of complex biological or social phenomena, or, which is basically the same, to the extent that biological or social sciences can employ physical theories or their analogs.

Mathematical theory of games is the theory of optimal decision models under conflict. A whole number of sciences is concerned almost exclusively with conflicts. Among such sciences are law, art of war, many divisions of economy, sociology, psychology and ethics. The list may include medicine and some branches of biology. Elaboration of formal, conceptual models, treated in the above sciences and their divisions with subsequent integration of the formal models obtained and their further mathematical development, represents an applied function of the game theory. The distinctive feature of the modern stage of development is the fact that mathematization is achieved not only and not so much by transferring mathematical apparatus by analogy from other models as by developing its own apparatus adapted specially for these purposes.

Mathematical modeling of biological and social systems necessarily involves investigation into the concept of behavior optimality. There is a sound background in consideration that, with some time having elapsed, the (mathematical) theory of games will be defined as the mathematical science of different optimality principles and their relations. Even today, among many forms of optimality investigated by the theory of games, we can discuss three of them as those having received the most attention. They may be called, somewhat arbitrarily, efficiency, stability and validity. The above forms of optimality have many features in common and the division suggested has no classification character. The simplest is the concept of efficiency which resides in the fact that the participants of a conflict should behave in such a way that the vectors of their gains would be maximal, or else they should attempt to reach the situations from which any deviation would involve a decrease in the gain even of one hand. The optimization, thus understood, is called the Pareto optimum. The concept of situation efficiency can be understood as the quest of the participants of a conflict for such strategy that the minimal gain obtained as a result of its application might be maximal. The principle of situation stability may be described, say, as the indifference of the conflict participants to violating the established situation which is known as the equilibrium. The concept of validity, as a form of optimization, can be interpreted as symmetry conditions: if the conflict is invariant with respect to a particular rearrangement of the conflict participants, then the set of all solutions to this conflict is also invariant with respect to the conflict. The above concepts of optimality prove to be symmetric, i.e. valid, in a sense. One of the aspects of validity, as a form of optimality, is the gambler's concept (that of the conflict participant) of obtaining the gain which he can surely get under any circumstances and the payment of additional sums to the conflict participant according to his participation in obtaining the overall gain. In some principles of optimality these conditions are introduced axiomatically.

The modern mathematical apparatus is unable to model people's intuition, feelings and will, in other words their creative capabilities. This insufficiency of the mathematical apparatus may be obviated, to a certain extent, by heuristic modeling. By using heuristics, we model inductive conclusions and develop programs for solving some problems by methods somewhat similar to those employed by human intellect. In our opinion, it is a promising combination of heuristic methods and mathematical modeling using not only deterministic formulations but chiefly stochastic statements of models.

The direct description of the original system, in which homomorphic correspondence between the structure of the model and that of the original is achieved, exists in simulation models. Such models are utilized in simulation games (here simulation is understood as the imitation of properties and functions of the original), machine simulation, simulation with the use of analog computers, etc. The undoubted advantage of the simulation models for direct description of the original system resides in their being more suitable for a researcher and a practicing engineer with respect to language. The language of such a model coincides with natural language. The latter allows for the inclusion of man in the process of modeling and the model itself as an element of utilizing human creative capabilities in modeling. In the development of models, particularly those of large systems, we employ the aggregation method when integrating the blocks of a system, specifically of a subsystem, the behavior of which is sufficiently monotonous. The approach of that kind allows in a simplified form the simulation of the original system as a whole. Experience in application shows that the aggregative simulation models prove to be most efficient particularly in developing long-run predictions when they yield sufficiently accurate results.

Concurrent with quantitative features, complex systems exhibit qualitative characteristics which do not always have to be formalized. The formalized model of the original system can also contain nonformalized elements expressed in natural language. Simulation models result from a synthesis of man and data processors.

In practice, each system can be mathematically described either as a particular transformation of input actions (stimuli) into output data (responses), which constitutes a phenomenological approach (identified sometimes at the causal or terminal approach), or from the standpoint of the system accomplishing a particular objective or performing a function, which constitutes an expedient or decision making approach.

§5. MATHEMATICAL LOGIC AS A MEANS OF COGNITION

About three centuries ago the versatile mind of Leibnitz explicitly expressed, for the first time in the history of science, the possibility of devising a symbolic calculus exhibiting all the advantages of mathematical procedures and, simultaneously, capable of expressing any logical statements and their combinations. Leibnitz thought that his idea of a "universal characteristic", having been implemented, would lead in the final analysis to a situation wherein any scientific debate would be easily and quickly settled by a simple symbolic calculus without leaving any doubt about falsity or validity of any one of the statements advanced by debaters. 150 years later, in the middle of the last century, a distinguished British mathematician G. Bool published his remarkable works, in which he actually managed to create logic calculus in a mathematized form, or symbolic logic – Boolean algebra, i.e. in a sense he realized Leibnitz's idea.

Boolean algebra did not immediately acquire established status (that is to say, initially it had not quite happy designations), the more so that it was not widely accepted immediately. To complete Boolean algebra much was done by Platon Sergeyevich Poretsky (1846–1907), a Russian astronomer-observer, private docent at the Kazan University. Couteur, Shreuder and other scientists relied further on his works. According to Shreuder, the Bool-Poretsky algebra in concept represented logic, and in form the section of mathematics being a letter calculus with two numerical values: 0 and 1, the former corresponding to falsity and the latter to validity.

At the border of 19th and 20th centuries, symbolic methods were widely applied in the study of logic bases of mathematics, which was prepared by a general trend of mathematical sciences at that time. The quest for logicizing mathematics arose over and above the mathematization of logic which was reflected in the creation of Boolean algebra. On the face of it, one might think that mathematics did not need such logicizing at all, being in itself a system of strictly logical statements. However, the study of the elementary geometry fundamentals particularly in the 19th century disclosed that Euclidean geometry possessed a number of statements containing considerable elements of intuition. Hence comes the axiomatic lead in the investigation of mathematical fundamentals trying to employ the techniques of mathematized logic in an effort to study the foundation of mathematical sciences, a purely logical aspect of their construction. This lead was often compelled to devise some special techniques which Boolean algebra did not need. And they were generated by specific features of mathematical sciences where the transfinite processes absent in the standard formal logic are employed. This produced differences from the latter and refers particularly to the principle of the eliminated third which cannot be always applied to transfinite operations, whereas in traditional logic it works without any limitations due to finiteness of all statements. The specific character of mathematical sciences also determined the circumstance that designations and initial concepts of the above lead of mathematical logic considerably differ from the Boolean, being more comprehensive and diverse. Boolean algebra represents a simple calculus of classes, and in compliance with this point all the concepts and techniques of this algebra are perfectly elementary although with its use we can form combinatorial logical constructions of any complexity. The principal simplest concepts of Boolean algebra, on the basis of which all the rest is constructed, are basically simple and clear; these involve negation, logical (disjunction) sum and logical (conjunction) product which can be regarded as elementary "algebraic" operations with classes of things or concepts, statements, etc.

In the branch of mathematical logic developed specifically for the needs of mathematical axiomatics there were available partly the same elementary concepts or those (say, implication and others) which only in appearance differed from Boolean basic definitions, whereas in actuality they were easily expressed through the latter. Nevertheless such logical relations as implication and the like are suitable precisely for mathematical purposes. Apart from elementary concepts and operations, mathematics incorporated those which could not be expressed in terms of Boolean symbols, e.g., universal and existential quantifiers were adapted for the needs of mathematical axiomatics and mathematical statements in general, and in this sense they are, so to say, "transcendental", since they are applied to infinite (and transfinite) statements and processes. It is this condition that produces considerable differences of the present lead in mathematical logic from Boolean algebra, and the principal difference is a limitation on employment of the principle of the eliminated third, since, apart from A or not-A, sometimes there may be the third, namely: we cannot basically establish with the use of the given means if it is "A or not-A". There are also other features differentiating symbolic logic, which investigates a logical structure of mathematics, from Boolean algebra, i.e. mathematized logic.

In the 1930s the symbolic methods of logical calculus were again developed, which constituted, so to say, the third branch of mathematical logic. Initially the mathematicians were concerned with fairly abstract subjects such as solvability or unsolvability of some qualitative mathematical problems of a general character, which was investigated by means of combinatorial techniques, best expressed by symbols of mathematical logic though in new concepts and designations with new initial definitions and operations which were substantially different in different authors. Here a variety of schools and individual trends differing from the two indicated above (Boolean and axiomatic) were formed in mathematical logic. They may include the following: calculus of problems (among the founders are Soviet mathematicians such as A. N. Kolmogorov, P. S. Novikov, *et al.*), theory of algorithms (normal algorithms and constructive logic were studied by A. A. Markov, N. A. Shanin, *et al.*), recursive functions of different types, so-called many-valued logics, etc. These trends are particularly abundant in USA. All of them may be integrated as the quest for qualitative study of solvability of mathematical problems of quite a general type by means of algorithmic procedures of a sufficiently definite character. The terms and designations in the works of this trend in mathematical logic are often considerably different from those of the former two trends. Finally, symbolic logic found extensive applications in constructing data processors, automatic constructions, etc.

None of the above three main trends in symbolic logic deals with separate statements and their general types, but they mostly give complex structures of a synthetic nature or, being engaged in analysis, they analyze exactly different combinatorial relations from the standpoint of their feasibility or unfeasibility, possibility or impossibility within a particular algorithm or a given system of specific operations. Hence it is clear that traditional and mathematical logics, though related, represent special, independent sciences, each of them with its own place and value in a general system of humanitarian and natural sciences.

We consider now the value of symbolic logic as a means of cognition and its power as a method for investigating various problems of natural and social sciences. In this respect, somewhat important is the fact that symbolic calculus, in general, provides a great convenience for carrying out intensive operations with the prescribed complex systems of objects, actions, premises, axioms, etc. We should recall that many branches of mathematics have been thoroughly developed since the introduction of proper symbols. Although K. F. Gauss said that mathematics derives its strength not from designations but from concepts, which is perfectly correct, it is proper symbolics that often makes some branches of science, so to say, technically more accessible. That was the case with algebra, which started to develop rapidly after Francise Viette introduced letter designations. It is much clearer in the case of mathematical logic since its problems are complex and combinatorial, i.e. they are such that we are interested not in the analysis of individual syllogisms in the spirit of Aristotle but in the study of a general character and properties of integral complex logical constructions connected into something uniform by a system of operations with objects which interact with one another according to particular rules. The theoretical combinatorial schemes obtain technical expression as data processors and other machines functioning unattended. In this situation the difficulty resides exactly in a combinatorial aspect of the matter and a large number of possibilities arising here, each of which has to be taken into consideration and which interact with each other creating a general picture often utterly boundless in its expanse if we fail to use mathematical notation procedures. Mathematical notation in association with rapid operation of modern data processors and other devices enables one to carry out such calculations and obtain such results of which recently we could not dream. Thus, one may observe Leibnitz's dreams partially come true. Of course, here only symbolic methods can be employed, formal traditional logic application being out of the question in such cases, since its scope is utterly different. It should be emphasized that operation of various "thinking" machines designed by means of symbolic logic basically differs from that of the human brain notwithstanding an external similarity. This involves one of the basic practical applications of mathematical logic and it could not be carried out with the use of unsymbolizable reasonings. However, the human brain comprises much more possibilities than any machine and logic-formal scheme, the latter being taken as a "completely mathematized" logical construction of any complexity. Human thought can thoroughly discuss and compare with other constructions any scheme in various aspects and from different standpoints. In other words, the human mind is capable of what may be called dialectical logic which does not confine itself to a particular symbolized scheme. However, the schemes of the latter type, i.e. the proper field of symbolical logic are vital, as we see, to our cognition of the surrounding world. Mathematical logic cannot claim a universal value and application as a comprehensive means of cognition, notwithstanding that its field of operations and applications is very wide and important. Being an extremely useful and even indispensable means of investigation in its field, which we tried above to characterize to a certain extent, mathematical logic and its associated branches of knowledge represent chiefly and only a combinatorial aspect of mental operations and processes, which is very important, though it does not embrace the whole variety of cognitive possibilities of the human brain.

Chapter 2

PROBLEMS OF QUANTITATIVE ANALYSIS IN NATURAL AND SOCIAL SCIENCES

A skilled master always takes care of his tools. The tools of research are the procedures of construction, analysis and application of models. The quality of these instruments predetermines to a large extent that of results of the entire scientific work. The modern stage of development of natural and social sciences is unthinkable without mathematics. Application of mathematical methods, primarily quantitative, in a specific field of knowledge requires solving a number of methodological problems. The present chapter is an attempt to fill, in a sense, the gaps in the literature on this subject. Here we continue to inquire into the part played by mathematics in investigation of the real world, considering different points of view. In so doing we proceed from the statement that mathematics is an effective means for constructing hypotheses consistent with the available knowledge of the real world. In the process of investigation, when developing theories and constructing models, we cannot avoid idealization which is not a result of application of mathematics but a requisite element of any science. The breakoff from reality is possible here, which may be called "production costs", but these costs are often too high. Control over these costs is accomplished by the rules of correspondence between ideal and empirical objects, specifically due to operationality of definitions.

The chapter focuses the reader's attention on the problems of the measurement theory and primarily the representation of variables in quantitative scales. The key point in the theory of measurements is the problem of adequacy, i.e. elucidating which of the operations on the numbers representing some values are intelligent. The dimensionality analysis which is a powerful and simple means for construction and logical verification of quantitative models is based on the theory of scales and solving the problems of adequacy. We consider the application of quantitative models including the optimization ones in planning, predicting and developing systems. Social sciences, as distinct from natural, feature the necessity to take account of goal-seeking behavior. This problem is the subject of inquiry in philosophy, psychology, biology and other sciences. We also discuss the problems of predicting the development of systems and processes. The principal possibility of prediction is provided by the availability of cause-and-effect relations in those systems and processes. The chapter ends in a comparative analysis of the model and expert methods intended for prediction.

§6. PROBLEMS OF SYSTEMS QUANTITATIVE ANALYSIS

The problems of possibility and necessity to apply mathematics in natural and social sciences were and still are debatable. Some authors think that mathematical methods are essentially appropriate only in quantitative analysis, and this analysis can ostensibly play an auxiliary part in investigating phenomena, processes, and systems. Because of this, it seems to be incorrect to believe that mathematical methods can substantially help in tackling important theoretical problems and devising more sophisticated systems. In general, the criticism of mathematics and its applications in other scientific disciplines is apt to exist as long as mathematics itself. Classical antiquity loyally supported mathematical studies, though not requiring and expecting from them any serious applications. When, after the Depression, an interest in science was awakened, R. Bacon had to prove the power of mathematics "... in sciences, secular matters and occupations".[1] The review of mathematical criticism as such and the analysis of this criticism do not enter into the scope of the present book, though of interest is even a brief survey of objections to mathematical applications in various fields of knowledge. Mathematical and, specifically, quantitative methods made it possible to obtain a number of fundamental results in natural sciences, primarily in physics from the beginning of the last century. These advances did not eliminate, however, doubts in the necessity and even possibility of applying mathematics in physics. The objections made to the application of mathematics in physics are rather instructive; moreover, it was at that time that, as a reaction to the advances of mathematics in physics, a serious substantiation arose with respect to the concept stating that even though mathematics is capable of treating physical problems, it cannot be used to tackle sociological, psychological, economic and other problems. However, "physics without mathematics" could not compete with "physics with mathematics". The matter is somewhat different with the sciences that treat animate nature. Here mathematization runs into considerable difficulties and its progress, as compared, say, with that in "biology without mathematics", is still moderate. This serves as an "empirical basis" to state that mathematics practically yields no results although it may be effective in physics.

Objections to applying mathematics in sciences concerned with nature and society rely on fairly stable philosophical traditions. Thus, A. Shopenhauer perceived in mathematical rationality a challenge to his view of the world as will and notion. G. Hegel pointed out insufficiency of the one-sided mathematical view of the world which "aims at comparing the subjects under consideration". It must be admitted that in this manner some important results have been achieved and thus we should particularly recall great progress in modern times in comparative anatomy and comparative linguistics. In this situation scientists went too far under the assumption that the comparative method can be applied with equal success in all fields of knowledge.... The comparison alone cannot completely satisfy scientific needs and the above results obtained by that method are to be regarded only as preliminaries for a really cognizing knowledge. However, since the comparison aims at reducing available differences to identity, mathematics must be regarded as a science in which this aim is attained most fully, and it achieved success exactly because the quantitative difference represents a completely external one.

Neither empirical sciences nor philosophy are to grudge mathematics this advantage.[2] The most developed substantiation of impossibility or, at least, minor efficiency of applying mathematics in sciences concerned with living beings (biology, psychology, economics, history, etc.) is proposed by V. Vindelbandt, G. Rickert and other representatives of the Baden School of Neo-Kantianism[3,4] according to which there is a strict line of demarcation between sciences concerned with animate and those dealing with inanimate nature. The starting point of establishing such a line between sciences is the

Kant's thesis of a distinction between the theoretical and practical mind as well as of the prime importance of utility,[5] from which Neo-Kantian philosophers have drawn a conclusion of two basically different ways of cognition and two basically different groups of sciences: nomothetic and ideographic sciences, or those based on the generalizing method of cognition, which are natural-mathematical sciences, and sciences based on the individualizing method of cognition, which are historical including economics, sociology, etc. Nomothetic or natural sciences discover something general, recurrent, regular. Application of mathematics is quite natural here. Ideographic sciences employing the individualizing method study something individual, singular, unique. Mathematics is unacceptable here since it is abstract and by its nature cannot consider singular, absolutely individual phenomena which do not recur in space and time. Nomothetic sciences are "those dealing with laws, while others with events".[3] In this case, as G. Rickert wrote, "the greater perfection we impart to our natural scientific theories, the further we get away from reality"[4] which is always individual, specific, unique, and immediately experienced. Natural scientific cognition is, thus, thought as a process of successive abstracting from reality, and not as an "ascent from abstract to specific". You won't find here a critical examination of the views of the Neo-Kantian Baden School with respect to the subject and theory of cognition. It should be noted, however, that estimation of these views will be one-sided and, therefore, incorrect if we ignore the fact that in the times of Vindelband and Rickert the idea of "the laws of phenomena in the social-economic field" was oversimplified by mechanistic materialism.

Incorrect and often incompetent application of mathematical apparatus, particularly that of the theoretical-probabilistic, to social, economic and other "human" problems produced appropriate spirits among mathematicians, the desire to estrange themselves from solving the applied problems in this field. The pros and cons with respect to application of mathematics in the sciences of society and nature always, explicitly or implicitly, take into consideration a particular concept of relations between the real world and the world of mathematical entities. We can isolate, however, few images of these concepts reflecting their most distinctive features, though not separated from each other with strict and clear bounds.

As far as the relation of mathematics to the real world is concerned, we may focus on the Pythagorean-Platonic approach according to which mathematical objects and, in the final analysis, only these objects exist in reality. The Pythagorean-Platonic thesis includes at least three components:

— the true reality, which is the world of ideas, i.e. eternal, invariable entities outside space and time;
— mathematical entities exist in reality;
— these two worlds are identical.

The second component plays an important part in discussions on possibility or impossibility of applying mathematics in sciences dealing with nature and society. At the same time, the acceptance of realism as mathematical philosophy does not mean that we accept the Pythagorean-Platonic thesis of the identity of the mathematical entities world and the real one.

According to another approach the real world and that of mathematical entities are isomorphic, which implies that mathematical theorems involve data on the real world. From the isomorphism of physical and mathematical worlds it follows that we can employ the knowledge of physical processes to validate mathematical statements. Since isomorphism prescribes the relation of equivalence, the real world plus the world of mathematics can be divided into the classes of equivalence. This idea underlies the "general theory of systems" by L. von Bertalanffy and leads in the final analysis to the Pythagorean-Platonic thesis if the world of mathematical entities is assumed to exist objectively, in reality.

As some scientists see it, mathematics is an experimental science describing a specific part of the real world, though it is not clear what part is precisely reflected in mathematical theorems and lemmas. According to other scientists the world of mathematical entities is that of pure forms of being. From this we may conclude the existence of two types of sciences dealing with the real world, which are the conceptual and the formal sciences.

The aphorism ascribed to G. Galilei and I. Newton says that mathematics is a language. If mathematics is the science of forms, then we may ask: the forms of what? From what content are these forms abstracted? The preceding thesis treated the forms of the real world. We would less violate language if we discussed not the content of the world but that of statements about the world and the form of these statements. But this is another thesis of the relation between the world of mathematical entities and the real world.

The thesis that mathematics is the grammar (or, to define it more precisely, the syntax) of the scientific language, focuses on the functions performed by mathematics in the developing system of scientific knowledge. At the same time, the results being obtained by means of mathematical and logical conclusions do not involve any ontological commitments, i.e. they do not immediately testify to the objective existence and objective truth of appropriate entities, as it is assumed by the ontology of the Kantian or materialistic type. Thus, the truths of mathematics and logic, though being uncontestable, are always relative. Compared to the Pythagorean-Platonic concept, the isomorphic concept of the real world and mathematics and the notion of mathematics as a science of the real world forms, the above notions look modest, though they constitute a firm foundation.

Here it is wise to point out the "everyday" concept of knowledge mathematization according to which mathematical methods are used if some mathematical symbols are employed, the latter is identified as representational mathematics. Actually it can be stated that "representational" mathematics is somewhat beneficial: the use of symbols, however minimal it may be, disciplines a scientist and contradictions, nonsense, gross mistakes, trivialities become obvious provided they are present in the text.

Logic and mathematics enable one to establish consistency of scientific theories and primarily serve as an efficient means to form hypotheses in line with the available knowledge. In this case the meaning of terms must be fixed since it is useless to carry out operations with indeterminate terms according to definitive rules. But fixation of one meaning of the term signifies separation, abstraction from other possible knowledge, i.e. a certain idealization. The application of mathematics is not the point here. The necessity to fix the meaning of terms is not the only reason of scientific abstraction. Science "without mathematics" as well as "mathematized" science do not dispense with abstraction and idealization of their objects. "Mathematized" and "nonmathematized" sciences are the sign systems, the difference between them is to be found in the nature of signs. Idealization and abstraction are the necessary elements of cognition which results in the creation of ideal objects in the nature of the simple pendulum or S. Karno's ideal steam engine, mathematical and analog models, theories.

The problem of determining the initial, primary, basic terms (concepts) which are not defined logically by means of other terms constitutes only part of a wider problem of establishing correspondence between the real world, particularly empirical data, and the results of the linguistic and cognitive activity of man. This problem has been an object of close attention of the philosophers of all trends throughout the entire history of philosophy. Isolation of an object, a system or a phenomenon, its separation from other objects and its definition constitute the most sophisticated and least formalizable part of scientific work which is rather an art than a science.

In general, description of any system does not become quantitative because the values of variables are presented as numbers. Description becomes quantitative if, concurrent with the variable values, the operationally defined operations on the modeled entities can be presented as those on numbers, specifically, as arithmetic operations. In constructing and analyzing the system models it is equally important to draw a distinction between the quantification, i.e. definition of a variable as numerical and the measurement, i.e. prescription of a proper numerical value to the variable. A nominalist approach to model construction rejects introducing variables into a model, these values being not amenable to measurement. Such an approach offers a number of advantages since the above requirement prevents construction of dummy models, i.e. the models which do not model anything. At the same time, by and large the nominalist methodology is not constructive. A distinction between formal description, numerical representation and measurement does not give rise to doubts, though it should be noted that in publications on the modeling of systems, objects, phenomena, processes and the like this distinction is not sufficiently clear, which is attested by an inconsistency in the terminology employed.

The measurement theory is usually taken as a theory the foundations of which have been laid by J. von Neumann and O. Morgenstern in their fundamental work,[6] and also in a more formalized form in the work by P. Suppes and G. Lines.[7] The above and many other papers study the problems of variable numerical representation, i.e. the problems of measurement in the sense the term is used in physics, technology, econometrics and other fields of research, are not analyzed in detail. In this case we employ both the term "measurement" and the term "representation in the form of numerical information" or the similar one. Suppes and Lines take the measurement actually as the solution of the following problems of the measurement theory: the problems of representing a relation empirical system, a relation numerical system, the problem of uniqueness of this representation (definition of the scale type) and that of the adequacy of such a representation.

The principles of the measurement theory have been adequately discussed in appropriate papers, nevertheless a few remarks are in order. The numerical representation of a variable is defined by the concept of a "relation system" or that of the n-term sequence $S = \langle P, Q_1, \ldots, Q_n \rangle$, here P is a nonempty set of the relation system; $Q_i\,(i = \overline{1,n})m$ are local relations in $P\,(m \geq 2)$. The type of the relation system S is identified as the n-term sequence $\Sigma = \langle m_1, \ldots, m_n \rangle$ if for each $i = \overline{1,n}$ of the system S the relation Q_i is m_i-local. The single type relation systems are called similar. The single type relation systems S_1 and S_2 are referred to as isomorphic if there exists such one-to-one representation $\bar{\varphi}$ mapping the domain (set) P_1 of the relation system S_1 into the domain P_2 of the relation system S_2 that for each $i = \overline{1,n}$ and each element sequence of the set $P_1 - \langle p_1, \ldots, p_{m_i} \rangle$ the relation $Q_i^{(1)}(p_1, \ldots, p_{m_i})$ prevails if and only if there is the relation $Q_i^{(2)}(\varphi(p_1), \ldots, \varphi(p_{m_i}))$, in other words, such relation systems are isomorphic provided they have the same structure.

The empirical and numerical relation systems are vital to the Suppes-Lines theory. The numerical relation system is termed the relation system, the domain of which is a set of real numbers. We say that the numerical system S represents the empirical system S_l if S_l is isomorphic (or at least homomorphic) to the numerical system S.

When examining the possibility of determining the truth in natural and artificial languages, A. Tarsky noted that the border between these two languages is relative since "... the objective reasons to establish a strict line of demarcation between these two groups of terms are unknown." Because of this, we find it possible to include in logical terms those usually regarded by logicians as extralogical, which is consistent with the practical employment of language.[8] Establishment of a strict dividing line between the language, by means of which nonlinguistics facts are discussed, and that is dominant in Tarsky's concept according to M. Taube.[9] Establishment of such distinction in the logical analysis of constructing a logic-deductive system means that the object of analysis is a formalized approximation of the conceptual theory. The model of constructing a model is the metamodel with respect to the latter, its elements being some formal systems and procedures. Therefore, all the three elements, which are an empirical system, a representation and a numerical system, are formal objects, i.e. in a broad sense they constitute the results of an operational definition because of their being the objects of logical analysis. Naturally, with mathematical theories being no exception there are no completely formalized theories, that is why metatheories study their idealized images.

§7. PROBLEMS OF SYSTEMS DESCRIPTION

Description of an empirical system aims at defining the elements of a subject region and finding their relations. In other words, we have to formulate some statements about these elements and establish their truth or falsity, thus the empirical relation system represents a formalized fragment of conceptual description, and a numerical system is only a part of formal description. Now the concept of a complete numerical system is employed. We say that a numerical system is complete if its domain is a set of all real numbers. Let us define the scale of measurement as the n-tuple $S = \langle \Omega', \Omega'', \varphi \rangle$, where Ω' is the empirical relation system; Ω'' — the complete numerical relation system; φ — isomorphism (or homomorphism) of Ω' in Ω''. The form of the scale is determined by the properties of uniqueness of the mapping φ. Let f be such mapping that $\langle \Omega', \Omega'', f \rangle$ would also be the scale, then the transformation ψ is identified as feasible for the given scale if $\varphi = \psi f$ is satisfied. A set of feasible transformations is a group. Definition of the scale or the group of feasible transformations is needed to solve the problem of adequacy. We assume that a numerical statement is adequate if its truth does not vary under feasible transformation of the numerical representation of the variable. The fact that we can operate with the quantitative values of variables just as with real numbers, here real operations with real objects being behind these operations, sometimes allows an experiment on real objects to be substituted by simple or complex calculations. Arithmetic properties of quantitative variables can be described by the Helder theorem on the ordered groups.

Operational definition of a variable, particularly that of the quantitative variable, is generally a result of the protracted process, the basis of which is provided by social practice. We do not run into difficulties with indicating who and when introduced such units as, say joule, second, meter, etc. whereas it is impossible to indicate who "invented" such fundamental physical quantities as force, time, and length.

In the relation scale, the names of units would designate a standard to which the objects measured are compared, and we operate with the names of units themselves just as with normal values. The theory of adequacy of numerical statements enables us to isolate feasible operations on dimensions. A dimension which is not expressed in terms of other ones, represents a fundamental dimension. As a rule, the choice of fundamental dimensions and their number is largely arbitrary and determined by convenience of their utilization. The analysis of dimensions in physics and engineering is employed not only as a means of control, but also as a fairly effective heuristic method of constructing the models of objects, processes, systems, and the like. In the social sciences and economics there is no standard system of units. Moreover, statements are made that it is impossible to isolate once and for all a set of fundamental dimensions, which can be basically agreed upon.

However trivial it may be, we take the liberty of reminding that mathematical models of phenomena, processes and systems are, by the employed variable values, those of other variables. Definition and explication of the concept of "connection" have acquired an ample philosophical literature, but in some fields of knowledge, particularly in social-economic sciences, the literature still admits confusion about the concept.

We isolate some types of dependences employed in mathematical models. Here, of course, we do not seek an exhaustive classification, but emphasize a methodological aspect of the matter.

Relations-definitions. We most commonly employ the equations-definitions characterized by the fact that the unknown value cannot be measured irrespective of the determining values.

Composition relations. Specifically, the equations of composition describe the possibility of developing an object, a system into a sum of components. It should be noted that in the general case a desired development may not exist.

Relations of distribution. The equations of distribution express distribution of a result in different directions. In this case they formally bear resemblance to the composition equations, but, as distinct from the composition equations which can or cannot be satisfied, the distribution equations represent identities. In other words, if all directions of distribution of a particular result are included they are valid by definition. Incidentally, the equations-definitions are identities too. The transformation relations, specifically the equations, as distinct from the above types of relations, reflect the cause-and-effect connections realized by a really functioning mechanism.

Relations of behavior and (or) decision making. They differ from the transformation relations in that a conscious subject or a group of subjects possessing the liberty of choosing among alternative actions respond to a stimulus. In the given type of relations it is important that decisions can be instantly changed, at least in principle, while, say, the properties of an abstract technology, cannot be arbitrarily changed. This condition distinguishes the present type of relations from that of transformation relations bearing a substantially stable character.

The above types of relations do not exhaust all potentialities but we think that any relation (equation), which is a part of a certain model of an object, process, system, may be reduced as a result of proper transformations to the relation of one and only one of the above types. The balance relations (equations), which actually represent a combination of equations of the indicated types, can be added to the above. These balance equations are frequently encountered in economic models.

The foregoing types of relations of mathematical models are isolated with the object of establishing criteria, which enable us to judge what can and what cannot be accomplished by the given type of relations. For example, the system containing only equations-definitions, composition equations and distribution equations is unsuitable for prediction, i.e. for calculus of variable values at a particular moment $t+\tau$ by their values at the moment of time t. Such a model cannot be employed for long-range planning. It is caused by the fact that the above equations (relations) describe connections between variables at the same moment of time. Serious problems arise with respect to definition of the relation type, but attempts can be made to formulate the test allowing equations-definitions, composition equations and distribution equations to be distinguished from equations of transformation and behavior. The latter group of equations (relations) contains constants the values of which are determined from the analysis of empirical material.

As stated above, one of the problems solved when constructing a mathematical model of a phenomenon or a system is to seek description of a connection between the isolated variables, assuming that such connections exist. As applied to relations-definitions such an assumption is always valid; it holds by definition. Relations-definitions do not describe connections between phenomena, they only introduce variables, hence to construct a model by employing only relations-definitions is impossible: the relations (equations) of behavior and (or) the equations of transformation are needed, they reflect the essence of the modeled phenomenon. Interrelations among variables of the mathematical model are generally deformed under the influence of various factors. Sometimes the deformation may be so great that from the analysis of specific empirical material it is possible to come to false conclusions with respect to the absence of the connection when it is available or to conclusions about the existence of the dependence when there is no such dependence. Of course, all these difficulties are encountered only in construction of transformation relations or behavior relations. In the case of relations-definitions, relations of composition and distribution there are no problems with the existence of dependences or the evaluation of parameters of these dependences. Real problems may be solved by employing the relations of transformation and behavior. To do this, we often need to solve the problems of existence of dependences and evaluation of their parameters based on statistical information. When considering the above problem we often employ the terms: "independence", "correspondence", "influence", "causality", "correlation", etc. Among these concepts, "independence" is one of the most fundamental and employed in many fields. Of vital importance is the establishment of causality among events or phenomena. It is said that one (or several) variable influences another (or several) variable.

An important indication of causality between variables is seen by the availability of the causing mechanism relating a cause to an effect. In this case the cause does not produce the effect instantly: in this sense, the well-known postulate of the constancy of the velocity of light is an immediate consequence of the causality principle. At the model level this means that ideal description of causal dependence must include time delays. In practice, to simplify description, these delays may be set sometimes equal to zero, i.e. it may be assumed that the speed of interaction is infinite, but such simplification does not imply disregard of delays. The cause-effect relations between the model variables show themselves properly under particular conditions; incompleteness of information on these conditions leads to errors in description of the relations mentioned. At the model level this involves incorporation of random variables into description, e.g. addition of a random error additively entering into the model. In this case the model does not obtain any "statistical" or "probabilistic" causality, the independent variables of the model can be merely determined (predicted, controlled) and stochastic. In the latter case we employ such terms as "correspondence", "correlation", etc. but these terms do not represent an alternative causality, though of course the availability of correspondence between the model variable values does not provide the cause-and-effect relation between appropriate phenomena or processes. In practice when modeling systems, among other things, we employ various quantitative measures for inadequacy of connection between variables (tests, elements phenomena, etc.). Efficient application of the measures requires the analysis of premises on which they are constructed, considering that the effect similar to that of false correlation is always possible. In this case, of vital importance is the theoretical-probabilistic model of the system, within the limits of which the function of distribution of measures for inadequacy of connection between variables.

The problem generally consists in constructing not just a logically consistent model of the system, but a model reflecting the real situation. Basically, models are developed with a view to decision making and to do this variables and parameters of the model must be assigned specific numerical values. To substantiate numerical values, they have to be correlated in some way with the true values of parameters and variables of the model. The situation of complete coincidence of the numerical values of parameters and variables employed in the models may prove to be unattainable due to errors of observation, "noise", etc. This however does not eliminate the problem of estimating the truth of measurements and parameters and values of the model variables. From the standpoint of Marxist-Leninist philosophy the problem of the criterion of truth has been solved: man has to prove the truth in practice. The problem of practice as a criterion of truth obtained a fairly ample philosophical literature. However, general formulation of the criterion of truth is adequate for solution of general-philosophical, ideological problems, but for evaluation of the quality of a specific scientific theory, model, and quantitative statement, the criterion of practice should be specified. The problem of specifying the criterion of truth as applied to particular fields of scientific knowledge is rather complicated. We know many more or less different approaches to the solution of the problem. However, we cannot claim that it has been adequately (sufficiently) developed in the methodological aspect.

We do not seek to review and analyze the criteria of choice (adaption, quality) of scientific theories or, more generally, scientific statements. Without claiming completeness we set forth a minimal list of criteria to be satisfied with quantitative statements, model, and theory. Concurrent with logical criteria such as consistency, completeness, etc. the list may include the following criteria: truth (T), verifiability (C), utility (U), self-descriptiveness (I), value (V). To a first approximation, it suffices to assume that each of the variables T, C, U, I, V takes on only two values. It is believed that truth, utility and verifiability do not represent independent characteristics of statements. Moreover, $I \subset T \cup U \cup C$; $V \subset I$ exists, i.e. of importance may be only a self-descriptive statement, and the statement can be self-descriptive only when it is true and verifiable (utility does not provide a necessary condition for self-descriptiveness). At the same time, e.g. an erroneous statement can be verifiable and useful, and the true one can be useless, etc. We do not seek here to discuss in detail these characteristics and relations.

To be noted is one more problem that arises when describing systems. We discuss the problem of goal-seeking behavior which distinguishes the elements of social-economic systems from those, say of physical, chemical and other systems. The problems of teleology are numerous and have a long history; it is believed that they arose in the pre-socratic times when science had not detached itself from myth. The term "goal-seeking behavior" has at least two meanings: "the behavior directed toward a goal" and "the behavior directed by a goal". The statement that an object is moving toward a goal or it is directed toward a goal, etc. does not contain any teleology; in the given case the talk is about the answer to the question of where the object is moving. However, it is different with the statement that the object is directed by the goal. Here the goal acts as some cause of movement of the object, which is external to it. Teleology forms the basis of the functional approach to description of the reality, in this case, phenomena, objects, systems, processes prove to be penetrated by mutual usefulness: one thing emerges for the sake of another. Teleological problems came to the foreground of scientific investigations in connection with the appearance of cybernetics, general theory of systems, systems approach and a number of other scientific and quasi-scientific disciplines containing in their names the word "system".

§8. DEFINITIONS OF SYSTEMS

Systems optimization methodology represents a scientific discipline which studies various phenomena, separating itself from their specific nature in an effort to optimize them, considers merely formal relations between their constituent factors and studies the nature of their changes determined by external effects. This must find its reflection in definitions of the concept "system" where the object under consideration is constituted not by a physical reality, but by a formal relation between the observed tests and properties. Definitions of the concept "system" should apply equally to the systems exhibiting behavioral purposefulness and employing the concepts of decision making and control as well as to their phenomenological description fixing the nature of cause-and-effect transformations of input actions into output values. The study of corollaries from the fact that systems display definite properties should constitute the content of any theory including the systems optimization theory which must never be confined to a mere descriptive classification of systems.

By virtue of the fact that the concept of system represents the starting point of our study, we will consider some approaches to definition of this concept. We will name only the approaches in which the basic system concepts are introduced by means of formalization, i.e. those in which proceeding from a verbal description of an intuitive notion there is an exact mathematical definition utilizing for this purpose a minimal mathematical structure, say a minimum of axioms admitting its correct interpretation. In some cases, relying on the basic concepts obtained from formalization, the authors of the approaches extend mathematical theory adding new mathematical structures needed for investigation into various properties of systems. Indeed, the procedure of that kind proves to be effective and enables one to study to what extent a specific property is fundamental and, simultaneously, what minimal set of assumptions is needed for the system to possess this property or a given relation to be satisfied with it.

The concept of system may be defined proceeding from special notions formulated on the sets with a rich mathematical structure, specifically, in terms of flows in topological spaces or e.g. mappings in functional spaces. Such approaches however poorly agree with the quest for the limiting generality and a desire to describe ill-structurized indefinite situations.

Attempts were repeatedly undertaken to define and construct a purely algebraic theory of systems. In some places, the theory thus constructed, proved to be successful since it enabled one to utilize appropriate algebraic structures for the solution of specific problems. However, when it comes to introduction of fundamental concepts, the language of algebraic structures does not turn out to be the most successful.

We are familiar with an extensive use of logic for the study of statements about systems and deductive analysis of their properties and behavior. This approach seems to be rather attractive. However, abstract mathematical structures employed in axiomatic logical systems are too special to serve as a basis for the definition of the fundamental concept of system.

Some authors, in their efforts to define the system concept not only by a relation, but also by describing the space of states, transition and output functions, encounter many difficulties. In this approach, systems are described by a family of their subsystems and relationships. However, even though the state transition functions are known for each of the subsystems, to define state transitions for the entire system as a whole turns out to be a tedious job, if possible at all. On the other hand, any system combination of that kind also constitutes a system for which the possibility of defining the state transition function is open to question. Furthermore, in nature are found the (biological, social) systems with purposeful behavior, admitting formal definition and they can be described only as such. As applied to such systems it is generally difficult, if possible, to introduce the concept of state, and it is impossible to describe transitions in the state space.

In our discussion, the system definition is given in theoretical-multiple terms. At this level, the system is rather simply and quite naturally defined as a relation in the language of the set theory.

Let a family of nonempty sets be $M = (m_k : k \in K)$, here K is a set of indices. Define a system specified on M as a proper subset of the Cartesian product $\times M\Sigma \subset \langle \times (m_k : k \in K) \rangle$. All components $m_k, k \in K$ of the Cartesian product $\times m_k$ will be identified as objects of the system Σ.

Definition 8.1: (Descriptive). A general system is identified as a relation on nonempty sets

$$\Sigma \subset \langle \times (m_k : k \in K) \rangle , \tag{8.1}$$

here $\times$ is a symbol for the Cartesian product, K a set of indices. The set m_k will be referred to as an object or element of the system.

Specifically, if the set K is finite, then the system can be represented as

$$\Sigma \subset \langle m_1 \times m_2 \times \cdots \times m_k \rangle . \tag{8.2}$$

In those cases when we are interested in cause-and-effect transformations of input actions into output values we employ the phenomenological description fixing the character of these transformations. Assume that the set K can be broken down into two sets $K_p \subset K$ and $K_q \subset K$. Here we complete $K_p \cap K_q = \phi, K_p \cup K_q = K$. The set $P = \times (m_k : k \in K_p)$ will be identified as the input object of the system, and the set $Q = \times (m_k : k \in K_q)$ as its output object, then the system may be defined as follows.

Definition 8.2: (Phenomenological). The "input-output" system is identified as the relation

$$\Sigma \subset \langle P \times Q \rangle . \tag{8.3}$$

Specifically, if Σ is the function $\Sigma : P \to Q$ it may be said that the system Σ is called functional.

If the set of indices K is linearly ordered, then it may be identified as the set of time moments, and the functions defined on such sets are referred to as the time functions. The objects, the elements of which are time functions, will be termed the time objects, and the systems defined on the time objects, – the time systems.

In M. Mesarevich and Y. Takahara (*General Theory of Systems: Mathematical Foundations*), we find a fairly comprehensive and developed investigation of systems in which it is possible to describe a transformation of input actions (stimuli) into output values (responses), and the phenomenological or cause-and-effect approach to describe systems is implemented. We consider systems proceeding from their achievement of an objective or based on performing a particular function which reflects the approach from the standpoint of purposefulness or decision-making.

The principal relations characterizing a purposeful system and distinguishing it from the systems of other nature show themselves in who and how they participate in the system (phenomenon), in what form and who are interested in the result (outcome). Any description (definition) of the purposeful system (phenomenon) should reflect specific conditions.

To begin with, we should fix participation in the system (phenomenon) of particular decision-making parties, specifically, subjects. These parties may be in different relations among themselves, in other words, a set of participants (parties) displays a structure and these parties will be identified as the coalitions of actions. Denote a set of all action coalitions in terms of $m_j : j \in J$. Next, we should isolate the coalitions of objectives (interests), i.e. among the action coalitions we need to point at the parties standing for particular objectives (interests). A set of interest coalitions will be denoted in terms of $m_i : i \in I$. We have to define potentialities of the participants of a phenomenon, i.e. indicate which actions may be carried out by each of the action coalitions. These actions will be termed the coalition strategies $r \in m_j$; $r \in R$, and the set of all action coalition strategies will be denoted by R_r. The result of all action coalitions choosing their strategies with respect to all restrictions is called a situation. Thus, the set S of all situations may be taken as a subset of the Cartesian product $\prod_{r \in m_j} R_r$.

Definition 8.3: (Constructive). A system is referred to as a collection of the form

$$\Sigma = \langle m_j; m_i; \{R_r\}_{r \in m_j}; S \rangle . \tag{8.4}$$

For purposeful systems and those of decision-making, we need to indicate, for each coalition of interests $r \in m_i$ on the set of situations S, the preference relation $>_r$, e.g. the binary relation of preference. The fact that the situation S_1 is more preferable for the coalition r than the situation S_2 is commonly denoted as $S_1 >_r S_2$.

Finally, we need to evaluate the results of the decisions being made, and denote the penalty coalition adopted in the phenomenon (system) by $\{\Delta_r\}_{r \in mj}$.

Now we can define the purposeful system.

Definition 8.4: (Normative). The purposeful system will be taken as the n-term sequence of the form

$$\Sigma = \langle m_j : j \in \mathcal{J}; m_i : i \in I; \{R_r\}_{r \in m_j}; S; \{>_r\}_{r \in m_i}; \{\Delta_r\}_{r \in m_j} \rangle , \qquad (8.5)$$

here $m_j : j \in \mathcal{J}; m_i : i \in T; \{R_2\}$ are arbitrary abstract nonempty sets; $S \subset \prod_{r \in m_j} R_r$, $>_r$ $(r \in m_i, i \in I)$ are arbitrary binary relations on S.

The classes of purposeful systems are isolated as a result of notion: coalitions of action, coalitions of objectives or interests and coalitions of strategies. Conditions of that kind may apply to each of these elements and restrict them according to different indications. Thus, of interest is the construction and study of the structure characterizing specific classes of purposeful systems.

To begin with, it should be noted that the very concept of the purposeful system or systems of decision-making (controls), and its definition (8.5) does not take into account a large number of complex features: dynamics (time dependence) inherent in informational states of the coalitions of interests, actions and, hence, in a set of situations, representations of strategies (behaviors, decisions) as time functions, opportunities arising with respect to exchange of information between the elements of the action coalition, participation or nonparticipation of particular elements in the system, etc.

It stands to reason that indication of new properties of the object described by the concept denotes specification of this concept, i.e. reduction of its logical body which simultaneously is accompanied by enrichment of this concept with content.

The foregoing undoubtedly holds when we discuss a specific system as something given and remaining identical with itself for the entire reasoning. Otherwise we can discuss the variables of purposeful systems the values of which may be the constants of the known kind system from a set of system constants. This remark is of practical value since the decision system constants have a variable structure, relations and objectives; in other words, all of the system elements may change with time, though formally the purposeful system will be represented in the same known form. Here we will not reconstruct a classification of the control systems, but set forth several concrete classes of systems with the only aim of making the content of this work formally less dependent on the papers previously published by other authors on decision systems.

Suppose we have a purposeful system where $m_j = m_i - m$. In other words, each element of the action coalition pursues only one objective (has only one interest), i.e. we have system elements, for each of which the behavior criteria are defined, so that there is mutually unambiguous correspondence between elements of the action coalition and those of the interest set. We have, as before, $S = \prod_{r \in m_j} R_r$, evaluated the situation preferability $>_r$ for each of the action coalition elements by the functions of result (outcome), each of which is defined on the set of all situations and assumes real values. In such an event, we have a coalitionless system of decision making which is specified in the form

$$\Sigma = \langle m, \{R_r\}_{r \in m}, \{\Delta_r\}_{r \in m} \rangle \,. \tag{8.6}$$

Specifically, we may have a coalitionless system for the two-element (person) control. It is natural to isolate, among coalitionless systems, the dynamic type of systems in which decision-making (adoption of controls, strategies) is expanded into the process proceeding with time. In the mathematical theory of games, such systems are modeled by differential or pursuit games. Exact definitions for the systems of these classes are rather intricate and thus are not mentioned here.

The coalitionless (noncooperative) purposeful systems may be exemplified by numerous systems constructed on the elements of two levels (say higher and subordinate) in control systems where only one interest (objective) of functioning is determined for each element. The coalitionless control system, especially the two-element systems, represent the class of purposeful systems rather simply from the standpoint of their construction. However, if relations between the elements of a system admit changes with time in strategies, especially interests (objectives), then such systems become extremely complex.

We isolate an important class of purposeful systems which contain merely one element of the action coalition. In a sense this case is the simplest since the systems with a nonempty set of action coalitions are of no interest. In such systems, a set of situations coincides with the strategies of a unique action coalition, which finds its reflection in the definition of the decision system where the strategies are not mentioned at all and only situations are noted. Such systems are naturally identified as the nonstrategic purposeful systems. We note three special classes among the nonstrategic control systems.

As a coalition of interests, we will consider the subset of a given set m, the elements of which, as in the coalitionless system case, are termed the action coalition elements. Each situation in the nonstrategic control system corresponds to the value of a particular characteristic which may be conventionally identified as the gain and is obtained by the action coalition elements. The set of all situations may then be taken as a subset of the Cartesian space E^m. Implementation of each situation x is interpreted as obtaining the ith component x_i of the vector x by the ith element of the action coalition. In the mathematical theory of games, similar situations are referred to as sharings. In the nonstrategic control system, each coalition corresponds to a real number $q(r)$, the set of sharings is a simplex of the form

$$\left\{x : x_i \geqq q(i) : i = m; \sum_{i \in m} x_i = q(m)\right\}. \tag{8.7}$$

In this case $x >_r y$ when and only when $\sum_{i \in r} x_i \leqq q(r)$, if $x_i > y_i$ for all $i \in r$, then by analogy with the mathematical theory of games, the nonstrategic control system is identified as the classical cooperative purposeful system.

In those cases where $m_j = m_i = m$, R is the closed and bounded subset E^R and $x^0 \in R$, and the set S of situations $S = \prod_{r \in m} R_r$ is defined as $\{x : x \geqq x^0, x \in R\}$, then the obtained control system is referred to as the arbitration scheme.

When the situation set is of arbitrary nature, specifically, it may be finite and the preference relation $>_r$ for each ith element of the action coalition is the arbitrary trichotomic relation such that of any two situations either one is more preferable for i than the other or they are equivalent for i, such decision-making systems are named the group decisions. The foregoing does not eliminate the possibility of the action coalition elements being integrated into coalitions which express the interests common to all members of the coalition.

Suppose we have a set of nodes of a certain oriented graph (X, G), and time is discretely distributed. As a strategy of each element in the action coalition m_j, we assume the function which assigns each pair (x, σ) to a node in G_x. Let us assume that there are the original node x_0 and the function determining the "queue of a move" for the action coalition element in the situation S in the position x at the moment of time τ, thus

$\psi : S_x X_x T \rightarrow m$. As the result of such an interpretation, each situation corresponds to a trajectory on the graph, for which we juxtapose the gain function of the action coalition element on the graph. As a consequence, we obtain a system of decision-making on graphs.

Next we consider the elements of the tree-like ordered set G. All positions immediately following the given one are identified as alternatives to this position, then all positions without alternatives prove to be conclusive. Let G' be a partition of the set of all inconclusive positions of the tree, thus: $\bigcup_{i \in m} g_i \cup g_0$, $g_{i_1} \cap g_{i_2} = \varnothing$ with $i_1 \neq i_2$. As applied to each position $g \in G_0$, we assign a probabilistic distribution P_g on the set of alternatives x. Consider the partition z_i of each of the sets G_i $(i \in m)$ into pairwise nonoverlapping subsets wherein there are no positions following each other, and the sets of alternatives have the same power in all positions. The indicated subsets of positions are referred to as the informational sets of the ith element in the action coalition. The associated alternatives of positions belonging to the informational set are named the alternative of this informational set. In such an event, it is natural that the strategy of the ith element in the action coalition is identified as any function specified on the family of its informational sets, the values of which on each informational set are the alternatives in this informational set. Each situation in this case determines the sequence of positions immediately following each other. Here it is natural to call the maximal sequence of positions of that kind the play of a game. Evidently, all final positions may also be assigned to the plays. Let B denote a set of all plays, in each play $b \in B$ the ith element in the action coalition obtains the payoff $h_i(b)$ then the collection

$$\Sigma = \langle G, G', \{P_g\}_{g \in G_0}, \{z_i\}_{i \in m}, \{h_i(b)\}_{i \in m}^{b \in B} \rangle \tag{8.8}$$

is called the positional system of decision making. Specifically, in the general positional system of decision making, the set of positions may be finite.

If each informational set consists of a unique position, then there is a system which may be referred to as the positional decision system with complete information. Such systems are of vital importance for mathematical investigations, but they are not so important when utilizing in reality the decision-making systems.

The above examples of the purposeful systems give an indication of the possible classification of such systems, employing a familiar scheme of the general notion of the decision (control) system. At present, however, it seems impossible to offer any harmonious classification of these systems, especially to find representatives of separate classes from the systems already developed.

We consider the concept of optimality as applied to the systems of decision-making (control).

To begin with, it should be noted that the mathematical theory of purposeful systems does not contain any assumptions of the optimal behavior of the action coalition elements. In general, they do not act optimally, and there is no reason to state that having obtained appropriate recommendations they will act optimally in real systems. The optimization problem itself is most likely to be the subject of investigation as viewed from the systems theory. In this case, one of the basic problems is to elucidate in what sense we have to understand optimality for the proper class of systems.

The systems theory states a set of different optimality principles, but not all of them are applicable in purposeful systems. Thus we will discuss three different forms of optimality which will be conventionally identified as efficiency, stability and validity.

The idea of efficiency, as a form of optimality, shows itself in that the action coalition elements should seek the situations (choose the situations), in which any deviation involves a decrease in the result (outcome) if only for one of them. Such ideas of optimality were likely found for the first time in the works of Pareto pertaining to 1896. Sometimes the idea of situation efficiency may be implemented as a tendency of the action coalition element to choose such a strategy that a minimal result (payoff) obtained from its application, would be maximal. In the mathematical theory of games, this principle is sometimes called the maximum principle or the principle of guaranteed result.

In purposeful systems, the situation stability principle can be implemented as a disinterested position of the action coalition elements with regard to violation of the established situation which, in this case, is called the equilibrium one. In some papers, the appropriate principle of optimality is referred to as the objective feasibility principle.

The principle of validity in the systems theory appears as conditions for symmetry: if the system is invariant with respect to a rearrangement of the action coalition elements, then the decision set of this system, in a sense, is also invariant with respect to it. This implies that the above forms and principles of optimality turn out to be symmetrical; in other words, they are valid in a sense.

In the general model for the purposeful system, it is difficult to say what alternative should be chosen by the action coalition element of decision-making or which of its strategies is optimal. From the foregoing it follows that optimality in the purposeful systems is taken as something expected, or feasible. The optimal behavior is to be called that which is possible under conditions of feasible actions of the action coalition elements and those accomplished according to their interests. Confusion exists in the literature with respect to optimality problems, which is in great part due to the very object of the theory or to be more exact, the object of its application — social and biological phenomena and systems. Owing to ambiguity, complexity, dynamism, integrity and other features of purposeful systems, we need a careful description and a strict line of demarcation drawn between theoretical game models of the systems. By and large the optimality principle from the standpoint of purposeful systems is the rule which is needed to solve the problem treated in the system.

An important point in the theory of purposeful systems is the demonstration of the existence for particular classes of situation systems satisfying the optimality conditions adopted for the systems of these classes. The problems arising in this respect are of vital importance since they apply to feasibility of the principles of optimal control (behavior). The practical problem of purposeful systems is to actually find the situations implementing particular principles of optimality. Here, concurrent with traditional methods of solving extremal mathematical problems, we need developments of a new apparatus adequate to purposeful systems.

The present work deals with the most important (in our opinion) problems of purposeful system optimization, that is, the problems of analysis and synthesis of systems and multi-objective optimization. It is this circumstance that explains the subsequent discussion of the theory of numerical optimization methods for large-dimension problems and methodology of multi-objective, multicriteria and vector optimization as applied to purposeful systems.

Chapter 3

DANTZIG-WULF DECOMPOSITION

§9. DANTZIG-WULF DECOMPOSITION METHOD

§9.1. Consider the problem of linear programming (LP), the constraint matrix which has a block-diagonal structure with connecting rows of the form

$$A = \begin{pmatrix} A_1 & A_2 & \cdots & A_p \\ B_1 & & & \\ & B_2 & & \\ & & \ddots & \\ & & & B_p \end{pmatrix},$$

where $p \geq 1$. The matrix of any LP problem can be reduced to that form with $p = 1$ as a result of proper partitioning of constraints into two subsets.

We write LP problem in the form:
minimize

$$\mathcal{L} = Cx \tag{9.1}$$

under conditions

$$A_1 x = b_1 \qquad (m_1 \text{ constraints}), \tag{9.2}$$

$$A_2 x = b_2 \qquad (m_2 \text{ constraints}), \tag{9.3}$$

$$x \geq 0,$$

where

$$C = (c_1, c_2, \dots, c_n) \qquad n\text{-dimensional vector,}$$
$$b_1, b_2 \qquad m_1, m_2\text{-dimensional vectors,}$$
$$A_1, A_2 \qquad (m_1 \times n), (m_2 \times n)\text{-dimensional matrices,}$$
$$x \qquad n\text{-dimensional vector.}$$

Assume that the set determined by conditions (9.2) – (9.3) is bounded (it will be shown below that this assumption is not necessary).

Denote

$$S = \{x | A_2 x = b_2,\ x \geq 0\}\,.$$

We set forth theorems on which, together with the column generation procedure, the Dantzig-Wulf decomposition principle is based.

Theorem 9.1[19]**:** Let $X = \{x | Ax = b, x \geq 0\}$ be a nonempty constrained set and x^i $(i = \overline{1,r})$ its boundary points. Then any element $x \in X$ can be represented in the form

$$x = \sum_{i=1}^{r} z_i x^i,\ z_i \geq 0,\ i = 1, \dots, r\,;$$

$$\sum_{i=1}^{r} z_i = 1\,.$$

Theorem 9.2[19]**:** Let $X = \{x | Ax = b, x \geq 0\}$ be a nonempty set. A point of $x \in X$ if and only if it can be represented as the sum of a convex combination of boundary points and a linear combination with nonnegative coefficients of extreme rays of homogeneous solutions of the set X i.e.

$$x = \sum_i z_i x^i,$$

$$\sum_i z_i \delta_i = 1,\ z_i \geq 0$$

$$\delta_i = \begin{cases} 1\,, & x^i \text{ is a boundary point of } X\,, \\ 0\,, & x^i \text{ is a ray of the set } X\,. \end{cases}$$

Let $x^1, x^2, \ldots, x^N$ be extreme points of the set S. Then by theorem 9.1, any element of S can be represented as $x = \sum_{i=1}^{N} z_i x^i$, where

$$\sum_{i=1}^{N} z_i = 1, \qquad z_i \geq 0, \qquad i = \overline{1, N}. \tag{9.4}$$

Substituting (9.4) in (9.1), (9.2) we get

$$\mathcal{L} = \sum_{i=1}^{N} (cx^i) z_i,$$

$$\sum_{i=1}^{N} (A_1 x^i) z_i = b_1 .$$

Denote

$$p_i = A_1 x^i \tag{9.5}$$

$$f_i = cx^i . \tag{9.6}$$

The original LP problem is rewritten in the form:
minimize

$$\mathcal{L} = \sum_{i=1}^{N} f_i z_i \tag{9.7}$$

under

$$\sum_{i=1}^{N} p_i z_i = b_1 , \tag{9.8}$$

$$\sum_{i=1}^{N} z_i = 1 , \tag{9.9}$$

$$z_i \geq 0 , \quad i = 1, \ldots, N . \tag{9.10}$$

Problem (9.7) – (9.10) is called coordinating. It has $(m_1 + 1)$ rows as compared to $(m_1 + m_2)$ rows of the original problem and a large number of columns. In order not to store all these columns in the computer memory, they are obtained as needed using for these purposes the column generation method. To put it briefly, we call problem (9.7) – (9.10) the Z-problem, and the original – the X-problem.

§9.2. Consider the Z-problem solution. Assume that basis vectors of the Z-problem basic solution are known. For each i, calculate characteristic differences $\Delta_i = f_i - \Lambda \binom{p^i}{1}$, where Λ is the dual variable vector. Represent Λ as $\Lambda = (\Lambda_1, \Lambda_0)$, where Λ_1 corresponds to constraints (9.8), and Λ_0 to constraint (9.9). Using formulae (9.5) and (9.6), we write

$$\Delta_i = (C - \Lambda_1 A_1)x^i - \Lambda_0\,. \tag{9.11}$$

In compliance with the simplex procedure for determining a variable say, z_s, introduced in the basis, one has to obtain

$$\min_i \Delta_i = \Delta_s = (C - \Lambda_1 A_1)x^s - \Lambda_0\,. \tag{9.12}$$

Since an optimal solution of the LP problem, the constraint set of which is bounded, is achieved at the extreme point of this set, performing operation (9.12) is equivalent to solving a problem:
minimize

$$\mathcal{L}_\Lambda(x) = (C - \Lambda_1 A_1)x \tag{9.13}$$

with constraints

$$A_2 x = b_2, \qquad x \geq 0\,. \tag{9.14}$$

Problem (9.13) – (9.14) is referred to as the X_Λ-problem. Calculating the x^s solution of this problem enables one to obtain the column to be introduced into the Z-problem basis:

$$p_s = \begin{pmatrix} A_1 x^s \\ 1 \end{pmatrix}. \tag{9.15}$$

An appropriate coefficient in the goal function is given by

$$f_s = Cx^s\,. \tag{9.16}$$

That kind of approach is very effective if the number of independent diagonal blocks is more than one, i.e. the original problem can be represented in the form:

minimize

$$\mathcal{L} = c_1x_1 + c_2x_2 + \cdots + c_px_p \tag{9.17}$$

with constraints

$$A_1x_1 + A_2x_2 + \cdots + A_px_p = b_0\,, \tag{9.18}$$

$$\begin{aligned} B_1x_1 \qquad\qquad\qquad\qquad &= b_1\,,\\ B_2x_2 \qquad\qquad\quad &= b_2\,,\\ \ddots \qquad\qquad &\ \ \vdots\\ B_px_p &= b_p\,, \end{aligned} \tag{9.19}$$

$$x_i \geq 0\,, \qquad i = 1, \ldots, p\,.$$

Denote

$$S_t = \{x | B_tx = b_t,\ \ x \geq 0\},\ \ t = 1, \ldots, p\,.$$

By theorem 9.1, any element of S_t can be represented as

$$x^{(t)} = \sum_{j=1}^{N_t} z_j^{(t)} x_j^{(t)}\,; \tag{9.20}$$

$$\sum_{j=1}^{N_t} z_j^{(t)} = 1\,, \tag{9.21}$$

where N_t is the number of extreme points of the set S_t. Using (9.20), (9.21) and considerations, §9.1, the Z-problem is written to problem (9.17) – (9.19):
minimize

$$\sum_{t=1}^{p}\sum_{j=1}^{N_t} f_j^{(t)} z_j^{(t)} \tag{9.22}$$

under conditions

$$\sum_{t=1}^{p}\sum_{j=1}^{N_t} p_j^{(t)} z_j^{(t)} = b_0\,, \tag{9.23}$$

$$\sum_{j=1}^{N_t} z_j^{(t)} = 1\,, \qquad t = 1, 2, \ldots, p\,, \tag{9.24}$$

$$z_j^{(t)} \geq 0\,, \quad j = 1, 2, \ldots, N_t\,; \quad t = 1, 2, \ldots, p\,, \tag{9.25}$$

where

$$f_j^{(t)} = c_t x_j^{(t)}, \tag{9.26}$$

$$p_j^{(t)} = A_t x_j^{(t)}. \tag{9.27}$$

If the numbers of $z_j^{*(t)}$ constitute a solution to the Z-problem then the vectors

$$x^{*(t)} = \sum_{j=1}^{N_t} z_j^{(t)} x_j^{(t)}$$

constitute the optimal program

$$x^* = \left(x^{*(1)},\ x^{*(2)}, \dots, x^{*(p)}\right)$$

of the original problem. The X_Λ-problems associated with each step of the Z-problem solution fall into $p(t = \overline{1,p})$ of the $X_\Lambda^{(t)}$ problems which take the form:
minimize

$$f_\Lambda^{(t)} = (c_t - \Lambda_1 A_t)x^{(t)} \tag{9.28}$$

with constraints

$$B_t x^{(t)} = b_t, \qquad x^{(t)} \geq 0. \tag{9.29}$$

§**9.3.** When constructing the Z-problem we proceed from the assumption that the set determined by the conditions $\{A_2 x = b_2,\ x \geq 0\}$ is bounded. Only a fair modification of the Z-problem allows rejecting this assumption.

Consider the original problem (9.1) – (9.3). In the general case the set $S = \{A_2 x = b_2,\ x \geq 0\}$ is convex polyhedral. By theorem 9.2, §9.1, any point x of the set can be represented as

$$x = \sum_{i=1}^{N} z_i x_i, \tag{9.30}$$

$$\sum_{i=1}^{N} z_i \delta_i = 1,\ z_i \geq 0, \tag{9.31}$$

where

$$\delta_i = \begin{cases} 1, & \text{if } x^i \text{ is the boundary point of } S, \\ 0, & \text{if } x^i \text{ is the directing vector of the unbounded edge of the set } S. \end{cases}$$

In order to retain equivalence of the X-problem and the Z-problem one should replace condition (9.9) of the Z-problem by condition (9.31). Thus, in the general case, the Z-problem takes the form:
minimize

$$\mathcal{L} = \sum_{i=1}^{N} f_i z_i \tag{9.32}$$

under conditions

$$\sum_{i=1}^{N} p_i z_i = b_1 , \tag{9.33}$$

$$\sum_{i=1}^{N} z_i \delta_i = 1 , \qquad z_i \geq 0 , \qquad i = 1, \dots , N . \tag{9.34}$$

For each i the characteristic difference Δ_i is calculated from the relation

$$\Delta_i = (C - \Lambda_1 A_1) x^i - \Lambda_0 \delta_i . \tag{9.35}$$

Case 9.1: It corresponds either to the Z-problem solution or requires the inclusion, in the next basis, of the vector $(p_i, \delta_i)^T$ corresponding to the node of the set S.

Let x_{i^*} be an optimal program to the X_Λ-problem.

The inequalities are

$$\mathcal{L}_\Lambda(x_{i^*}) \leq \mathcal{L}_\Lambda(x_i), \qquad i = 1, 2, \dots , N_1 .$$

N_1 is the number of nodes of the set S, which together with relation (9.35) lead to the condition

$$\Delta_{i^*} \leq \Delta_i, \qquad i = 1, 2, \dots , N_1 .$$

We show that

$$\mathcal{L}_\Lambda(x_i) \geq 0, \qquad i = 1, 2, \dots , N , \tag{9.36}$$

where $x_{N_1+1}, \dots, x_N$ are the directing vectors of the unbounded edges of the set S. Indeed, let the unbounded edge corresponding to x_i come from the node x_{i^0}. If $\mathcal{L}_\Lambda(x_i) < 0$, then the value of the linear form $\mathcal{L}_\Lambda(x_{i_0} + \mu x_i) = \mathcal{L}_\Lambda(x_0) + \mu \mathcal{L}_\Lambda(x_i)$ on the program $(x_{i_0} + \mu x_i)$ unboundedly decreases as μ increases. Thus, violation of even one of the inequalities (9.36) contradicts solvability of the X_Λ-problem. Therefore for $i = N_1 + 1, \dots, N$; $\Delta_i \geq 0$. Consequently, in this case the vectors of conditions (p_i, δ_i) corresponding to the unbounded edges of the polyhedral set S should not be introduced into the basis. The equality $\Delta_{i^*} = 0$ signifies that the Z-problem is solved. If $\Delta_{i^*} < 0$, then the vector (p_{i^*}, δ_{i^*}) corresponding to the node x_{i^*} of the set S is introduced into the next basis.

Case 9.2: On some iteration the estimate Δ_j^Λ of a vector of the X_Λ-problem conditions proves to be negative and as it does so, all the constituents x_{ij} of a composition of this vector into basis vectors are nonpositive. Assume that the preceding basic program x_{i^0} has as basis vectors the first $\bar{p}$ of the X_Λ-problem vectors ($\bar{p}$ is the number of independent conditions for the X_Λ-problem), so that

$$x_{i^0} = (x_{10}, x_{20}, \dots, x_{\bar{p}0}, 0, \dots, 0)\,.$$

Consider the vector

$$x_i = (-x_{1j}, -x_{2j}, \dots, -x_{\bar{p}j}, 0, \dots, 1, \dots, 0)\,.$$

By constructing x_i,

$$A_2 x_i = 0\,.$$

Owing to the conditions $x_{ij} \leq 0$ it follows that $x_i \geq 0$. Therefore, the vector $x = x_{i0} + \mu x_i$ is the X_Λ-problem program with any $\mu \geq 0$. It means that the vector x_i is the directing vector of the unbounded edge of the set S.

Calculate the value of the functional $\mathcal{L}_\Lambda(x)$ on the vector x_i

$$\mathcal{L}_\Lambda(x_i) = c_j^\Lambda - \sum_{i=1}^{\bar{p}} c_i^\Lambda x_{ij}\,,$$

where c_i^Λ are the vector components

$$C_\Lambda = C - \Lambda_1 A_1\,.$$

On the other hand, by defining the vector estimate of the X_Λ-problem conditions we have

$$\Delta_j^\Lambda = -\sum_{l=1}^{P} c_i^\Lambda x_{ij} + c_j^\Lambda .$$

Therefore

$$\mathcal{L}_\Lambda(x_i) = \Delta_j^\Lambda .$$

Using relation (9.35) and the fact that for the directing vector x_i of the unbounded set $\delta_i = 0$ we obtain

$$\Delta_i = \mathcal{L}_\Lambda(x_i) = \Delta_j^\Lambda .$$

And since in case 9.2 the estimate Δ_j^Λ is negative, then it is possible to introduce into the next basis of the Z-problem the vector $\{A_1 x_i, 0\}$ corresponding to the unbounded edge of the X_Λ-problem definition domain.

§**9.4.** Formulate the two-level algorithm for solving problems (9.17) – (9.19) where at the first level, problems (9.28) – (9.29) are solved whereas at the second, the coordinating, i.e. the Z-problem is solved.

Step 0: Suppose that there exists the feasible basis solution of the Z-problem (9.22) – (9.25). If there is agreement between the basis matrix B and the vector of estimates of the constraints (Λ_1, Λ_0).

Step 1: Using the estimates Λ_1 we solve problems (9.28) – (9.29) for each $t = 1, 2, \dots, p$, derive their solutions $x^{(t)}(\Lambda_1)$ and optimal values of the goal functions $\mathcal{L}_\Lambda^{(t)}(x^{(t)}) = z_t^0$ $(t = 1, 2, \dots, P)$.
Let

$$x^*(\Lambda_1) = \left(x^{*(1)}(\Lambda_1), \dots, x^{*(p)}(\Lambda_1)\right).$$

Step 2: Calculate characteristic differences of the vectors $p_j^{(t)}, \delta_j^{(t)} l_t$ for the fixed t:

$$\Delta_{i*}^{(t)} = \begin{cases} \mathcal{L}_\Lambda^{(t)}\left(x^{*(t)}\right) - \Lambda_{0(t)} , \\ \mathcal{L}_\Lambda^{(t)}(x^*(t)) , \end{cases}$$

the $X_\Lambda^{(t)}$-problem is solvable, the linear form of the $X_\Lambda^{(t)}$-problem is unbounded.

Step 3: If

$$\min_{t=1,\dots,p} \Delta_{i*}^{(t)} \geq 0 ,$$

then the Z-problem basic program understudy constitutes its solution (because here all the condition vectors have nonnegative estimates). Go to step 5.

<u>*Step 4*</u>: If $\min\limits_{t=1,\dots,p} \Delta_{i^*}^{(t)} < 0$, then we introduce into the Z-problem basis the vector $\left(p_{i^*}^{(k)}, \delta_{i^*}^{(k)} l_k\right)$ for which

$$\Delta_{i^*}^{(k)} = \min_{t=1,\dots,p} \Delta_{i^*}^{(t)} .$$

The vector components are calculated with the formula

$$\begin{pmatrix} p_{i^*}^{(k)} \\ \delta_{i^*}^{(k)} l_k \end{pmatrix} = \begin{pmatrix} A_k x^{*(k)} \\ \delta_{i^*}^{(k)} l_k \end{pmatrix} .$$

The column obtained is multiplied by B^{-1} and with the use of the standard leading operation we define a new inverse matrix and the vector of constraint estimates. Go to step 1.

<u>*Step 5*</u>: Define the solution of problem (9.17) – (9.19)

$$x^* = \left(x^{*(1)}, \dots, x^{*(p)}\right),$$

where

$$x^{*(t)} = \sum_{j=1}^{N_t} z_j^{(t)} x_j^{*(t)} .$$

$z_j^{(t)}$ are the Z-problem solutions.

Terminate.

If the coordinating problem is nondegenerate, then the goal function value decreases at each iteration. Since there is only a finite number of possible bases and none of them is used twice, employment of the Dantzig-Wulf decomposition method will lead to an optimal solution for a finite number of iterations.

§10. DUAL APPROACH IN BLOCK PROGRAMMING

§10.1. In §9 we considered one of the possible approaches (the Dantzig-Wulf decomposition method) to solving block structure problems. The present paragraph sets forth the dual formulation of the block programming problem and reveals a sufficiently general dual approach to the analysis of block problem.[11]

Consider the problem:
maximize

$$\mathcal{L} = Cx \tag{10.1}$$

with constraints

$$A_1 x = b_1\,, \tag{10.2}$$

$$A_2 x = b_2\,, \tag{10.3}$$

$$x \geq 0\,, \tag{10.4}$$

where C, x are n-dimensional vectors, A_1, A_2 – $(m_1 \times n)$, $(m_2 \times n)$-dimensional matrices, b_1, $b_2 - m_1$, m_2-dimensional vectors.

As in §9, S is referred to as a polyhedral set of points defined by conditions (10.3), (10.4).

Problem (10.1) – (10.4) generates the Z-problem:
maximize

$$\mathcal{L} = \sum_{i=1}^{N} f_i z_i \tag{10.5}$$

under

$$\sum_{i=1}^{N} p_i z_i = b_1\,, \tag{10.6}$$

$$\sum_{i=1}^{N} \sigma_i z_i = 1\,, \tag{10.7}$$

$$z_i \geq 0\,, \qquad i = 1, \ldots, N\,, \tag{10.8}$$

where p_i, f_i are defined by formulae (9.5) – (9.6), §9,

$$\sigma_i = \begin{cases} 1, & \text{if } x_i \text{ is the node of the set } S, \\ 0, & \text{if } x_i \text{ is the directing vector of the unbounded edge of the set } S. \end{cases}$$

Enumerate the elements of x_i so that the first N_1 of vectors x_i might correspond to the nodes of the set S whereas the remaining $(N - N_1)$ to

the unbounded edges of this set. Introduce for consideration the $\tilde{Z}$-problem dual with respect to the Z-problem (10.1) – (10.4):
minimize

$$b_1\Lambda_1 + \Lambda_0 \tag{10.9}$$

under conditions

$$p_i\Lambda_1 + \Lambda_0 \geq f_i, \quad i = 1, 2, \ldots, N_1\,, \tag{10.10}$$

$$p_i\Lambda_1 \geq f_i, \quad i = N_1 + 1, \ldots, N\,, \tag{10.11}$$

where $\Lambda_1 = (\lambda_1, \lambda_2, \ldots, \lambda_{m_1})$.
Since linear form (10.9) is subject to minimization, then, with the fixed Λ_1 satisfying (10.11), the number Λ_0 has to be selected as small as possible.

Replace the system of conditions (10.10) by the condition

$$\Lambda_0 = \max_{i=\overline{1,N_1}} \{f_i - p_i\Lambda_1\}\,. \tag{10.12}$$

From the definitions of the number f_i and the vector p_i we have

$$f_i - p_i\Lambda_1 = Cx^i - A_1x^i\Lambda_1 = (c - A_1\Lambda_1, x^i) = c_\Lambda x^i\,,$$

where $C_\Lambda = C - \Lambda_1 A_1$.

Rewrite relation (10.12) as

$$\Lambda_0 = \max_{i=\overline{1,N_1}} C_\Lambda x^i\,. \tag{10.13}$$

Thus, the problem of function minimization

$$b_1\Lambda_1 + \max_{i=\overline{1,N_1}} C_\Lambda x^i \tag{10.14}$$

under conditions (10.11) is equivalent to the $\tilde{Z}$-problem.

§**10.2.** Let R be the set of points Λ_1 for which

$$\max_{x\in S} C_\Lambda x < \infty\,. \tag{10.15}$$

It will be shown that R is a convex polyhedral set to which a system of conditions (10.11) corresponds. Condition (10.11) is written in the equivalent form:

$$C_\Lambda x^i \leq 0, \quad i = N_1 + 1, \ldots, N\,. \tag{10.16}$$

By theorem 9.2, §9, any point of $x \in S$ can be represented as

$$x = \sum_{i=1}^{N} z_i x^i, \quad \sum_{i=1}^{N_1} z_i = 1, \quad z_i \geq 0, \quad i = 1, \ldots, N. \tag{10.17}$$

Because of this, under conditions (10.16),

$$C_\Lambda x \leq \max_{i=\overline{1,N_1}} C_\Lambda x^i < \infty. \tag{10.18}$$

If for a number of $i_0 \in \{N_1 + 1, \ldots, N\}$

$$C_\Lambda x^{i_0} > 0,$$

then assuming that $x(\theta) = x^1 + \theta x^{i_0}$, we have

$$\lim_{\theta \to \infty} C_\Lambda x(\theta) = \infty$$

and since $x(\theta) \in S$ for any $\theta \geq 0$, then

$$\sup_{x \in S} C_\Lambda x = \infty.$$

Thus, the set R is defined by a system of inequalities (10.11). If S is a convex polyhedron, then $N_1 = N$, i.e. conditions (10.11) are absent. In this case R coincides with the whole m_1-dimensional space of points Λ_1. If R contains none of the points, then this case is equivalent to the inconsistency of the $\tilde{Z}$ problem conditions. Below R is assumed to be a nonempty set.

Introduce into consideration the function $\varphi(\Lambda_1)$ defined on the set R by the relation

$$\varphi(\Lambda_1) = b_1 \Lambda_1 + \max_{x \in S} C_\Lambda x. \tag{10.19}$$

Let $\Lambda_1 \in R$. From relation (10.17) we get

$$\varphi(\Lambda_1) - b_1 \Lambda_1 = \max_{x \in S} C_\Lambda x = \max_{i=\overline{1,N_1}} C_\Lambda x^i. \tag{10.20}$$

Therefore problem (10.14), under conditions (10.11) being equivalent to the $\tilde{Z}$-problem, reduces to minimization of the function $\varphi(\Lambda_1)$ defined on the set R.

If $\inf\limits_{\Lambda_1 \in R} \varphi(\Lambda_1) = -\infty$, then the Z-problem (10.5) – (10.8) has no program, i.e. conditions (10.2) – (10.4) of the original problem are inconsistent. If $\inf\limits_{\Lambda_1 \in R} \varphi(\Lambda_1) = \varphi(\Lambda_1^*) > -\infty$, then by (10.13) and (10.20) the vector

$$\bar{\Lambda}^* = (\Lambda_1^*, \Lambda_0^*),$$

where $\Lambda_0^* = \varphi(\Lambda_1^*) - b_1\Lambda_1^*$ is the solution of the $\tilde{Z}$-problem. In the (m_1+1)-dimensional space of the points $(\Lambda_1, \lambda_{m_1+1})$ each node x^i $(i = 1, 2, \ldots, N_1)$ of the set S generates the hyperplane H_i defined by the equation

$$\lambda_{m_1+1} + (A_1 x_i - b_1)\Lambda_1 = Cx^i \tag{10.21}$$

or, which is the same,

$$\lambda_{m_1+1} = C_\Lambda x^i + b_1\Lambda_1 . \tag{10.22}$$

In the same space the directing vector x^i $(i = N_1 + 1, \ldots, N)$ of each unbounded edge of the set S generates the hyperplane H_i defined by the equation

$$(A_1 x^i)\Lambda_1 = Cx^i \tag{10.23}$$

or, which is the same,

$$C_\Lambda x^i = 0 . \tag{10.24}$$

Define in the space under consideration the set $\bar{R}$ composed of the points $(\Lambda_1, \lambda_{m_1+1})$ for which

$$\lambda_{m_1+1} \geq \varphi(\Lambda_1), \qquad \Lambda_1 \in R . \tag{10.25}$$

By virtue of (10.20) the first condition from (10.25) can be written as

$$\lambda_{m_1+1} \geq C_\Lambda x^i + b_1\Lambda_1 , \qquad i = 1, 2, \ldots, N_1 . \tag{10.26}$$

The second condition of (10.25) is equivalent to conditions (10.11) or which is the same as (10.16).

From the conditions of (10.25) it follows that the problem of minimizing the function $\varphi(\Lambda_1)$ on the set R is equivalent to selecting the "lowest" (in the sense of the axis λ_{m_1+1}) point of the set $\bar{R}$. The surface

$$\lambda_{m_1+1} = \varphi(\Lambda_1)$$

coincides with part of the boundary of the convex polyhedral set $\bar{R}$. Therefore $\varphi(\Lambda_1)$ is a piecewise linear convex function.

Theorem 10.3: Analyzing the $\tilde{Z}$-problem is equivalent to:

(1) minimizing the piecewise linear convex function $\varphi(\Lambda_1)$ specified on the convex polyhedral set R where $\varphi(\Lambda_1)$, R are defined by conditions (10.19) and (10.16); or

(2) selecting the "lowest" (in the sense of the axis λ_{m_1+1}) point of the polyhedral set $\bar{R}$.

The problem of minimizing the function $\varphi(\Lambda_1)$ on the set R or, which is the same, the problem of seeking for the lowest point of the set $\bar{R}$ is the dual formulation of problem (10.1) – (10.4).

§**10.3.** Consider the dual formulation of block problem (10.1) – (10.4).

The difficulty of finding a minimum of the function $\varphi(\Lambda_1)$ resides in the fact that:

(1) $\varphi(\Lambda_1)$ is given as an algorithm (finding its value at the point Λ_1 requires solving the X_Λ-problem),

(2) $\varphi(\Lambda_1)$ is differentiable not at all points and, consequently, smooth function minimization techniques cannot be applied here.

Consider a general scheme of minimizing the function $\varphi(\Lambda_1)$. Select a point of $\Lambda' \in R$ and try to find a possible direction so that:

(1) $\Lambda' + \varepsilon l \in R$ with a sufficiently small $\varepsilon > 0$,

(2) the value of the function $\varphi(\Lambda_1)$ decreases along l.

The set of such directions is denoted by $\mathcal{L}_\Lambda$. Assume that there exists a particular technique T of selecting $l \in \mathcal{L}_\Lambda$ for a nonempty set $\mathcal{L}$: With the aid of the technique T we isolate the direction of $l \in \mathcal{L}_\Lambda$ or make sure that the set $\mathcal{L}_\Lambda$ is empty. In the latter case Λ' is the desired point of the minimum of the function $\varphi(\Lambda_1)$. If the direction of $l \in \mathcal{L}_\Lambda$ is found, shift to the point

$$\Lambda = \Lambda' + \Theta l\,.$$

The choice of the number of $\Theta > 0$ will obey the requirement: the function of the scalar argument t

$$\psi(t) = \varphi(\Lambda' + \Theta l)$$

must monotonously decrease with $0 \le t \le \Theta$. The technique of selecting the number of Θ is denoted by E. Each pair of the techniques T, E determines a block programming method. On the basis of these methods we can also formulate combined procedures for block programming when in the process of solving a problem, different techniques T, E are employed.

§10.4. By way of example we consider the $\varphi(\Lambda_1)$ minimization procedure based on the employment of the generalized gradient lowering.

Thus, we have problem (10.1) – (10.4). Assume that the set $S = \{x | A_2 x = b_2,\ x \ge 0\}$ is bounded. If this is not the case, it is always possible to add to system (10.3) the inequality of the form $\sum\limits_{i=1}^{n} x_i \le M$ and select M so large that even one of the optimal solutions of problem (10.1) – (10.4) will satisfy this inequality. Form the Lagrange function $\mathcal{L}(\Lambda, x)$, $\Lambda = (\lambda_1, \dots, \lambda_{m_1})$ for problem (10.1), (10.2) and consider the following problem:

$$\min_{\Lambda \ge 0} \max_{x \in S} \mathcal{L}(\Lambda, x) = \min_{\Lambda \ge 0} \max_{x \in S} \left[\sum_{i=1}^{n} c_i x_i + \sum_{j=1}^{m_1} \lambda_j \left(b_j^1 - \sum_{i=1}^{n} a_{ij}^1 x_i \right) \right], \quad (10.27)$$

where $(b_1^1, \dots, b_{m_1}^1) = b_1$; a_{ij}^1; $i = \overline{1,n}$; $j = \overline{1,m_1}$ are the elements of the matrix A_1. Study the function separately

$$\mathcal{L}^*(\Lambda) = \max_{x \subset S} \left[\sum_{i=1}^{n} c_i x_i + \sum_{j=1}^{m_1} \lambda_j \left(b_j^1 - \sum_{i=1}^{n} a_{ij}^1 x_i \right) \right]. \quad (10.28)$$

Evidently, the function $\mathcal{L}^*(\Lambda)$ is the function $\varphi(\Lambda)$ described above. Inasmuch as the polyhedron S is bounded, the function $\mathcal{L}^*(\Lambda)$ is defined for any Λ. Owing to applying maximum operation to the convex set of linear forms the function $\mathcal{L}^*(\Lambda)$ is convex. And since S is a polyhedron, $\mathcal{L}^*(\Lambda)$ is a piecewise linear function. For $\Lambda \ge 0$ and $x = \bar{x}$ where $\bar{x} = (\bar{x}_1, \dots, \bar{x}_n)$ is the feasible solution of problem (10.1) – (10.4)

$$\mathcal{L}^*(\Lambda) \ge \mathcal{L}(\Lambda, \bar{x}) = \sum_{i=1}^{n} c_i \bar{x}_i + \sum_{j=1}^{m_1} \lambda_j \left(b_j^1 - \sum_{i=1}^{n} a_{ij}^1 x_i \right) \ge \sum_{i=1}^{n} c_i \bar{x}_i .$$

Thus, $\mathcal{L}^*(\Lambda)$ is bounded below for $\Lambda \ge 0$, and since $\mathcal{L}^*(\Lambda)$ is piecewise linear, there exists Λ^* such that

$$\mathcal{L}^*(\Lambda^*) = \min_{\Lambda \ge 0} \mathcal{L}^*(\Lambda) .$$

Consider the problem dual to problem (10.1) – (10.4)

$$\min\left(\sum_{i=1}^{m_1} b_i^1\lambda_i+\sum_{k=1}^{m_2} b_k^2\sigma_k\right), \tag{10.29}$$

$$\sum_{i=1}^{m_1} a_{ij}^1\lambda_i+\sum_{k=1}^{m_2} a_{kj}^2\sigma_k \geq c_j, \quad j=\overline{1,n}\,, \tag{10.30}$$

$$\lambda_i \geq 0, \quad i=\overline{1,m_1}; \quad \sigma_k \geq 0, \quad k=\overline{1,m_2}\,. \tag{10.31}$$

Lemma: An optimal solution to problem (10.29) – (10.31) of the form

$$(\Lambda^*,\Sigma^*)=(\lambda_1^*,\dots,\lambda_{m_1}^*,\sigma_1^*,\dots,\sigma_{m_2}^*)$$

corresponds to any optimal solution of the problem (10.27). Proof.[41]
From this lemma it follows that the solution of problem (10.29) – (10.31) reduces to minimization of the function $\mathcal{L}^*(\Lambda)$ with the constraints $\Lambda \geq 0$. If $X^*(\Lambda)=(x_1^*(\Lambda),\dots,x_n^*(\Lambda))$ is the vector on which a maximum in (10.28) is achieved, then

$$\mathcal{L}^*(\Lambda)=\sum_{i=1}^{n} c_i x_i^*(\Lambda)+\sum_{j=1}^{m_1}\lambda_j\left(b_j^1-\sum_{i=1}^{n} a_{ij}^1 x^*(\Lambda)\right). \tag{10.32}$$

From expression (10.32) it follows that the generalized gradient lowering direction at the point Λ coincides with the direction opposite to that of the vector

$$\bar{\sigma}_\Lambda=\left\{b_j^1-\sum_{i=1}^{n} a_{ij}^1 x_i^*(\Lambda)\right\}, \quad j=\overline{1,m_1}\,. \tag{10.33}$$

Formulate an algorithm for solving the problem (10.29) – (10.31) dual with respect to the original problem (10.1) – (10.4).

Step 1: Let the iteration number $l=1$, fix the value $\Lambda^{(1)}$, in particular, $\Lambda^{(1)}$ can be taken as equal to 0.

Step 2: Solve problem (10.28), obtain the vector $X^*(\Lambda^{(l)})$ and the vector of an optimal solution to the dual problem $\sum(\Lambda^{(l)})$.

Step 3: Calculate the vector $\bar{\sigma}_l\ \{(b_j^1-\sum\limits_{i=1}^{n} a_{ij}^1 x_i^*(\Lambda^{(l)})\}, \quad j=\overline{1,m_1}$.

Step 4: Derive $\Lambda^{(l+1)}$ from the formula

$$\Lambda^{(l+1)}=p\left\{\Lambda^{(l)}-k_l\frac{\bar{\sigma}_l}{|\bar{\sigma}_l|}\right\},$$

where $k_l > 0$ is the value of the step along the lowering direction, p is the operator leaving nonnegative coordinates without any change and transforming negative coordinates into 0. Go to step 2.

If k_l is selected so that $k_l \underset{l\to\infty}{\longrightarrow} 0$, $\sum_{l=1}^{\infty} k_l = \infty$, then, in this case, converging $\mathcal{L}(\Lambda^{(l)})$ to an optimal value of problem (10.27) is guaranteed:

$$\mathcal{L}^*(\Lambda^{(l)}) \underset{l\to\infty}{\longrightarrow} \mathcal{L}^*(\Lambda^*)\,.$$

Here the pair of vectors $(\Lambda^{(l)}, \Sigma(\Lambda^{(l)}))$ converges to an optimal solution of problem (10.29) – (10.31).

Terminate.

§11. TRANSPORTATION PROBLEM SOLUTION BY THE DECOMPOSITION METHOD

§11.1. One of the most interesting special problems of linear programming is the transportation problem, the classical formulation of which takes the form:

$$\sum_{j=1}^{n}\sum_{i=1}^{m} c_{ij}x_{ij} \to \min\,, \tag{11.1}$$

$$\sum_{j=1}^{n} x_{ij} = a_i\,, \qquad i = \overline{1,m}\,, \tag{11.2}$$

$$\sum_{i=1}^{m} x_{ij} = b_j\,, \qquad j = \overline{1,n}\,, \tag{11.3}$$

$$x_{ij} \geq 0\,, \quad i = \overline{1,m}\,; \quad j = \overline{1,n}\,. \tag{11.4}$$

We consider the transportation problem (11.1) – (11.4) as the Block programming problem in which conditions are partitioned into two blocks. In the notation in §9, conditions (11.2) correspond to the submatrix A_1, and conditions (11.3) to the submatrix A_2. The decomposition method was used by Williams[8] to solve the transportation problems of large dimensionality. Following the scheme of the decomposition method, consider the set S prescribed by conditions (11.3) – (11.4). This set represents a polyhedron with the nodes that are obtained if for each j one of x_{ij} is assumed to be equal to b_j, the rest being equal to null. Denote the quantity of nodes $N = m^n$ by

$$x^k = \{x_{ij}^k\}, \qquad k = \overline{1, N}.$$

An arbitrary point of the polyhedron S is represented as:

$$x = \sum_{k=1}^{N} z_k x^k, \tag{11.5}$$

$$\sum_{k=1}^{N} z_k = 1, \qquad z_k \geq 0, \qquad k = \overline{1, N}. \tag{11.6}$$

Substitute expression (11.5) into conditions (11.1) and (11.2). Obtain the coordinating or Z-problem:

$$\begin{gathered} \sum_{k=1}^{N} f_k z_k \to \min, \\ \sum_{k=1}^{N} p_k z_k = a, \\ z_k \geq 0, \quad k = \overline{1, N}, \end{gathered} \tag{11.7}$$

where

$$a = (a_1, a_2, \dots, a_m)^T,$$

$$f_k = \sum_{j=1}^{n} \sum_{i=1}^{m} c_{ij} x_{ij}^k, \qquad k = \overline{1, N}$$

and the p_k vector components are determined from the equality

$$p_k^i = \sum_{j=1}^{n} x_{ij}^k, \qquad k = \overline{1, N}, \qquad i = \overline{1, m}. \tag{11.8}$$

The convexity condition $\sum_{k=1}^{N} z_k = 1$ for the transportation problem is the consequence of equalities in (11.7). Indeed, summing equalities in (11.7) by rows we get

$$\sum_{i=1}^{m} \sum_{k=1}^{N} p_k^i z_k = \sum_{i=1}^{m} a_i.$$

Employing (11.8), we have

$$\sum_{k=1}^{N}\sum_{i=1}^{m}\sum_{j=1}^{n} x_{ij}^{k} z_k = \sum_{i=1}^{m} a_i$$

and since x_{ij}^k satisfies conditions (11.3) and (11.4), we derive the equality

$$\sum_{k=1}^{N}\Big(\sum_{i=1}^{m} a_i\Big) z_k = \sum_{i=1}^{m} a_i$$

from which the condition for convexity follows.

The X_Λ-problem is constructed according to the scheme of the decomposition method. The linear form appears as:

$$\mathcal{L}_\Lambda(x) = \sum_{j=1}^{n}\sum_{i=1}^{m} (c_{ij} - \lambda_i) x_{ij} , \qquad (11.9)$$

where $\Lambda = (\lambda_1, \dots, \lambda_m)$ is a vector of dual variables. The X_Λ-problem definition domain is determined by conditions (11.3) and (11.4).

From the goal function inseparability (11.9) and independence of constraints (11.3) it follows that the X_Λ-problem falls into n independent subproblems of the form:

$$\begin{aligned} &\sum_{i=1}^{m} (c_{ij} - \lambda_i) x_{ij} \to \min , \\ &\sum_{i=1}^{m} x_{ij} = b_j , \\ &x_{ij} \geq 0 . \end{aligned} \qquad (11.10)$$

The complete optimal solution of the X_Λ-problem constitutes a vector in which for each j only one component is different from null and equal to b_j. The number S of the nonzero component is calculated for each j from the relation

$$c_{Sj} - \lambda_S = \min_{i=\overline{1,m}} (c_{ij} - \lambda_i) .$$

The vector p_{k^*} of the Z-problem conditions corresponds to the solution x^{k_*} of the X_Λ-problem. The estimate of this vector with respect to the current basis is equal to the value of linear form (11.9) calculated on the optimal solution of the X_Λ-problem:

$$\Delta_{k^*} = \sum_{j=1}^{n}\sum_{i=1}^{m}(c_{ij} - \lambda_i)x_{ij}^{k_*} . \tag{11.11}$$

If $\Delta_{k^*} = 0$, then the Z-problem basis under study is optimal, and an optimal solution to the original transportation problem is expressed in the form

$$x^* = \sum_{k=1}^{N} x^k z_k^* ,$$

where z_k^* is the optimal solution of the Z-problem. If $\Delta_{k^*} < 0$, the iteration process should be continued, here the vector p_{k^*} with the minimal characteristic difference Δ_{k^*} is introduced into the Z-problem basis. The p_{k^*} vector components are calculated by the formula

$$p_{k^*}^i = \sum_{j=1}^{n} x_{ij}^{k_*} , \qquad i = \overline{1,m} . \tag{11.12}$$

The algorithm described on a basis of the decomposition method has proved to be particularly effective for transportation problems under condition $n \ll m$.

§**11.2.** Formulate the solution algorithm for problem (11.1) – (11.4).

<u>*Step 0*</u>: Assume that there exists a feasible basic solution to the Z-problem (11.7). It corresponds to the basic matrix and the estimate vector of constraints $\Lambda = (\lambda_1, \ldots, \lambda_m)$.

<u>*Step 1*</u>: When employing the estimate vector Λ we solve problems (11.10) for each $j = \overline{1,n}$, i.e. for each j calculate the number S_j:

$$c_{S_j j} - \lambda_S = \min_{i=\overline{1,m}} (c_{ij} - \lambda_i) .$$

The complete optimal solution of the X_Λ-problem is given by

$$(\underbrace{0,\ldots,\overset{s_1}{b_1},\ldots,0}_{m},\underbrace{0,\ldots,\overset{s_2}{b_2},\ldots,0}_{m},\ldots,\underbrace{0,\ldots,\overset{s_n}{b_n},\ldots,0}_{m}) .$$

Step 2: From formula (11.11) calculate the estimate Δ_{k^*} with respect to the current basis of the vector p_{k^*} corresponding to the solution of the X_Λ-problem.

Step 3: If $\Delta_{k^*} \geq 0$, then the current basic program z_k^* of the Z-problem is optimal. Go to step 5.

Step 4: If $\Delta_{k^*} < 0$, then we introduce into the Z-problem basis the vector p_{k^*} the components of which are calculated with formula (11.12). The column obtained is multiplied by the inverse basic matrix, and a new inverse matrix and a vector of dual variables Λ are determined with the use of a standard leading operation. Go to step 1.

Step 5: Define the solution of problem (11.1) – (11.4):

$$x^* = \sum_{k=1}^{N} x^k z_k^* ,$$

where z_k^* is an optimal solution to the Z-problem.

Terminate.

§11.3. The decomposition method scheme considered above has been applied for the multi-index transportation problem.[10] For example, the three-index transportation problem takes the form:

$$\sum_{i_3=1}^{n_3} \sum_{i_2=1}^{n_2} \sum_{i_1=1}^{n_1} c_{i_1 i_2 i_3} x_{i_1 i_2 i_3} \to \min , \tag{11.13}$$

$$\sum_{i_2=1}^{n_2} \sum_{i_1=1}^{n_1} x_{i_2 i_1 i_3} = a_{i_3}^3 , \quad i_3 = \overline{1, n_3} , \tag{11.14}$$

$$\sum_{i_3=1}^{n_3} \sum_{i_1=1}^{n_1} x_{i_1 i_2 i_3} = a_{i_2}^2 , \quad i_2 = \overline{1, n_2} , \tag{11.15}$$

$$\sum_{i_3=1}^{n_3} \sum_{i_2=1}^{n_2} x_{i_1 i_2 i_3} = a_{i_1}^1 , \quad i_1 = \overline{1, n_1} , \tag{11.16}$$

$$x_{i_1 i_2 i_3} \geq 0 , \quad i_1 = \overline{1, n_1} ; \quad i_2 = \overline{1, n_2} ; \quad i_3 = \overline{1, n_3} . \tag{11.17}$$

Suppose that $n_3 \ll n_2$ and $n_3 \ll n_1$. Isolate constraints (11.15) – (11.17) and consider a polyhedron with their corresponding nodes $x^k_{i_1 i_2 i_3}$. The coordinating problem is formed on conditions (11.14), and the X_Λ-problem resides in minimizing the linear form

$$\mathcal{L}_\Lambda(x) = \sum_{i_3=1}^{n_3} \sum_{i_2=1}^{n_2} \sum_{i_1=1}^{n_1} (c_{i_1 i_2 i_3} - \lambda_{i_3}) x_{i_1 i_2 i_3}$$

under conditions (11.15) – (11.17). Solving the X_Λ-problem reduces in turn, to finding an optimal solution to the classical transportation problem of $n_1 \times n_2$ dimensionality. Thus, the k-index transportation problem reduces to the linear programming problem of a general form and the sequence of $(k-1)$-index transportation problems.

§12. DECOMPOSITION FOR PROBLEMS WITH A BLOCK-STAIRCASE STRUCTURE

§12.1. Consider the algorithm for solving LP problem with a block-staircase structure of the matrix of conditions[20] using the Dantzig-Wulf decomposition scheme. Consider the problem

$$\begin{gathered} \max\, Cx_n\,, \\ A_1 x_1 \le b\,, \\ A_j x_j \le B_j x_{j-1}\,, \quad j = 2, \dots, n\,, \\ x_j \ge 0\,, \quad j = 1, 2, \dots, n\,, \end{gathered} \tag{12.1}$$

where x_i are variable vectors, dimensions of matrices A_j, B_{j_1} ($j = \overline{1,n}$; $j_1 = \overline{2,n}$) correlate well.

Construct the problem dual to (12.1)

$$\begin{gathered} \min\, bu_1\,, \\ u_j A_j \ge u_{j+1} B_j\,, \quad j = \overline{1, n-1}\,, \\ u_n A_n \ge C\,, \\ u_j \ge 0\,, \quad j = \overline{1,n}\,. \end{gathered} \tag{12.2}$$

Assume that problem (12.1) has a feasible solution. We set forth the algorithm for constructing an optimal solution to problem (12.1) by the given feasible solution. The algorithm comprises n steps. Let $x_1^0, \dots, x_n^0$ be a feasible solution of (12.1). The algorithm step k results in constructing feasible solution (1) $x_1^k, \dots, x_n^k$, that $x_j^k = x_j^0$, $j = \overline{1, n-k}$; x_j^k, $j = \overline{n-k+1, n}$

constitute the optimal solution of the problem

$$\max\ Cx_n\,,$$
$$A_{n-k+1}x_{n-k+1} \le B_{n-k+1}x^0_{n-k}, \dots, \tag{12.3}$$
$$\dots, A_n x_n \le B_{n-1}x_{n-1}$$

as well as the optimal solution u^k_j, $j = \overline{n-k+1, n}$ of the problem dual to (12.3).

§**12.2.** Describe in detail separate steps of the algorithm and their connections. Each step involves a finite number of iterations.

<u>*Step 1*</u>: Solve the problem

$$\max\ Cx_n\,,$$
$$A_n x_n \le B_n x^0_{n-1}\,,$$
$$x_n \ge 0\,.$$

Let $\bar{x}_n$ be a basis solution of the problem, and $\bar{u}_n$ a solution to the dual problem. Let $x^1_j = x^0_j$, $j = \overline{1, n-1}$, $x^1_n = \bar{x}_n$, $u^1_n = \bar{u}_n$ and go to step 2.

Before formally describing step 2, we formulate the problem to be solved at this step:

$$\max\ Cx_n\,,$$
$$A_{n-1}x_{n-1} \le B_{n-1}x^1_{n-2}\,,$$
$$A_n x_n \le B_n x_{n-1}\,, \tag{12.4}$$
$$x_n, x_{n-1} \ge 0\,.$$

We apply the Dantzig-Wulf decomposition method to solve this problem as well as a similar problem at step k.

Let x^1_{n-1}, x^1_n be a feasible solution of (12.4). Assume that all the polyhedral sets under consideration are bounded. Let $x^{1,i}_{n-1}$ $(i = \overline{1, N})$ be all the extreme points of the polyhedron $A_{n-1}x_{n-1} \le B_{n-1}x'_{n-2}$. Its arbitrary point can be represented then as

$$x_{n-1} = \sum_{i=1}^{N} z^{1,i}_{n-1} x^{1,i}_{n-1}\,,$$

$$\sum_{i=1}^{N} z^{1,i}_{n-1} = 1\,, \quad z^{1,i}_{n-1} \ge 0\,, \quad i = \overline{1, N}\,.$$

(Here the first upper index belongs to the step number.) We shall present the point x_{n-1} in a somewhat different way

$$x_{n-1} = \sum_{i=0}^{N} z_{n-1}^{1,i} x_{n-1}^{1,i} ,$$

where $x_{n-1}^{1,0} = x_{n-1}^{1}, \sum_{i=0}^{N} z_{n-1}^{1,i} = 1$

or

$$x_{n-1} = x_{n-1}^{1} - \sum_{i=1}^{N} z_{n-1}^{1,i} (x_{n-1}^{1} - x_{n-1}^{1,i}) , \sum_{i=1}^{N} z_{n-1}^{1,i} = 1 .$$

Thus, the coordinating problem takes the form

$$\begin{gathered} \max \, Cx_n , \\ A_n x_n + B_n \sum_{i=1}^{N} z_{n-1}^{1,i} (x_{n-1}^{1} - x_{n-1}^{1,i}) \leq x_{n-1}^{1} , \\ \sum_{i=1}^{N} z_{n-1}^{1,i} = 1 , \\ x_n \geq 0, \quad z_{n-1}^{1,i} \geq 0, \quad i = \overline{1,N} . \end{gathered} \tag{12.5}$$

The vector $x_n = x_n^1$ is the basis solution (5), u_n^1 – its corresponding vector of dual variables. The optimality criterion for a feasible solution is constituted by fulfillment of inequalities.

$$\Delta_i = (x_{n-1}^{1} - x_{n-1}^{1,i}) u_n^1 \geq 0, \quad i = \overline{1,N}$$

or

$$\min_{i=\overline{1,N}} \Delta_i = \min_{i=\overline{1,N}} (x_{n-1}^{1} - x_{n-1}^{1,i}) u_n^1 \geq 0 .$$

Testing the latter condition by the Dantzig-Wulf method can be reduced to solving an LP problem of the form

$$\begin{gathered} \max \, x_{n-1} u_n^1 , \\ A_{n-1} x_{n-1} \leq B_{n-1} x_{n-2}^{1} , \\ x_{n-1} \geq 0 . \end{gathered} \tag{12.6}$$

The objective function value obtained is compared to $x^1_{n-1}u^1_n$.

Continue describing algorithm steps. The number enclosed in parentheses corresponds to the number of iteration at the step.

Step 2(0): Obtain the optimal solution $x^{1,1}_{n-1}$ of problem (12.6) and the dual variables $u^{1,1}_{n-1}$. If $x^{1,1}_{n-1}u^1_n \leq x^1_{n-1}u^1_n$, then let $x^2_j = x^1_j$; $j = \overline{1,n}$; $u^2_n = u^1_n$; $u^2_{n-1} = u^{1,1}_{n-1}$ and go to step 3, otherwise return to step 1, i.e. continue solving the coordinating problem (12.5) and introduce into its basis a new vector $(x^1_{n-1} - x^{1,1}_{n-1})$. Returning to step 1 results in solving problem (12.5) $\bar{x}^1_n$, $\bar{z}^{1,1}_{n-1}$ and obtaining its corresponding dual variables $u^{1,1}_n$, $v^{1,1}_n$ ($v^{1,1}_n$ is a new scalar dual variable associated with the constraint $\sum\limits_{i=1}^{N} z^{1,i}_{n-1} = 1$). After the jth return to step 1 go to iteration $2(j)$.

Step 2(j): Obtain the optimal solution $x^{1,j+1}_{n-1}$ of the problem

$$\max x_{n-1}u^{1,j}_n\,,$$
$$A_{n-1}x_{n-1} \leq B_{n-1}x^1_{n-2}\,,$$
$$x_{n-1}$$

and the dual variables corresponding to $x^{1,j+1}_{n-1}$. If

$$x^{1,j+1}_{n-1}u^{1,j}_n \leq x^1_{n-1}u^{1,j}_n + v^{1,j}_n\,,$$

then set $x^2_j = x^1_j$, $j = \overline{1, n-2}$;

$$x^2_{n-1} = x^1_{n-1} - \sum_{i=1}^{j} \bar{z}^{1,i}_{n-1}(x^1_{n-1} - x^{1,i}_{n-1})\,,$$

$x^2_n = \bar{x}^1_n$, $u^2_n = u^{1,j}_n$, $u^2_{n-1} = u^{1,j+1}_{n-1}$.

Go to step 3. If

$$x^{1,j+1}_{n-1}u^{1,j}_n > x^1_{n-1}u^{1,j}_n + v^{1,j}_n\,,$$

then return to step 1 and insert into the basis of problem (12.5) the column $(x^1_{n-1} - x^{1,j+1}_{n-1})$. As a result of performing the iteration in (12.5) we have $\bar{x}_n^1$, $\bar{z}_{n-1}^{1,i}$, $i = 1, \dots, j+1$ and the dual variables $u_n^{1,j+1}$, $v_n^{1,j+1}$. Now turn to iteration $2(j+1)$.

Evidently, the number of iterations at step 2 is finite and limited by the number of basic solutions (12.5).

Step k(0): Obtain the optimal solution $x_{n-k+1}^{k-1,1}$ of the problem

$$\max x_{n-k+1} u_{n-k+2}^{k-1},$$
$$A_{n+1-k} x_{n-k+1} \leq B_{n-k+1} x_{n-k}^{k-1},$$
$$x_{n-k+1} \geq 0$$

and the dual variables $z_{n-k+1}^{k-1,1}$ corresponding to $x_{n-k+1}^{k-1,1}$. If

$$x_{n-k+1}^{k-1,1} u_{n-k+2}^{k-1} \leq x_{n-k+1}^{k-1} u_{n-k+2}^{k-1},$$

then set $x_j^k = x_j^{k-1}$, $j = \overline{1,n}$; $u_j^k = u_j^{k-1}$, $j = \overline{n-k+2,n}$; $u_{n-k+1}^k = u_{n-k+1}^{k-1,1}$.

Go to step $(k+1)$. If

$$x_{n-k+1}^{k-1,1} u_{n-k+2}^{k-1} > x_{n-k+1}^{k-1} u_{n-k+2}^{k-1},$$

then go to step $(k-1)$ with regard to the inserted column $(x_{n-k+1}^{k-1} - x_{n-k+1}^{k-1,1})$. After the jth return to step $(k-1)$ go to step $k(j)$. Prior to it, we have an optimal solution to the problem of step $(k-1)$ $\bar{x}_j$, $j = \overline{n-k+2,n}$; $\bar{z}_{n-k+1}^{k-1,i}$, $i = \overline{1,j}$ and the dual variables $\bar{u}_j$, $j = \overline{n-k+3,n}$, $u_{n-k+2}^{k-1,j}$, $v_{n-k+2}^{k-1,j}$.

Step k(j): Obtain the optimal solution $x_{n-k+1}^{k-1,j+1}$ of the problem

$$\max x_{n-k+1} u_{n-k+2}^{k-1,j},$$
$$A_{n-k+1} x_{n-k+1} \leq B_{n-k+1} x_{n-k}^{k-1},$$
$$x_{n-k+1} \geq 0$$

and the dual variables $u_{n-k+1}^{k-1,j+1}$ corresponding to $x_{n-k+1}^{k-1,j+1}$. If

$$x_{n-k+1}^{k-1,j+1} u_{n-k+2}^{k-1,j} \leq x_{n-k+1}^{k-1} u_{n-k+2}^{k-1,j} + v_{n-k+2}^{k-1,j},$$

then set $x_j^k = \bar{x}_j$, $j = \overline{n-k+2,n}$; $x_j^k = x_j^{k-1}$, $j = \overline{1,n-k}$;

$$x_{n-k+1}^k = x_{n-k+1}^{k-1} - \sum_{i=1}^{j} \bar{z}_{n-k+1}^{k-1,i} \left(x_{n-k+1}^{k-1} - x_{n-k+1}^{k-1,i} \right),$$

$u_j^k = \bar{u}_j,\ i = \overline{n-k+3, n};$

$$u^k_{n-k+2} = u^{k-1,j}_{n-k+2}\,; \qquad u^k_{n-k+1} = u^{k-1,j+1}_{n-k+1}\,,$$

go to step $(k+1)$. Otherwise go to step $(k-1)$ and add the column $(x^{k-1}_{n-k+1} - x^{k-1,j+1}_{n-k+1})$ to the matrix A_{n-k+2}.

Evidently, the number of iterations at the step k is finite and restricted by the number of basis solutions to the problem at step k.

§13. SOLUTION OF THE INTERVAL PROGRAMMING PROBLEM

§13.1. Consider the interval linear programming problem

$$\begin{gathered} \max\ Cx \\ b^- \le Ax \le b^+\,, \end{gathered} \tag{13.1}$$

where A is an $(m \times n)$ matrix, b^-, b^+ are m-dimensional vectors, c is an n-dimensional vector.

The interval linear programming problem can be interpreted as the $\mathcal{L}p$ general problem provided the two-side constraints are regarded as a pair of individual conditions, and each free variable is represented as a difference of nonnegative variables. This procedure, however, enlarges the problem dimensionality. On the other hand, the interval programming problem finds numerous applications, say, in production planning, i.e. special methods for solving the problem are of interest.

Describe an approach to solving problem (13.1) based on results[2] using the Dantzig-Wulf decomposition method as the base.

Introduce the following notation: $S^{m\times n}$ – an $m \times n$ matrices space, $S_r^{m\times n}$ – a set of matrices in $S^{m\times n}$ with the rank r, $R(A) = \{y \in R^m \,|\, y = Ax$ for a particular $x \in R^n\}$ – the range of A matrix values, $N(A) = \{x \in R^n | Ax = 0\}$ – the A matrix zero subspace, $M = \{x \in R^n | b^- \le Ax \le b^+\}$ – a set of feasible points for problem (1).

Definition 1: Problem (13.1) is called feasible if $M \ne \varnothing$. If $A \in S_m^{m\times n}$, then problem (13.1) is feasible with any $b^- \le b^+$ since here $R(A) = R^m$.

Definition 2: Problem (13.1) is called constrained if $M \neq \varnothing$ and $\max\limits_{x \in M} Cx < \infty$.

Lemma 13.1: Let $M \neq \varnothing$. In this case problem (13.1) is constrained if and only if

$$Cx = 0 \quad \forall x \in N(A) . \tag{13.2}$$

Condition (13.2) can be tested if there is the basis $N(A)$. The following lemma provides a procedure for constructing the basis.

Lemma 13.2: Suppose $A \in S^{m\times n}$ and $\bar{E} \in S_m^{m\times n}$ satisfy the condition

$$\bar{E}A = \left(\frac{E_r|\Delta}{O_{(m-r)n}}\right) p ,$$

where p is a permutation matrix. Then the rank of $(A) = r$ and that of matrix column

$$N = p^T \begin{pmatrix} -\Delta \\ E_{n-r} \end{pmatrix}$$

form the basis $N(A)$.

In the following lemma it is established that problem (13.1), for which (13.2) holds, can be transformed into the interval problem with a full rank, i.e. with the matrix $A \in S_r^{m\times r}$.

Lemma 13.3: Let $A \in S_r^{m\times n}$ and $D \in S_r^{r\times n}$ be such matrices that $R(D^T) = R(A^T)$. Then $AD^T \in S_r^{m\times r}$. Here the proof is evident.

Lemma 13.4: Assume that problem (13.1) has the matrix $A \in S_r^{m\times n}$ and (13.2) is fulfilled. Let D be the matrix satisfying the conditions for lemma 13.3. Then an optimal solution to problem (13.1) is represented as

$$x^* = D^T y^* + N(A) , \tag{13.3}$$

where y^* is any optimal solution to the interval programming problem with the full rank matrix

$$\max\ CD^T y , \tag{13.4}$$

$$b^- \leq AD^T y \leq b^+ . \tag{13.5}$$

Proof: Using (13.2) and the fact that $R(A^T) = N(A)$ we find that an optimal solution to problem (13.1) is equal to

$$x^* + N(A), \tag{13.6}$$

where x^* is any optimal solution to the problem

$$\begin{aligned} &\max Cx, \\ &b^- \le Ax \le b^+ \\ &x \in R(A^t). \end{aligned} \tag{13.7}$$

From the condition $R(D^T) = R(A^T)$ for lemma 13.3 it follows that any element $x \in R(A^T)$ can be presented as

$$x = D^t y \tag{13.8}$$

for a particular $y \in R^r$. By substituting (13.8) in (13.6) and problem (13.1) in (13.4) we obtain the desired result. The following lemma will be utilized in constructing a decomposition procedure for problem (13.1).

Lemma 13.5: Assume that the interval programming problem is prescribed by the matrix $A \in S_n^{n\times n}$. Its optimal solution is

$$x^* = A^{-1}w^*, \tag{13.9}$$

where

$$w_i = \begin{cases} b_i \\ \theta_i a_i + (1-\theta_i)b_i, \\ a_i \end{cases} \quad \text{if } (CA^{-1})_i = \begin{cases} >0 \\ =0 \\ <0 \end{cases} \tag{13.10}$$

$i = \overline{1,n}$ and $0 \le \theta_i \le 1$.

Proof: Substituting $z = Ax$ in (1), we obtain the equivalent problem

$$\begin{aligned} &\max CA^{-1}z, \\ &b^- \le z \le b^+, \end{aligned}$$

the optimal solutions of which are prescribed by (13.10). A backward substitution gives (13.9).

§**13.2.** At this point we will show that any constrained problem of interval programming is transformed to the equivalent decomposable form

$$\begin{aligned} &\max\ Cx\,, \\ \bar{b}^- \le &\ \bar{A}x \le \bar{b}^+\,, \\ \hat{b}^- \le &\ \hat{A}x \le \hat{b}^+\,, \end{aligned} \tag{13.11}$$

where $\bar{A}$, $\hat{A}$ are invertible matrices $\bar{A}$, $\hat{A} \in S_p^{p\times p}$.

Thus, we have the interval problem

$$\begin{aligned} &\max\ Cx\,, \\ b^- \le &\ Ax \le b^+\,, \end{aligned}$$

where, by lemma 13.4, we can state, without violating generality, that $A \in S_r^{m\times r}$.

Partition the problem constraints as follows

$$\begin{aligned} b_1^- &\le A_1x \le b_1^+\,, \\ b_2^- &\le A_2x \le b_2^+\,, \\ b_3^- &\le A_3x \le b_3^+\,, \end{aligned} \tag{13.12}$$

where $A_1 \in S_r^{r\times r}$, $A_2 \in S_q^{q\times r}$, $A_3 \in S^{(m-r-q)\times r}$. Here A_1 is an invertible matrix, and the submatrix A_2 has a full rank. To be noted is that q is defined in an ambiguous way, but as shown below, q should be taken as large as possible ($q = 0$ can always be taken). It is possible to introduce new constraints of the form

$$b_4^- \le Bx \le b_4^+\,,$$

without violating the conditions of (13.12) and $\begin{pmatrix} A_2 \\ B \end{pmatrix} \in S_r^{r\times r}$. The original problem (13.1) takes the form

$$\begin{aligned} &\max\ Cx \\ b_1^- &\le A_1x \le b_1^+\,, \\ b_2^- &\le A_2x \le b_2^+\,, \\ b_3^- &\le A_3x \le b_3^+\,, \\ b_4^- &\le Bx \le b_4^+\,. \end{aligned} \tag{13.13}$$

Note that x^* is the optimal solution of (13.1) if and only if (x^*, y^*) is an optimal solution to the following interval problem written in the decomposable form:

$$\begin{pmatrix} b_1^- \\ 0 \end{pmatrix} \le \begin{pmatrix} A_1 & 0 \\ 0 & E_{m-r-q} \end{pmatrix} \begin{pmatrix} x \\ y \end{pmatrix} \le \begin{pmatrix} b_1^+ \\ 0 \end{pmatrix},$$

$$\begin{pmatrix} b_2^- \\ b_4^- \\ b_3^- \end{pmatrix} \le \begin{pmatrix} A_2 & 0 \\ B & 0 \\ A_3 & E_{m-r-q} \end{pmatrix} \begin{pmatrix} x \\ y \end{pmatrix} \le \begin{pmatrix} b_2^+ \\ b_4^+ \\ b_3^+ \end{pmatrix}.$$

§**13.3.** We set forth the decomposition method for (13.1). The decomposed interval problem (13.11) is written as the equivalent formula

$$\begin{gathered} \max Cx\,, \\ \bar{x} = \hat{x}\,, \\ \bar{b}^- \le \bar{A}\bar{x} \le \bar{b}^+\,, \\ \hat{b}^- \le \hat{A}\hat{x} \le \hat{b}^+\,. \end{gathered} \tag{13.14}$$

To solve (13.14), the Dantzig-Wulf decomposition principle is applied, the coordinating problem being constructed on the equality $\bar{x} = \hat{x}$. Assume that $\bar{S} = \{x \in R^p | \bar{b}^- \le \bar{A}x \le \bar{b}^+\}$ and $\bar{x}^i$, $i = \overline{1, \bar{N}}$ is a set of the nodes $\bar{S}$. Any point of $\bar{x} \in \bar{S}$ can then be represented in the form

$$\bar{x} = \sum_{i=1}^{\bar{N}} \bar{z}_i \bar{x}_i, \quad \sum_{i=1}^{\bar{N}} \bar{z}_i = 1, \quad \bar{z}_i \ge 0, \quad i = \overline{1, \bar{N}}\,.$$

The same is true for the polyhedron

$$\hat{S} = \{x \in R^p | \hat{b}^- \le \hat{A}x \le \hat{b}^+\}\,.$$

Write the coordinating problem for (13.14)

$$\begin{gathered} \max \sum_{i=1}^{\bar{N}} (C\bar{x}_i)\bar{z}_i\,, \\ \sum_{i=1}^{\bar{N}} \bar{z}_i \bar{x}_i - \sum_{i=1}^{\hat{N}} \hat{z}_i \hat{x}_i = 0\,, \\ \sum_{i=1}^{\bar{N}} \bar{z}_i = 1\,, \quad \sum_{i=1}^{\hat{N}} \hat{z}_i = 1\,, \\ \bar{z}_i \ge 0, \quad i = \overline{1, \bar{N}}, \quad \hat{z}_i \ge 0, \quad i = \overline{1, \hat{N}}\,. \end{gathered} \tag{13.15}$$

Problem (13.15) is the LP problem with many variables and unknown columns. According to the Dantzig-Wulf scheme, iteration of columns is performed by solving local problems. Introduce into consideration the $(p+2)$-dimensional vector of Lagrange multipliers

$$(\lambda_1, \dots, \lambda_p, \sigma_1, \sigma_2) = (\lambda, \sigma_1, \sigma_2)\,.$$

The vector $(\bar{x}^i, 1, 0)^T$ or $(-\hat{x}^i, 0, 1)^T$ can be introduced into the basis if it has a nonnegative estimate Δ_i with regard to this basis

$$\bar{\Delta}_i = (\lambda, \sigma_1, \sigma_2)\begin{pmatrix} \bar{x}^i \\ 1 \\ 0 \end{pmatrix} - C\bar{x}^i = (\lambda - C)\bar{x}^i + \sigma_1 < 0$$

or

$$\hat{\Delta}_i = (\lambda, \sigma_1, \sigma_2)\begin{pmatrix} -\hat{x}^i \\ 0 \\ 1 \end{pmatrix} = -\lambda\hat{x}^i + \sigma_2 < 0\,.$$

According to the simplex method, the vector with a minimized estimate Δ is introduced into the basis

$$\Delta = \min\{ \min_{i=\overline{1,\bar{N}}} ((\lambda - C)\bar{x}^i + \sigma_1),\ \min_{i=\overline{1,\hat{N}}} (-\lambda\hat{x}^i + \sigma_2)\}\,.$$

If $\Delta \geq 0$, then we have an optimal solution.

Define extremal points $\bar{x}^*$, $\hat{x}^*$ such that

$$(\lambda - C)\bar{x}^* = \min_{i=\overline{1,\bar{N}}} (\lambda - C)\bar{x}^i\,,$$

$$-\lambda\hat{x}^* = \min_{i=\overline{1,\hat{N}}} (-\lambda\hat{x}^i)\,.$$

Since

$$\min_{i=\overline{1,\bar{N}}} (\lambda - C)\bar{x}^i = \min_{x\in\bar{M}} (\lambda - C)x\,,$$

then, using lemma 13.5, obtain

$$\bar{x}^* = \sum_{i\in\bar{k}^+} \bar{b}_i^+ \bar{a}_i + \sum_{i\in\bar{k}^-} \bar{b}_i^- \bar{a}_i\,, \tag{13.16}$$

where $\bar{A}^{-1} = (\bar{a}_1, \dots, \bar{a}_p)$,

$$\bar{k}^+ = \{i | (\lambda - c)\bar{a}_i \geq 0\}\,, \qquad \bar{k}^- = \{i | (\lambda - c)\bar{a}_i < 0\}\,.$$

Similarly we have

$$\hat{x}^* = \sum_{i \in \hat{k}^+} \hat{b}_i^+ \hat{a}_i + \sum_{i \in \hat{k}^-} \hat{b}_i^- \hat{a}_i \,, \tag{13.17}$$

where $\hat{A}^{-1} = (\hat{a}_1, \dots, \hat{a}_p)$,

$$\hat{k}^+ = \{i | -\lambda \hat{a}_i \geq 0\}, \qquad \hat{k}^- = \{i | -\lambda \hat{a}_s < 0\} .$$

If $\Delta < 0$, then either $(\bar{x}^*, 1, 0)^T$ or $(\hat{x}^*, 0, 1)^T$ are introduced into the basis according to which of these vectors has a smaller characteristic estimate.

If $\Delta \geq 0$, then we have an optimal solution to problem (13.14) in the form

$$x^* = \sum_{i \in k_1} \bar{z}_i \bar{x}^i + \sum_{i \in k_2} \hat{z}_i \hat{x}^i \,,$$

where $k_1 = \{i | \bar{z}_i > 0\}$, $k_2 = \{i | \hat{z}_i > 0\}$.

The initial basis is chosen in the following way. Consider the problem equivalent to (13.14)

$$\begin{gathered} \max \sum_{i=1}^{\bar{N}} (C\bar{x}^i)\bar{z}_i - \theta \sum_{i=1}^{p} v_i \,, \\ \begin{pmatrix} \bar{D} & \hat{D} & -E_p \\ e & 0 & 0 \\ 0 & e & 0 \end{pmatrix} \begin{pmatrix} \bar{z} \\ \hat{z} \\ v \end{pmatrix} = \begin{pmatrix} 0_p \\ 1 \\ 1 \end{pmatrix} , \end{gathered} \tag{13.18}$$

where v is the p-dimensional vector of artificial variables, $\bar{D}$, $\hat{D}$ are matrices with columns $\bar{x}^i$, $\hat{x}^i$, respectively, and θ is a sufficiently large positive number.

First find optimal solutions to the subproblems

$$\begin{gathered} \max\ Cx, \ \ x \in \bar{S} \,, \\ \max\ Cx, \ \ x \in \hat{S} \,. \end{gathered} \tag{13.19}$$

If $\bar{x}$, $\hat{x}$ are solutions to problems (13.19), then the initial basis is taken in the form

$$\begin{pmatrix} \bar{x} \\ 1 \\ 0 \end{pmatrix}, \begin{pmatrix} \hat{x} \\ 0 \\ 1 \end{pmatrix}, \pm \begin{pmatrix} e_1 \\ 0 \\ 0 \end{pmatrix}, \dots, \pm \begin{pmatrix} e_p \\ 0 \\ 0 \end{pmatrix} . \tag{13.20}$$

The plus or minus sign in the column depends on the component sign $(\bar{x}_i - \hat{x}_i)$, $(i = \overline{1,p})$.

§**13.4.** We set forth the algorithm of the above procedure for solving the interval problem LP. Assume that problem (13.1) is reduced to the decomposable form (13.11).

Step 0: Finding the initial basis. To do this, subproblem (13.19) is solved and the initial basis B is taken in the form (13.20). Thereupon the basic feasible solution $(\bar{z}, \hat{z}, v)$ and its corresponding dual variables $(\lambda, \sigma_1, \sigma_2)$ are sought.

Step 1: Calculate the value $(\lambda - C)$ and with formulae (13.16) and (13.17) determine the extreme points $\bar{x}^*$, $\hat{x}^*$ delivering a minimum to the relative estimates $\bar{\Delta}$, $\hat{\Delta}$ of the relative basis vectors $(\bar{x}, 1, 0)^T$, $(\hat{x}, 0, 1)^T$. Let $\Delta = \min(\bar{\Delta}, \hat{\Delta})$. If $\Delta \geq 0$, then the basic solution under study is optimal. Go to step 4.

Step 2: One of the vectors $(\bar{x}, 1, 0)^T$, $(\hat{x}, 0, 1)^T$ possessing a smaller relative estimate Δ, is introduced into the basis of the coordinating problem.

Step 3: The vector introduced into the basis is multiplied by B^{-1} and, with the use of the standard leading operation, new inverse matrix and vector of dual estimates are defined. Go to step 1.

Step 4: Define a solution to problem (13.11)

$$x^* = \sum \bar{z}_i \bar{x}^i + \sum \hat{z}_i \hat{x}^i ,$$

where $\bar{z}_i$, $\hat{z}_i$ is the solution of the coordinating problem.

Terminate.

§14. EXTENSION OF THE DANTZIG-WULF DECOMPOSITION PRINCIPLE TO NONLINEAR PROGRAMMING PROBLEMS

§**14.1.** Since linear programming is an adequately developed means for solving problems, it is natural to try to use its methods in nonlinear problems with the help of linearization.

Consider the mathematical programming problem

$$\min\ f(x)\,, \\ g_i(x) \le 0,\ \ i = \overline{1,m}\,, \tag{14.1}$$

where f, g_i are convex functions, $x \in E^n$.

Select a set of mesh points $x_1, x_2, \dots, x_T$. The convex hull of this set has an element of the form

$$x = \sum_{t=1}^{T} z_t x_t,\ \ \sum_{t=1}^{T} z_t = 1,\ \ z_t \ge 0,\ \ t = \overline{1,T}\,.$$

Replacement of f, g_i $(i = \overline{1,m})$ by their linearization on the lattice leads to an LP problem with respect to the variables z_t:

$$\min \sum_{t=1}^{T} z_t f(x_t)\,, \\ \sum_{t=1}^{T} z_t g_i(x_t) \le 0\,,\quad i = \overline{1,m}\,, \\ \sum_{t=1}^{T} z_t = 1,\ \ z_t \ge 0\,. \tag{14.2}$$

Obtaining exact approximation of the nonlinear functions f, g_i $(i = \overline{1,m})$ calls for many nodal points. Therefore it is desirable to refine approximation only in the neighborhood of a minimum, and it can be done by employing the simplex method for choosing new mesh points when solving appropriate auxiliary problems. Assume that at the beginning of the iteration we introduce the lattice with the nodal points $x_1, \dots, x_T$. Let $\{z_t^k\}$ be an optimal solution to (14.2), and λ_i^k $(i = \overline{1,m})$, λ_0^k dual variables. The current approximation of an optimal solution to problem (14.1) is:

$$x^k = \sum_{t=1}^{T} z_t^k x_t\,.$$

From all the new nodal points that might be introduced for improving the approximation the simplex method would have chosen the point corresponding to the column with the smallest characteristic difference, i.e. the point x_t which minimizes

$$\bar{f}_t = f(x_t) - \sum_{i=1}^{m} \lambda_i^k g_i(x_t) - \lambda_0^k .$$

Calculating that point reduces to solving the nonlinear problem without constraints

$$\min \left\{ f(x) - \sum_{i=1}^{m} \lambda_i^k g_i(x) \right\} . \tag{14.3}$$

The problem can be solved by the proper unconstrained minimization method. If the minimum is attained at the point $\hat{x}^k$, then problem (14.2) is supplemented with a new column

$$p^k = (g_1(\hat{x}^k), \dots , g_m(\hat{x}^k), 1)$$

and $f(\hat{x}^k) z_{T+1}$ is introduced into the objective function. Thereupon problem (14.2) is newly solved. If the objective function and constraints in (14.1) are linear and defined on the convex polyhedron, then any point inside it can be represented as a convex combination of extreme points forming a finite lattice, and approximation can be made exact. The process of linearization on the lattice is then transformed into the Dantzig-Wulf decomposition algorithm for the LP problem. In this case retaining all the generation columns in the coordinating problem is not necessary. In the nonlinear case the existing proof of convergence[11] requires retaining all these columns.

If

$$\min \left\{ f(x) - \sum_{i=1}^{m} \lambda_i g_i(x) \right\} - \lambda_0^k \geq 0, \ \ \lambda_i \leq 0, \ \ i = \overline{0, m} \tag{14.4}$$

then the current approximation is optimal for problem (14.1), which is shown in the following theorem.

Theorem 14.1: Let (λ^k, λ_0^k) be the dual estimates corresponding to the optimal solution of the coordinating problem (14.2) on iteration k. If (14.4) is fulfilled, then the point $x^k = \sum_t z_t^k x_t$ is optimal, i.e.

$$\min_{g_i(x) \leq 0, \ i=\overline{1,m}} f(x) = f(x^k) .$$

Proof: Characteristic differences for basic variables are equal to zero:

$$\bar{f}_t^k = f(x_t) + \sum_{i=1}^{m} u_i^k g_i(x_t) + u_0^k = 0\,,$$

where $u_i^k = -\lambda_i^k \geq 0$ multiplying these equalities by z_t^k and summing them, we get

$$\sum_t z_t^k f(x_t) + \sum_{i=1}^{m} u_i^k \sum_t z_t^k g_i(x_t) + u_0^k = 0\,.$$

Under conditions for complementary nonhardness we have

$$\sum_{i=1}^{m} u_i^k \sum_t z_t^k g_i(x_t) = 0\,.$$

Thus, $u_0^k = -\sum_t z_t^k f(x_t) = -z^k$.

Consider the inequality

$$\min_{x \in E^n} \mathcal{L}(x, u) \leq \min_{x \in S} f(x)\,, \tag{14.5}$$

where

$$S = \{x | g_i(x) \leq 0, \;\; i = \overline{1, m}\}\,,$$

$\mathcal{L}(x,u)$ is the Lagrangian, $\mathcal{L}(x,u) = f(x) + \sum_i u_i g_i(x)$. Since f is convex and x^k are feasible, then

$$\min_{x \in S} f(x) \leq f(x^k) \leq \sum_t z_t^k f(x_t) = z^k\,. \tag{14.6}$$

Add the value $u_0^k + z^k$ equal to zero to the right-hand sides of (14.5) and, considering (14.6), we obtain

$$\min_{x \in E^n} \{\mathcal{L}(x, u^k) + u_0^k\} + z^k \leq \min_{x \in S} f(x^k) \leq z^k\,. \tag{14.7}$$

Under conditions for the theorem

$$\min_{x \in E^n} \{\mathcal{L}(x, u^k) + u_0^k\} \geq 0\,,$$

so that this value can be removed from the left-hand side of (14.7)

$$z^k \le \min_{x\in S} f(x) \le f(x^k) \le z^k\,,$$

i.e.

$$\min_{x\in S} f(x) = f(x^k) = z^k\,.$$

The proof is completed.

§14.2. Consider the case when f, g_i are additively separable, i.e.

$$f = \sum_j f_j(x_j);\quad g_i = \sum g_{ij}(x_j);\quad j = \overline{1,n};\quad i = \overline{1,m}\,.$$

In this case the advantages manifest themselves.

A set of the nodal points $\{x_{jt}\}$ can be defined for each variable x_j and each of the functions f_j, g_{ij} is linearized separately so that the problem arises

$$\min \sum_{j,t} z_{jt} f_j(x_{jt})\,,$$

$$\sum_{j,t} z_{jt} g_{ij}(x_{jt}) \le 0,\quad i = \overline{1,m}\,, \tag{14.8}$$

$$\sum_t z_{jt} = 1,\quad j = \overline{1,n};\quad z_{jt} \ge 0\,.$$

Auxiliary problems (14.3) are divided here into independent single-dimensional problems

$$\min\left\{ f_j(x_j) - \sum_{i=1}^{m} \lambda_i g_{ij}(x_j)\right\}.$$

§14.3. We will set forth a nonlinear variation of the Dantzig-Wulf decomposition principle.[34]

Consider the problem

$$\min\left\{ f(x) = \sum_{i=1}^{p} f_i(x_i)\right\}, \tag{14.9}$$

where f_i ($i = \overline{1,p}$) are convex functions, and linear connecting constraints are specified

$$\sum_{i=1}^{p} A_i x_i = b\,, \tag{14.10}$$

and a series of possibly nonlinear constraints is given with respect to each x_i:

$$x_i \in S_i, \quad i = \overline{1,p}\,, \tag{14.11}$$

where S_i are convex sets, not necessarily polyhedrons, A_i ($i = \overline{1,p}$) are m-dimensional matrices. Inasmuch as the constraints on each x_i are independent, it is possible to choose an individual lattice of nodal points for each of them. Each function $f_i(x_i)$ is substituted by its linearization on the lattice $\{x_i^t\}$:

$$f_i(x_i) = \sum_t z_i^t f_i(x_i^t)\,, \tag{14.12}$$

where

$$\sum_t z_i^t = 1, \quad z_i^t \geq 0\,. \tag{14.13}$$

Its associated values of x_i are as follows

$$x_i = \sum_t z_i^t x_i^t\,. \tag{14.14}$$

Write the approximating linear problem for (14.9) – (14.11):

$$\min\left\{z = \sum_{i,t} z_i^t f_i(x_i^t)\right\}, \tag{14.15}$$

$$\sum_{i,t} (A_i x_i^t) z_i^t = b\,, \tag{14.16}$$

$$\sum_t z_i^t = 1, \quad i = \overline{1,p}\,, \tag{14.17}$$

$$z_i^t \geq 0\,. \tag{14.18}$$

Inasmuch as $x_i^t \in S_i$ and S_i is convex, the point x_i in (14.14) is also an element of S_i. If z_i^t satisfies (14.16), then x_i satisfies (14.10) and represents a feasible solution to the original problem. As before, it is undesirable to select many nodal points in advance. It can be avoided with the use of the simplex method for generating them as needed. The column related to z_i^t is

$$p_i^t = (A_i x_i^t, l_i)^T . \tag{14.19}$$

Assume that a feasible basis in (14.15) – (14.18) exists, and let (λ_1, λ_2) be dual variables. Then the characteristic difference corresponding to the column p_i^t is this:

$$\Delta_i^t = f_i(x_i^t) - \lambda_1 A_i x_i^t - \lambda_{2i} . \tag{14.20}$$

The problem of determining the point of $x_i^t \in S_i$, transferring a minimum to this expression, takes the form

$$\min\{f_i(x_i) - \lambda_1 A_i x_i\} , \tag{14.21}$$

$$x_i \in S_i, \quad i = \overline{1,p} . \tag{14.22}$$

Let $\hat{x}_i$ be a solution to the subproblem i. Assume that if only one of the characteristic differences is negative:

$$f_i(\hat{x}_i) - \lambda_1 A_i \hat{x}_i - \lambda_{2i} < 0 \quad \text{for an } i . \tag{14.23}$$

Form the columns

$$\hat{p}_i = (A_i \hat{x}_i, l_i)^T, \quad i = \overline{1,p} . \tag{14.24}$$

The columns are utilized in constructing a reduced coordinating problem. The problem has, as variables, a set of basic variables $\{z_i^t\}$ plus p new variables $\hat{z}_t$, one for each column $\hat{p}_i$:

$$\min\left\{z = \sum_i \sum_{t_{baz}} z_i^t f_i(x_i^t) + \sum_i \hat{z}_i f_i(\hat{x}_i)\right\} ,$$

$$\sum_i \sum_{t_{baz}} (A_i x_i^t) z_i^t + \sum_i (A_i \hat{x}_i) \hat{z}_i = b ,$$

$$\sum_{t_{baz}} z_i^t + \sum_i \hat{z}_i = 1, \quad i = \overline{1,p}, \quad z_i^t \geq 0, \quad \hat{z}_i \geq 0 .$$

The problem has as the initial feasible basis associated with the variables z_i^t and p columns $\hat{p}_i$ of form (14.24) as nonbasic columns.

§**14.4.** We set forth the algorithm of the nonlinear analog of the Dantzig-Wulf decomposition method under assumption that the coordinating problem (14.15) – (14.18) has $(m+1)$ constraints in the form of an equality, i.e. conditions (14.11) are regarded as a single whole.

Step 1: Obtain linearly independent solutions to $x_i^t \in S_i$, $i = \overline{1,p}$, $t = \overline{1, m+1}$ and calculate the following values

$$\alpha_t = \sum_{i-1}^{p} A_i x_i^t, \quad \gamma_t = \sum_{i=1}^{p} f_i(x_i^t), \quad t = p_1, \dots, p_{m+1},$$

where $(p_1, \dots, p_{m+1})$ is the current basis of the coordinating problem.

Step 2: Obtain the dual variables (λ_1, λ_2) from the formula

$$(\lambda_1, \lambda_2) \begin{pmatrix} \alpha_{p_1} & \alpha_{p_2} & \dots & \alpha_{p_{m+1}} \\ 1 & 1 & \dots & 1 \end{pmatrix} = \left(\gamma_{p_1}, \dots, \gamma_{p_{m+1}}\right).$$

Step 3: Obtain solutions to the following p problems

$$\min \Delta_i = f_i(x_i) - \lambda_1 A_i x_i; \quad x_i \in S_i \tag{14.25}$$

Denote the optimal solutions of these problems by $\hat{x}_i$ $(i = \overline{1,p})$ and the corresponding Δ_i by

$$\hat{\Delta}_i = f_i(\hat{x}_i) - \lambda_1 A_i \hat{x}_i, \quad i = \overline{1,p}.$$

Step 4: Calculate

$$\Delta^* = \sum_{i=1}^{p} \hat{\Delta}_i - \lambda_2.$$

If $\Delta^* = 0$, go to step 6.

Step 5: Calculate

$$\bar{\alpha} = \sum_{i=1}^{p} A_i \hat{x}_i; \qquad \bar{\gamma} = \sum_{i=1}^{p} f_i(\hat{x}_i),$$

introduce the column $\binom{\bar{\alpha}}{1}$ into the coordinating problem and carry out one operation of the simplex method.

Let $z_i, i = \overline{1, m+1}$ be the current solution of the coordinating problem.

Go to step 2.

Step 6: Calculate the optimal solution of the original problem (9) – (11):

$$x^* = (x_1^*, \dots, x_p^*),$$

where $x_i^* = \sum_{t=p_1}^{p_{m+1}} z_t \hat{x}_i^t, \quad i = \overline{1,p}.$

Terminate.

In conclusion it should be noted that convergence of this algorithm to an optimal solution of problem (14.9) – (14.11) was proved by Sekain.[34]

Chapter 4

PARAMETRIC DECOMPOSITION

§15. KORNAI-LIPTACK METHOD

§15.1. Consider the linear programming problem

$$\max cx\,,$$
$$A^0 x \le b^0\,, \tag{15.1}$$
$$x \ge 0\,,$$

where c, x are n-dimensional vectors, b^0 is an m-dimensional vector, and A^0 an $(m \times n)$ matrix.

Partition the matrix A^0 into submatrices

$$A^0 = (A_1^0, \ldots, A_k^0)\,,$$

where with each $j \in [1:k]$ the matrix A_j^0 has $m \times n_j$ dimensionality. Then partition also the vectors c, x respectively:

$$c = (c_1, \ldots, c_k), \quad x = (x_1, \ldots, x_k)\,,$$

the dimensions $c_j, x_j\,(j = \overline{1,k})$ being equal to n_j. With due regard to partitioning problem (15.1) is transformed to the form

$$\max \sum_{j=1}^{k} c_j x_j\,,$$

$$\sum_{j=1}^{k} A_j^0 x_j \le b^0 \,, \qquad x_j \ge 0, \quad j = \overline{1,k} \,. \tag{15.2}$$

Introduce the m-dimensional column vectors $y_j = (j - \overline{1,k})$ satisfying the condition

$$\sum_{j=1}^{k} y_j \le b^0 \,. \tag{15.3}$$

Formulate k problems LP:

$$\max c_j x_j \,, \qquad A_j^0 x_j \le y_j \,, \qquad x_j \ge 0 \,. \tag{15.4}$$

Introduce the vector $y = (y_1, \dots, y_k)$ and denote by M_y a set of the vectors y such that (15.3) is fulfilled and problems (15.4) have solutions. Optimal values of the functionals of problems (15.4) are functions from the values of y_j:

$$f_j = f_j(y_j) \,.$$

Then problem (15.2) reduces to the following problem

$$\max \mathcal{F}(y) = \sum_{i=1}^{k} f_j(y_j) \,, \qquad \sum_{j=1}^{k} y_j \le b^0 \,. \tag{15.5}$$

If decomposing the matrix A^0 is interpreted as its partitioning into k subsystems, and the vector b^0 is regarded as a common resource of the system, then problem (15.5) resides in finding optimal distribution of the common resource. Here the constraint on the sign of the values y_j, $j = \overline{1,k}$ is not presupposed since the subsystem can also consume resources. The analysis of problem (15.5) is complex in that the function is specified algorithmically. To find the values of this function we have to solve LP problem (15.4). Application of one or another scheme, when maximizing the function $\mathcal{F}(y)$, generates an appropriate decomposition method based on the Kornai-Liptack decomposition principle. In the initial work by Kornai-Liptack,[14] problem (15.1) was reduced to the maximum problem and to obtain an optimal solution it was suggested to employ the Brown-Robinson iterative scheme. But, as noted in Ref. 28, numerical experiments show that the convergence rate of the Brown-Robinson method is basically unsatisfactory, which results in its modification and construction of essentially different schemes.

§15.2. We set forth the iterative scheme for solving problem (15.5) based on application of the method of feasible directions in the set of the variables.[28]

Return to problem (15.5) where the original problem (15.2) takes the form

$$\begin{aligned} &\max \sum_{j=1}^{k} c_j x_j\,, \\ &\sum_{j=1}^{k} A_j^0 x_j \le b^0\,, \\ &A_j x_j \le b_j\,, \\ &x_j \ge 0, \quad j = \overline{1,k}\,, \end{aligned} \tag{15.6}$$

where with each $j = \overline{1,k}$, b_j is an m_j-dimensional vector, A_j is an $(m_j \times n_j)$ matrix. As before, introduce the m-dimensional column vectors $y_j/j = \overline{1,k}$, and formulate k problems LP.

$$\begin{aligned} &\max\, c_j x_j\,, \\ &A_j^0 x_j \le y_j\,, \\ &A_j x_j \le b_j\,, \\ &x_j \ge 0\,. \end{aligned} \tag{15.7}$$

In Ref. 29 it has been shown that the functions $f_j(y_j)$, $j = \overline{1,k}$ are concave and piecewise linear, and break points correspond to replacement of the basis in problems (15.7).

Consider the scheme of feasible directions for the case where the unique optimal dual variables u_j, $j = \overline{1,k}$ correspond to some given vectors y_{j0}, $j = \overline{1,k}$. By the first duality theorem, we have

$$f_j(y_{j0}) = \sum_{i=1}^{m} u_j^i y_{j0}^i + \psi_j; \quad \psi_j = \sum_{i=1}^{m_j} b_j^i u_j^{m+i}\,. \tag{15.8}$$

According to the scheme of the method of feasible directions, subsequent approximation is defined with the formula

$$y_{j_1} = y_{j_0} + \alpha z_j, \quad j = \overline{1,k}, \tag{15.9}$$

where z_j is the m vector of feasible directions satisfying a particular condition for valuation, e.g. $-1 \le z_j^i \le 1$, $j = \overline{1,k}$, $i = \overline{1,m}$ and α is the nonnegative parameter, the magnitude of the step along a direction.

Substituting (15.9) in (15.5) with regard to (15.8), we obtain the following LP problem:

$$\begin{gathered} \sum_{j=1}^{k}\sum_{i=1}^{m} u_j^i z_j^i \to \max, \\ \sum_{j=1}^{k} z_j^i \le 0, \quad i = \overline{1,m}, \\ -1 \le z_j^i \le 1, \quad j = \overline{1,k}, \quad i = \overline{1,m}, \end{gathered} \tag{15.10}$$

which falls into m independent subproblems of the form

$$\begin{gathered} \max \sum_{j=1}^{k} u_j^i z_j^i, \\ \sum_{j=1}^{k} z_j^i \le 0, \\ -1 \le z_j^i \le 1, \quad j = \overline{1,k} \end{gathered}$$

the solution of which is evident so that

$$z_j^i = \begin{cases} -1, & j = l_2, \\ 1, & j = l_1, \\ 0, & j \ne l_1, l_2, \quad i = \overline{1,m}, \end{cases}$$

where the indices l_1, l_2 are prescribed by the conditions

$$u_{l_1}^i = \max_j u_j^i, \quad u_{l_2}^i = \min_j u_j^i.$$

Thus, optimal rearrangement coincides with that flowing from elementary economic considerations; a resource is removed from the subsystem, for which it is least valuable and transferred to the subsystem where it is most valuable. To complete the description of the algorithm, indicate distinctive characteristics of selecting the α parameter determining the length of step along the direction selected.

We turn to problem (15.5). The magnitude of α is defined as $\alpha = \min\{\alpha_1, \alpha_2\}$, where α_1 is the maximal value of α satisfying the constraint of (15.5)

$$\sum_{j=1}^{k} y_{j0} + \alpha \sum_{j=1}^{k} z_j < b^0 ,$$

where α_2 is derived from the maximization condition for the single variable function

$$\mathcal{F}(\alpha) = \sum_{j=1}^{k} f_j(y_j(\alpha)) ,$$

$$A_j^0 x_j \le y_{j0} + \alpha z_j, \quad A_j x_j \le b_j, \quad x_j \ge 0, \quad j = \overline{1,k} .$$

Consecutive application of the scheme of the method of feasible directions, as shown in Ref. 31, ensures monotone convergence to an extremum for the finite number of steps.

§16. SOLUTION TECHNIQUE FOR BLOCK-SEPARABLE NONLINEAR PROBLEMS

§16.1. Consider the following block-separable problem

$$\min \sum_{i=1}^{k} f_i(x_i) ,$$
$$\sum_{i=1}^{k} g_i(x_i) \le b , \tag{16.1}$$
$$x_i \in S_i, \quad i = \overline{1,k} .$$

Rewrite (16.1) in the equivalent form, introducing the m-dimensional vectors y_i $(i = \overline{1,k})$

$$\max \sum_{i=1}^{k} f_i(x_i)\,,$$
$$g_i(x_i) \le y_i, \quad i = \overline{1,k}\,,$$
$$\sum_{i=1}^{k} y_i \le b\,, \tag{16.2}$$
$$x_i \in S_i, \quad i = \overline{1,k}\,.$$

With the fixed $y = (y_1, \ldots, y_k)$ problem (16.2) falls into k local problems of the form

$$\min f_i(x_i)\,,$$
$$g_i(x_i) \le y_i, \quad x_i \in S_i \tag{16.3}$$

It is natural that y_i should be chosen so that subproblems (16.3) might have feasible solutions, i.e.

$$y_i \in V_i = \{y_i | \exists x_i \in S_i : g_i(x_i) \le y_i\}\,. \tag{16.4}$$

Introduce into consideration the functions

$$w_i(y_i) = \min\{f_i(x_i) | g_i(x_i) \le y_i, \quad x_i \in S_i\}, \quad i = \overline{1,k}\,.$$

Evidently, to solve the original problem as a whole one should choose y_i minimizing the sum of these functions. Thus, the original problem (16.1) is equivalent to the following coordinating problem

$$\min w(y) = \sum_{i=1}^{k} w_i(y_i)\,, \tag{16.5}$$

$$\sum_{i=1}^{k} y_i \le b\,, \tag{16.6}$$

$$y_i \in V_i, \quad i = \overline{1,k}\,. \tag{16.7}$$

For LP problems such an approach to solving the original problem is described in the preceding paragraph and is called the Kornai-Liptack method.

Introduce the following propositions:

(1) S_i are nonempty, compact, convex;

(2) $f_i(x_i)$ are convex and differentiable on S_i;

(3) Each component of the vectors g_i is convex and differentiated on S_i;

(4) Problem (16.1) has a feasible solution.

Evidently, the convexity of the set V_i and the function w_i on the set follows from the convexity of f_i, g_i, S_i so that the coordinating problem (16.5) – (16.7) constitutes the convex programming problem.

The coordinating problem feasibility follows from the feasibility of the original problem. From compactness of S_i together with continuity of f_i, g_i, it follows that the ith subproblem has an optimal solution since it has the feasible one.

If $y^0 = (y_1^0, \dots, y_k^0)$ constitutes a solution to the coordinating problem (16.5) – (16.7) and x_i^0 is a solution to problem (16.3) with $y_i = y_i^0$, then $x^0 = (x_1^0, \dots, x_k^0)$ is a solution to the original problem (16.1).

§**16.2.** As in the preceding paragraph, we will employ the Loitendake method of feasible directions to solve problem (16.5) – (16.7).

We begin solving the coordinating problem with a feasible vector y, then construct the problem for determining the direction S or establishing the optimality of y. If y is not optimal then carry out a step in the direction S, determining a new distribution, and the process is repeated.

Denote

$$\mathcal{F} = \left\{ y \mid \sum_{i=1}^{k} y_i \le b,\ y_i \in V_i, \quad i = \overline{1,k} \right\}.$$

Assume that a feasible point of $y = (y_1, \dots, y_k) \in \mathcal{F}$ is known. To perform a small step in the direction S_i from the point y_i here $y_i + \alpha S_i \in V_i$, suppose that $y_i \in \operatorname{int} V_i$. Let $S = (S_1, \dots, S_k)$ be the assumed direction of search. To be useful, S should possess the following features:

(1) to be feasible at the point of $y \in \mathcal{F}$, i.e. $\varepsilon > 0$ must exist such that

$$y + \alpha S \in \mathcal{F}, \quad 0 < \alpha \le \varepsilon\,; \tag{16.8}$$

(2) the feasible direction S has to be suitable for w at the point of $y \in \mathcal{F}$, i.e.

$$w(y + \alpha S) < w(y) - \alpha\varepsilon, \quad 0 < \alpha \le \varepsilon\,. \tag{16.9}$$

Note that the parameter ε for (1), (2) is the same. Thus, for S to be a feasible direction, it is necessary to fulfill the following conditions in a field $\alpha \ge 0$

$$y_i + \alpha S_i \in V_i, \quad i = \overline{1,k} \tag{16.10}$$

$$\sum_{i=1}^{k}(y_i + \alpha S_i) \le b \tag{16.11}$$

(16.11) is rewritten as

$$\alpha \sum_{i=1}^{k} S_i \le b - \sum_{i=1}^{k} y_i \,. \tag{16.12}$$

Since under our assumption $y_i \in \operatorname{int} V_i$, then (16.10) is satisfied for any S direction. If the kth component of the vector $b - \sum_{i=1}^{k} y_i$ is positive, then there exists the region of such $\alpha > 0$ that the kth inequality is satisfied with any choice of the vectors $\{S_i\}$ so that the corresponding constraints may not be considered. If the kth component is equal to null, then a shift along some directions violates this constraint. Let

$$B = \left\{ j | b_j - \sum_{i=1}^{k} y_{ij} = 0 \right\}. \tag{16.13}$$

Then from (16.12) it follows that for feasibility of the direction one should fulfill the conditions

$$\sum_{i=1}^{k} S_{ij} \le 0, \quad j \in B \,. \tag{16.14}$$

Evidently, finding a locally better suitable direction requires finding the feasible direction that minimizes linear increment of an objective function, i.e. one has to solve the following problem

$$\min \Delta w(y, s) = \sum_{i=1}^{k} \Delta w_i(y_i, s_i) \,, \tag{16.15}$$

$$\sum_{i=1}^{k} s_{ij} \le 0, \quad j \in B \,, \tag{16.16}$$

$$\|s\| \le 1 \,. \tag{16.17}$$

Establish the optimality criteria for y and show that (16.15), (16.17) always yields a feasible direction, if any.

Theorem 16.1: Let y be feasible for the coordinating problem. If $s = 0$ is an optimal solution to (16.15) – (16.17), then y represents a solution to the coordinating problem.

Proof: Since $\Delta w(y, 0) = 0$, then

$$\Delta w(y, s) \geq 0 \tag{16.18}$$

for all s satisfying (16.16). If w is additionally defined by the condition

$$w(y) = \infty, \quad \text{if} \quad \sum_{i=1}^{k} y_i \nleq b\,, \tag{16.19}$$

then for the s vectors not satisfying (16.16), there exists $w(y + \alpha s) = \infty$, $\forall \alpha > 0$ so that $\Delta w(y, s) = \infty$. Thus (16.19) holds for all $s \in E^m$. There exists a theorem (say, Ref. 19) according to which y is the point of minimum for w if and only if

$$\Delta w(y, s) \geq 0, \quad \forall s \in E^m\,.$$

The proof is completed.

Theorem 16.2: Let s be a solution to (16.15) – (16.17) for a particular $y \in \mathcal{F}$. If $\Delta w(y, s)$ is negative, then s is a suitable direction for w at the point y.

Proof: Since, according to the condition, s is a feasible direction, then $\bar{\alpha}$ exists such that

$$y + \alpha s \in \mathcal{F}, \quad 0 \leq \alpha \leq \bar{\alpha}\,. \tag{16.19$'$}$$

To prove that s is suitable, employ the fact that

$$\Delta w(y, s) = \lim_{\lambda \to 0+} \frac{w(y + \lambda s) - w(y)}{\lambda} < 0\,. \tag{16.20}$$

There exists theorem (19) according to which if the function $w(y)$ is finite the difference relation

$$\alpha(\lambda) = \frac{w(y + \lambda s) - w(y)}{\lambda}$$

is nondecreasing for all $\lambda > 0$. Together with (16.20) this statement leads to the existence of a particular $\varepsilon > 0$, $\bar{\lambda} > 0$, thus

$$\alpha(\lambda) < -\varepsilon, \quad 0 < \lambda < \bar{\lambda}\,. \tag{16.21}$$

Denote $\delta = \min(\varepsilon, \bar{\lambda})$. Show that

$$\alpha(\lambda) < -\delta, \quad 0 < \lambda \le \delta\,. \tag{16.22}$$

(1) $\varepsilon > \bar{\lambda}$. (16.21) implies

$$\alpha(\lambda) < -\varepsilon < -\bar{\lambda} = -\delta, \quad 0 < \lambda \le \delta\,.$$

(2) $\varepsilon \le \bar{\lambda}$. Due to (16.21)

$$\alpha(\lambda) < -\varepsilon = -\delta\,.$$

Since the latter holds for all $0 < \lambda \le \bar{\lambda}$ it also holds for all $0 < \lambda \le \delta$. Suppose $\lambda^* = \min(\bar{\alpha}, \delta)$. Then due to (16.19′)

$$y + \alpha s \in \mathcal{F}, \quad 0 < \alpha \le \lambda^* \tag{16.23}$$

and owing to (16.22)

$$\alpha(\lambda) < -\lambda^*, \quad 0 < \lambda \le \lambda^*\,.$$

The latter is rewritten as

$$w(y + \lambda s) < w(y) - \lambda\lambda^*, \quad 0 < \lambda \le \lambda^*\,. \tag{16.24}$$

Since (16.23), (16.24) constitute the determinative inequality for the feasible direction s at the point y, then the proof of the theorem is completed. Return to the problem of obtaining an explicit expression $\Delta w(y, s)$.

Definition: The row vector x^* is identified as a subgradient of the function f at the point x if

$$f(z) \ge f(x) + x^*(z - x), \quad \forall z \in E^n\,. \tag{16.25}$$

Theorem 16.3: Let $\mathcal{L}(x, u) = f(x) + u(g(x) - y)$ be the Lagrange function for problems (16.3) and x^0 the solution of these problems. Then $-u^0 =$

$\{-u_i^0\}$; u_i^0 is the solution of the duality problem of (16.3) in the subgradient of w at the point y if and only if (x^0, u^0) is the saddle point of the Lagrange function, i.e.

$$\mathcal{L}(x^0, u) \leq \mathcal{L}(x^0, u^0) \leq \mathcal{L}(x, u^0).$$

Proof:
Necessity: Since (x^0, u^0) is the saddle point $\mathcal{L}(x, u)$, then

$$f(x^0) + u^0(g(x^0) - y) \leq f(x) + u^0(g(x) - y), \quad \forall x \in S$$

or, since $f(x^0) = w(y)$, $u^0(g(x^0) - y) = 0$, then

$$w(y) \leq f(x) + u^0(g(x) - y). \tag{16.26}$$

Select $z \in E^m$. If there exists $x \in S$ such that $g(x) \leq z$ then for such x, $g(x)$ in (16.26) can be replaced by z:

$$w(y) \leq f(x) + u^0(z - y).$$

Calculate inf of the right-hand side of this inequality by $x \in S$ so that $g(x) \leq z$. Then

$$w(y) < w(z) + u^0(z - y). \tag{16.27}$$

If there is no $x \in S$ such that $g(x) \leq z$, then $w(z) = +\infty$. Comparing that relation with the definition of subgradient (16.25), we see that $-u^0$ is the subgradient of the function w at the point y.

Sufficiency: Assume that (16.27) holds for all $z \in E^m$. Choose

$$z = g(x^0), \tag{16.28}$$

hence

$$w(z) = f(x^0) = w(y). \tag{16.29}$$

Substituting (16.28), (16.29) in (16.27), obtain

$$u^0(g(x^0) - y) \geq 0.$$

And since $u^0 \geq 0$ and $g(x^0) - y \leq 0$, then

$$u^0(g(x^0) - y) = 0. \tag{16.30}$$

Condition (16.30), combined with the relation $g(x^0) \leq y$, yields two from three conditions for the saddle point. It remains to show that x^0 minimizes $\mathcal{L}(x, u^0)$ on S. Substituting the equalities $f(x^0) = w(y)$, $u^0(g(x^0) - y) = 0$ in (16.27) obtain

$$f(x^0) + u^0(g(x^0) - y) \leq w(z) + u^0(z - y), \quad \forall z \in E^m . \tag{16.31}$$

For an arbitrary $x \in S$ set $z = g(x)$. Since x is a feasible point for problems (16.3) with $y = z = g(x)$, then

$$w(z) \leq f(x) . \tag{16.32}$$

Using condition (16.32) in (16.31), obtain

$$f(x^0) + u^0(g(x^0) - y) \leq f(x) + u^0(g(x) - y) ,$$

which proves the third condition. The proof of the theorem is completed.

There exists a theorem according to which a convex function is differentiable at a point if and only if it has a unique subgradient at this point. Since, with x^0 being optimal for problems (16.3), (x^0, u^0) is the saddle point α if and only if u^0 is the solution of the problem dual to (16.3), then the following statement flows from theorem 16.3.

Theorem 16.4: The function is differentiable at y if and only if the problem dual with respect to (16.3) has a unique optimal solution.

Theorem 16.3 can be employed for obtaining the explicit expression $\Delta w_i(y_i, s_i)$ at the points $y_i \in \text{int } V_i$.

Theorem 16.5: Assume that the convex function f has the nonempty region S of its finite values. Then

$$\Delta f(x, s) = \max\{x^* s | x^* - \text{subgradient } f\} ,$$

where $x \in \text{int } s$.

According to theorem 16.5 there exists

$$\Delta w_i(y_i, s_i) = \max_{u_i \in V_i} (-u_i s_i) , \tag{16.33}$$

where V_i is the set of all optimal dual variables for problem (16.3). Let each set s_i be defined with the use of the vector function c_i

$$s_i = \{x_i | c_i(x_i) \leq 0\}.$$

Suppose that each component of c_i is differentiable and convex. Then, if x_i^0 is an optimal solution to (16.3), u_i is an optimal dual variable for problem (16.3) if and only if x_i^0, u_i and a particular λ_i satisfy the Kuhn-Tucker conditions for (16.3). Form the Lagrange for (16.3)

$$\tilde{\mathcal{L}}_i = f_i + u_i(g_i - y_i) + \lambda_i c_i .$$

At its saddle point, the following conditions are fulfilled

$$I_{g_i'} u_i + I_{c_i'} \lambda_i = -\nabla f_i , \tag{16.34}$$
$$u_i(g_i - y_i) = 0 , \tag{16.35}$$
$$\lambda_i c_i = 0 , \tag{16.36}$$
$$u_i \geq 0; \quad \lambda_i \geq 0 , \tag{16.37}$$

where I_{g_i}, I_{c_i} are Jacobian for g_i, c_i i.e. the matrices, the rows of which are the gradients of the functions constituting g_i, c_i: Due to (16.33)

$$\Delta w_i(y_i, s_i) = \max\{-u_i s_i | (u_i, \lambda_i) \text{ satisfies (16.34) – (16.37)}\}. \tag{16.38}$$

Since $g_i \leq y_i, c_i \leq 0$, conditions (16.35), (16.36) are equivalent to the following

$$u_{ij} = 0, \quad \text{if} \quad g_{ij} - y_{ij} < 0 \tag{16.39}$$
$$\lambda_{ij} = 0, \quad \text{if} \quad c_{ij} < 0 . \tag{16.40}$$

Thus, (16.38) can be rewritten as

$$\max \left\{ -\sum_j u_{ij} s_{ij} \right\}, \tag{16.41}$$
$$\sum_j \nabla g_{ij} u_{ij} + \sum_j \nabla c_{ij} \lambda_{ij} = -\nabla f_i , \tag{16.42}$$
$$u_{ij} \geq 0, \quad \lambda_{ij} \geq 0 , \tag{16.43}$$

where summation is done by such j that

$$g_{ij} - y_{ij} = 0, \quad c_{ij} = 0\,.$$

Write the dual problem to (16.41) – (16.43) with the provisionally fixed s_i:

$$\min\{-\nabla f_i z_i\}\,, \tag{16.44}$$

$$\nabla g_{ij} z_i \geq -s_{ij}\,, \tag{16.45}$$

$$\nabla c_{ij} z_i \geq 0\,. \tag{16.46}$$

Thus, for the fixed $s = (s_1, \ldots, s_k)$ the directional derivative $w(y)$ is equal

$$\Delta w(y, s) = \sum_{i=1}^{k} \min\{-\nabla f_i z_i | (z_i, s_i) \text{ satisfy (16.45), (16.46)}\}\,. \tag{16.47}$$

In the problem of direction selection we had to find the vector that was minimized (16.47), and this is equivalent to solving the problem:

$$\min_{z,s}\left\{-\sum_{i=1}^{k} \nabla f_i z_i\right\}, \tag{16.48}$$

$$\sum_{i=1}^{k} s_{ij} \leq 0, \quad j : \sum_{i=1}^{k} y_{ij} = b_j\,, \tag{16.49}$$

$$\nabla g_{ij} z_i \geq -s_{ij}, \quad j : g_{ij}(x_i^0) - y_{ij} = 0, \quad i = \overline{1,k} \tag{16.50}$$

$$\nabla c_{ij} z_i \geq 0, \qquad j : c_{ij}(x_i^0) = 0, \qquad i = \overline{1,k}\,. \tag{16.51}$$

The problem constructed will be linear if we use the norm

$$\|s\| = \max_j |s_j|\,.$$

Then requirements for normalization are written as

$$-1 \leq s_{ij} \leq 1\,. \tag{16.52}$$

Remark 16.1: If all the functions w_i are differentiable, the problem of direction selection reduces to the following (see (16.15) – (16.17), (16.33)):

$$\min\left\{-\sum_{i=1}^{k} u_i^0 s_i\right\},$$

$$\sum_{i=1}^{p} s_{ij} \leq 0, \quad j \in B, \tag{16.53}$$

$$-1 \leq s_{ij} \leq 1.$$

Remark 16.2: Problem (16.48) – (16.52) has an angle structure, (16.49) connecting constraints, (16.50) – (16.51) angle blocks. It remains to consider the problem of step length selection. If a suitable direction s has been found, then a new point y is found via selection of the step $\alpha > 0$ in the expression

$$y^{n+1} = y^n + \alpha s.$$

We set forth one of the existing schemes for selecting α.

Solve the convex single-dimensional problem

$$\min\{w(y + \alpha s) | y + \alpha s \in \mathcal{F}\}. \tag{16.54}$$

To solve the problem, we can apply any single-dimensional minimization technique for the case of constraint absence when it is considered that the constraints $y + \alpha s \in \mathcal{F}$ in the explicit form appear as

$$\sum_{i=1}^{k} (y_i + \alpha s_i) \leq b, \tag{16.55}$$

$$y_i + \alpha s_i \in V_i, \quad i = \overline{1,k}. \tag{16.56}$$

Constraint (16.55) is equivalent to calculating the upper bound of α. Violation of constraints (16.56) can be detected by the absence of feasible solutions in a subproblem.

§**16.3.** We set forth the algorithm for solving problem (16.1)

<u>*Step 1*</u>: Select y^0 so that $\exists\, x_i : g_i(x_i) \leq y_i^0$, $i = \overline{1,k}$. Set $n = 0$.

<u>*Step 2*</u>: Solve the problem of the s direction selection for (16.48) – (16.51) or, in the case of differentiability of all w_i, $i = \overline{1,k}$ problem (16.53).

<u>*Step 3*</u>: Determine the value of the step α in the direction s from solving (16.54).

Step 4: Test optimality criteria. (Theorem 16.1) If $s = 0\,(\|s\| < \varepsilon)$, go to step 6.

Step 5: Determine a new point y with the formula

$$y^{n+1} = y^n + \alpha s\,.$$

Setting $n = n + 1$, go to step 2.

Step 6: Terminate.

§17. ON PARAMETRIC DECOMPOSITION OF THE RESOURCES ALLOCATION PROBLEM

The preceding paragraph was concerned with one method of parametric decomposition for what might be termed the resources allocation problem.

$$\begin{gathered} \min \sum_{i=1}^{k} f_i(x_i)\,, \\ \sum_{i=1}^{k} g_i(x_i) \le b\,, \\ x_i \in s_i, \quad i = \overline{1,k}\,, \end{gathered} \tag{17.1}$$

where x_i are the n_i-dimensional vector, g_i the m-dimensional vectors ($i = \overline{1,k}$), $m \le k$.

A choice of subproblems and a coordinating problem was suggested for the problem.

Here we shall consider some other possibilities of selecting subproblems and appropriate conditions for coordination[38] so that they might constitute a unified system equivalent to the original problem (17.1).

The present paragraph is rather of a survey nature since the approaches to solve local and coordinating problems are not considered here.

§17.1. We expand problem (17.1) into the following subproblems ($i = 1, \ldots, k$)

$$\begin{gathered} \min f_i(x_i)\,, \\ g_i(x_i) + \alpha_i \le b\,, \\ x_i \in s_i\,, \end{gathered} \tag{17.2}$$

where $\alpha_i = (\alpha_{i_1}, \ldots, \alpha_{m_i})$ is the vector of coordinating parameters.

Denote $\alpha = (\alpha_1, \dots, \alpha_k)$.

Assume that there exist the Lagrange function saddle points corresponding to problem (17.1) and subproblems (17.2) with the fixed α_i. For problem (17.1) define the so-called interaction functions[25]

$$k_i(x) = \sum_{\substack{j=1 \\ j\neq i}}^{k} g_j(x_j), \quad i = \overline{1,k}$$

and represent the conditions for coordination as

$$\alpha_i = k_i(x(\alpha)) = \sum_{\substack{j=1 \\ j\neq i}}^{k} g_j(x_j(\alpha_j)), \quad i = \overline{1,k}\,, \tag{17.3}$$

$$\lambda_1(\alpha_1) = \lambda_2(\alpha_2) = \cdots = \lambda_n(\alpha_n) = \hat{\lambda}\,, \tag{17.4}$$

where $x_i(\alpha_i)$ is a solution, and $\lambda_i(\alpha_i)$ is a vector of the Lagrange optimal multipliers of the ith subproblem (17.2).

(1) Show that system (17.3), (17.4) has a solution, i.e. there exists the vector $\bar{\alpha} = (\bar{\alpha}_1, \dots, \bar{\alpha}_k)$ satisfying the condition for coordinating (17.3), (17.4). Assume that we know a solution to problem (17.1) $\bar{x} = (\bar{x}_1, \dots, \bar{x}_k)$ and a vector of the Lagrange optimal multipliers $\bar{\lambda} = (\bar{\lambda}_1, \dots, \bar{\lambda}_m)$. In this case

$$\sum_{i=1}^{k} f_i(\bar{x}_i) + \left(\lambda, \sum_{i=1}^{k} g_i(\bar{x}_i) - b\right) \le \sum_{i=1}^{k} f_i(\bar{x}_i) + \left(\bar{\lambda} \sum_{i=1}^{k} g_i(\bar{x}_i) - b\right)$$

$$\le \sum_{i=1}^{k} f_i(x_i) + \left(\bar{\lambda}, \sum_{i=1}^{k} g_i(x_i) - b\right),$$

$$\forall \lambda \ge 0,\ \forall x_i \in s,\ i = \overline{1,k}\,. \tag{17.5}$$

Setting in inequalities (17.5) $x_i = \bar{x}_i$, $i \neq \bar{k}$, obtain

$$f_{\bar{k}}(\bar{x}_{\bar{k}}) + \left(\lambda, \sum_{i=1}^{k} g_i(\bar{x}_i) - b\right) \le f_k(\bar{x}_{\bar{k}}) + \left(\bar{\lambda}, \sum_{i=1}^{k} g_i(\bar{x}_i) - b\right)$$

$$\le f_{\bar{k}}(x_{\bar{k}}) + \left(\bar{\lambda}, g_{\bar{k}}(x_{\bar{k}}) + \sum_{\substack{i=1 \\ i\neq\bar{k}}}^{k} g_i(\bar{x}_i) - b\right),\ \forall \lambda \ge 0,\ \forall x_{\bar{k}} \in s_{\bar{k}}\,.$$

Select $\bar{\alpha} = (\bar{\alpha}_1, \ldots, \bar{\alpha}_k)$ so that

$$\bar{\alpha}_i = \sum_{\substack{j=1 \\ j \neq i}}^{k} g_j(\bar{x}_j), \quad i = \overline{1,k}. \tag{17.6}$$

Considering (17.6), rewrite the above inequalities

$$f_{\bar{k}}(\bar{x}_{\bar{k}}) + (\lambda, g_{\bar{k}}(\bar{x}_{\bar{k}}) + \bar{\alpha}_{\bar{k}} - b) \leq f_{\bar{k}}(\bar{x}_{\bar{k}}) + (\bar{\lambda}, g_{\bar{k}}(\bar{x}_{\bar{k}}) + \bar{\alpha}_{\bar{k}} - b)$$
$$\leq f_{\bar{k}}(\bar{x}_{\bar{k}}) + (\bar{\lambda}, g_{\bar{k}}(x_{\bar{k}}) + \bar{\alpha}_{\bar{k}} - b), \quad \forall \lambda \geq 0, \forall x_{\bar{k}} \in s_{\bar{k}}\,.$$

Thus, we obtain the conditions of the saddle point for subproblem (17.2) with $\alpha_{\bar{k}} = \bar{\alpha}_{\bar{k}}$; therefore

$$\bar{x}_{\bar{k}} = x_{\bar{k}}(\bar{\alpha}_{\bar{k}}), \quad \bar{\lambda} = \lambda^k(\bar{\alpha}_{\bar{k}})\,. \tag{17.7}$$

Since the index $\bar{k}$ is arbitrary, then equalities (17.7) hold for $\bar{k} = \overline{1,k}$.

As a result

$$\bar{\alpha}_{\bar{k}} = \sum_{\substack{j=1 \\ j \neq k}}^{k} g_j(x_j(\bar{\alpha}_j))\,,$$
$$\lambda^k(\bar{\alpha}_{\bar{k}}) = \lambda^j(\bar{\alpha}_j), \quad j, \bar{k} = \overline{1,k}\,,$$

which required to be proved.

(2) It remains to be seen whether it follows from fulfilling coordination conditions (17.3), (17.4) that $(x_1(\bar{\alpha}_1), \ldots, x_k(\bar{\alpha}_k))$ is a solution to problem (17.1).[24]

We write the conditions of the saddle point for subproblem (17.2) with $\alpha_i = \bar{\alpha}_i$:

$$f_i(x_i(\bar{\alpha}_i)) + (\lambda^i, g_i(x_i(\bar{\alpha}_i)) + \bar{\alpha}_i - b) \leq f_i(x_i(\bar{\alpha}_i))$$
$$+ (\lambda^i(\bar{\alpha}_i), g_i(x_i(\bar{\alpha}_i)) + \bar{\alpha}_i - b) \leq f_i(x_i)$$
$$+ (\lambda^i(\bar{\alpha}_i), g_i(x_i) + \bar{\alpha}_i - b), \forall \lambda^i \geq 0, \forall x_i \in s_i, i = \overline{1,k}\,. \tag{17.8}$$

Considering conditions (17.3), (17.4), from inequalities (17.8) we get

$$
\begin{aligned}
f_i(x_i(\bar{\alpha}_i)) + \left(\lambda^i, \sum_{j=1}^{k} g_j(x_j(\bar{\alpha}_j)) - b\right) &\le f_i(x_i(\bar{\alpha}_i)) \\
+ \left(\hat{\lambda}, \sum_{j-1}^{k} g_j(x_j(\bar{\alpha}_j)) - b\right) &\le f_i(x_i) \\
&+ (\hat{\lambda}, g_i(x_i) + \bar{\alpha}_i - b)\,, \\
&\forall \lambda^i \ge 0,\ \forall x_i \in s_i,\ i = \overline{1,k}\,.
\end{aligned}
\tag{17.9}
$$

Perform summation of inequalities (17.9) by the index i. Denote

$$
\sum_{i=1}^{k} \lambda^i = \lambda\,.
$$

Owing to arbitrariness and non-negativity of the vectors λ^i $(i = \overline{1,k})$ the vector λ is also arbitrary and nonnegative. Obtain

$$
\begin{aligned}
\sum_{i=1}^{k} f_i(x_i(\bar{\alpha}_i)) + \left(\lambda, \sum_{j=1}^{k} g_j(x_j(\bar{\alpha}_j)) - b\right) &\le \sum_{i=1}^{k} f_i(x_i(\bar{\alpha}_i)) \\
+ n\left(\bar{\lambda}, \sum_{j=1}^{k} g_j(x_j(\bar{\alpha}_j)) - b\right) &\le \sum_{i=1}^{k} f_i(x_i) \\
+ \left(\bar{\lambda} \sum_{i=1}^{k} [\, g_i(x_i) + \bar{\alpha}_i - b]\right), &\ \forall \lambda \ge 0,\ \forall x_i \in s_i,\ i = \overline{1,k}\,.
\end{aligned}
\tag{17.10}
$$

From the saddle point conditions for subproblem (17.2) (with $\alpha_i = \bar{\alpha}_i$) it follows that

$$
n\left(\hat{\lambda}, \sum_{j=1}^{k} g_j(x_j(\bar{\alpha}_j)) - b\right) = \left(\hat{\lambda}, \sum_{j=1}^{k} g_j(x_j(\bar{\alpha}_j))\right) = 0\,.
$$

Considering this fact, we get

$$\left(\hat\lambda, \sum_{i=1}^{k}[\,g_i(x_i)+\bar\alpha_i - b]\right)$$
$$= \left(\hat\lambda, \sum_{i=1}^{k} g_i(x_i)+\sum_{i=1}^{k}\sum_{j=1}^{k} g_j(x_j(\bar\alpha_j)) - nb\right)$$
$$= \left(\hat\lambda, \sum_{i=1}^{k} g_i(x_i) - a\right) + \left(\hat\lambda, (n-1)\left[\sum_{j=1}^{k} g_j(x_j(\bar\alpha_j))-b\right]\right)$$
$$= \left(\hat\lambda, \sum_{i=1}^{k} g_i(x_i)-b\right).$$

Thus, inequalities (17.10) can be represented as

$$\sum_{i=1}^{k} f_i(x_i(\bar\alpha_i)) + \left(\lambda, \sum_{i=1}^{k} g_i(x_i(\bar\alpha_i)) - b\right) \le \sum_{i=1}^{k} f_i(x_i(\bar\alpha_i))$$
$$+ \left(\hat\lambda, \sum_{i=1}^{k} g_i(x_i(\bar\alpha_i)) - b\right) \le \sum_{i=1}^{k} f_i(x_i)$$
$$+ \left(\hat\lambda, \sum_{i=1}^{k} g_i(x_i) - b\right), \ \forall\lambda \ge 0;\ \forall x_i \in s_i,\ i = \overline{1,k} \qquad (17.11)$$

Inequalities (17.11) are the conditions for the saddle point of (17.1), and, consequently, the vector $x(\bar\alpha) = (x_{_i}(\bar\alpha_1), \ldots, x_k(\bar\alpha_k))$ constitutes a solution to problem (1), i.e. $x(\bar\alpha) = \bar x$, whereas the vector of the Lagrange optimal multipliers $\hat\lambda$ common to all subproblems (17.2) is the same also for problem (17.1). Note that conditions (17.3), (17.4), taken singly, are not sufficient for the above statements but if subproblems are chosen as (17.2), then condition (17.4) is the necessary one for coordinating problem (17.1).[24]

§**17.2.** Choose subproblems in the following way: for $i = \overline{1,m}$

$$\min\left\{ f_i(x_i) + \sum_{\substack{j=1\\ j\ne i}}^{m} (\beta_j, g_{ij}(x_i)) \right\}, \qquad (17.12)$$
$$g_{ii}(x_i) + \alpha_i \le b_i, \quad x_i \in s_i\,,$$

for $i = \overline{m+1,k}$

$$\min\left\{f_i(x_i) + \sum_{j=1}^{m}(\beta_j, g_{ji}(x))\right\}, \qquad (17.13)$$

$$x_i \in s_i$$

where $\alpha = (\alpha_1, \ldots, \alpha_m)$, $\beta = (\beta_1, \ldots, \beta_m)$ are the coordinating parameter vectors.

Introduce the following coordination conditions:

$$\alpha_i = \sum_{j=1}^{m} g_{ij}(x_j(\alpha_j, \beta)) + \sum_{j=m+1}^{k} g_{ij}(x_j(\beta)), \qquad (17.14)$$

$$\beta_i = \lambda_i(\alpha_i, \beta), \quad i = \overline{1,m},$$

where $x_i(\alpha_i, \beta)$, $\lambda_i(\alpha_i, \beta)$ $(i = \overline{1,m})$ are respectively a solution and a vector of the Lagrange optimal multipliers in the ith subproblem (12), and $x_i(\beta)$ $(i = \overline{m+1,k})$ is a solution to the ith subproblem (17.13).

On the assumption that the saddle points of the Lagrange functions of problems (17.1), (17.12) – (17.13) exist we have proved that the statements similar to statements (1) and (2), §17.1.[38]

§17.3. Consider the following expansion of problem (17.1) for $i = \overline{1,m}$

$$\min_{x_i, u_i}\left\{f_i(x_i) + \sum_{\substack{j=1\\ j\neq i}}^{m}(\beta_j, g_{ji}(x_i)) - (\beta_i, u_i)\right\} \qquad (17.15)$$

$$g_{ii}(x_i) + u_i \le b_i, \quad x_i \in s_i, \qquad (17.16)$$

for $i = \overline{m+1,k}$

$$\min\left\{f_i(x_i) + \sum_{j=1}^{m}(\beta_j, g_{ji}(x_i))\right\}, \quad x_i \in s_i$$

where $u_1, \ldots, u_m$ are the connecting variables, $\beta = (\beta_1, \ldots, \beta_m)$ is the coordinating parameter vector. In this case the following theorems[37] are valid.

Theorem 17.1: Assume that there exists the vector $\bar{\beta} = (\bar{\beta}_1, \ldots, \bar{\beta}_m)$ such that the solutions $(\bar{x}_i, \bar{u}_i), i = \overline{1,m}$, $\bar{x}_i$ $(i = \overline{m+1,k})$ of subproblems

(17.15), (17.16) satisfy the coordination conditions

$$\bar{u}_i = \sum_{\substack{j=1 \\ j \neq i}}^{k} g_{ij}(\bar{x}_j), \quad (i = \overline{1, m}) .$$

Then $\bar{x} = (\bar{x}_1, \ldots, \bar{x}_n)$, $\bar{\beta} = (\bar{\beta}_1, \ldots, \bar{\beta}_m)$ satisfy the following saddle point conditions for subproblem (17.1)

(1) $\bar{x}$ minimizes the function

$$\sum_{i=1}^{k} f_i(x_i) + \left(\bar{\beta}, \sum_{i=1}^{k} g_i(x_i) - b \right)$$

on the set $s_1 \times \cdots \times s_k$;

(2) $\bar{\beta} \geq 0$;

(3) $\sum_{i=1}^{k} g_i(\bar{x}_i) \leq b$,

(4) $\left(\bar{\beta}, \sum_{i=1}^{k} g_i(x_i) - b \right) = 0.$

Thus, from theorem 17.1 it follows that $\bar{x}$ is a solution to problem (17.1).

Theorem 17.2: Assume that $\hat{x} = (\hat{x}_1, \ldots, \hat{x}_k)$, $\hat{\beta} = (\hat{\beta}_1, \ldots, \hat{\beta}_m)$ satisfies conditions (1) – (4) of theorem 17.1, i.e. $(\hat{x}, \hat{\beta})$ is the saddle point for problem (17.1) and

$$\hat{u}_i = \sum_{\substack{j=1 \\ j \neq i}}^{k} g_{ij}(\hat{x}_j), \quad (i = \overline{1, m}) .$$

Then $(\hat{x}_i, \hat{u}_i)\, (i = \overline{1, m})$ and $\hat{x}_i\, (i = \overline{m+1, k})$ constitute solutions to proper subproblems (17.15), (17.16) with $\beta = \hat{\beta}$.

The theorem implies that the vector $\bar{\beta} = (\bar{\beta}_1, \ldots, \bar{\beta}_m)$ described in theorem (17.1) exists. From theorem (17.2) it follows that such a decomposition technique for solving problem (17.1) is possible only when the saddle point of problem (17.1) exists.

In conclusion, it might be noted that in decomposition (17.2) the global constraints of problem (17.1) are expanded into components in a "vertical" way, whereas in decompositions (17.12), (17.13), (17.15), (17.16) expansions are implemented in a "horizontally diagonal" way.

§18. ONE METHOD OF PARAMETRIC DECOMPOSITION FOR LINEAR AND SEPARABLE PROGRAMMING PROBLEMS

§18.1. The present paragraph deals with the method of parametric decomposition[12] associated with introducing the parameters that enable one to partition at one's discretion the original problem into independent blocks. It is wise to employ the method when the constraint matrices possess a large number of zero elements, although not having a marked block structure.

We will set forth the main idea of the method using the following example.

It is required to find

$$\begin{aligned} \min(c,x) &= (c_1x_1 + \cdots + c_rx_r) + (c_{r+1}x_{r+1} + \cdots + c_nx_n)\,, \\ &(a_1x_1 + \cdots + a_rx_r) + (a_{r+1}x_{r+1} + \cdots + a_nx_n) \le b\,, \\ &x_1, \ldots, x_n \ge 0. \end{aligned}$$

Introduce the variables y_1, y_2 so that

$$a_1x_1 + \cdots + a_rx_r \le y_1;\; a_{r+1}x_{r+1} + \cdots + a_nx_n \le y_2\,,$$
$$y_1 + y_2 \le b\,.$$

The (c,x) function minimization with the fixed y_1, y_2 is expanded into two independent subproblems and the (c,x) minimum is the function from y_1, y_2. Denote $\mathcal{F}(y) = \min(c,x)$, where y is the domain of definition of $\mathcal{F}(y)$; $y^* = (y_1^*, y_2^*)$ is the point of minimum of $\mathcal{F}(y)$ in the region Y, x^* is a solution to the original problem LP. Since x^* satisfies the original problem constraints, the $(c,x^*) \ge (c,x^{**})$. On the other hand,

$$y_1^{**} = \sum_{j=1}^{r} a_jx_j^{**}; \quad y_2^{**} = \sum_{j=r+1}^{n} a_jx_j^{**}; \quad y_1^{**} + y_2^{**} \le b\,,$$

therefore $\mathcal{F}(y^{**}) \geq \mathcal{F}(y^*)$, i.e. $(c, x^*) \leq (c, x^{**})$. Thus, the LP original problem reduces to finding the $\mathcal{F}(y)$ minimum and solving the problems of smaller dimensions.

§**18.2.** Consider now the general case. Find

$$\min(c, x) = \sum_{j=1}^{n} c_i x_i\,,$$
$$\sum_{j=1}^{n} a_{ij} x_j \leq b_i, \quad i = \overline{1, m}\,, \tag{18.1}$$
$$x_j \geq 0, \quad j = \overline{1, n}\,.$$

Represent

$$\sum_{j=1}^{n} a_{ij} x_j = \sum_{k=0}^{r_i - 1} \sum_{j=n_{i_k}}^{n_{i_{k+1}}} a_{ij} x_j, \qquad n_{i0} = 1$$

and rewrite problem (18.1) in the form

$$\min(c, x)\,, \tag{18.2}$$

$$\sum_{j=n_{i_k}}^{n_{i_{k+1}}} a_{ij} x_j \leq y_{i_k}, \quad k = 1, \ldots, r_i, \quad i = \overline{1, m}\,, \tag{18.3}$$

$$x_j \geq 0, \quad j = \overline{1, n}\,, \tag{18.4}$$

where the parameters y_{i_k} are such that

$$\sum_{k=1}^{r_i} y_{i_k} \leq b_i, \quad i = \overline{1, m}\,. \tag{18.5}$$

Problems (18.2) – (18.4) are called the subproblems of problem (18.1). The optimal value of the objective function in (18.2) – (18.4) is the function from $y = \{y_{i_k}, k = \overline{1, r_i}, i = \overline{1, m}\}$ and we denote it by $\mathcal{F}(y)$ with the domain of definition Y. We set forth the problem equivalent to (18.2) – (18.4):

$$\min \mathcal{F}(y)\,,$$
$$\sum_{k=1}^{r_i} y_{i_k} \leq b_i, \quad i = \overline{1, m}\,. \tag{18.6}$$

Let y^* be a solution to (18.6), x^* a solution to the subproblem with $y = y^*$, x^{**} a solution to the original problem.

Theorem 18.1: $\mathcal{F}(y^*) = (c, x^*) = (c, x^{**})$. The proof of the theorem fully repeats the proof of the corresponding statement in §18.1.

§18.3. Thus, we have reduced the solving of the original problem to solving a nonlinear equivalent problem (18.6). The functions of the form $\mathcal{F}(y)$ are convex, but not smooth, therefore, for $\mathcal{F}(y)$ minimization it is possible to employ, say, the method of subgradient projection.[4] Let $\hat{\mathcal{F}}(y) = \{\hat{\mathcal{F}}_{i_k}(y)\}$, $k = \overline{1, r_i}$, $i = \overline{1, m}$, D is the feasible region of problem (18.6).

$$\hat{\mathcal{F}}_{i_k}(y) = \lambda_{i_k}(y)\,, \tag{18.7}$$

where $\lambda_{i_k}(y)$ are the dual variables complying with an optimal solution to subproblem (18.2) – (18.4) (with the specified y). Then, according to the subgradient projection method, a search for the solution y^* of the equivalent problem and its corresponding solution of the subproblem amounts to the following procedure. Let $y^0 = \{y^0_{i_k}\}$ be an arbitrary point of the region D, $y^s = \{y^s_{i_k}\}$ the point obtained after the sth step, x^s the solution of (18.2) – (18.4) with $y = y^s$, $\lambda^s = \{\lambda_{i_k}(y^s)\}$ are the corresponding dual variables. Then

$$y^{s+1} = p(y^s - \rho_s \lambda(y^s)), \quad s = 0, 1, \ldots,$$

where p is the operation of projecting onto the region D, ρ_s the step descent value which, for the convergence of the sequence $\{y^s\}\, y^*$ is chosen so that $\rho_s \geq 0$, $\sum_{s=0}^{\infty} \rho_s = \infty$, e.g. $\rho_s = (s+1)^{-1}$, $s = 0, 1, \ldots$. The subgradient projection method does not ensure fulfillment of the monotonicity condition $\mathcal{F}(y^s) > \mathcal{F}(y^{s+1})$ at each iteration and, in general, slowly converges, but if the points on the set D and the subgradient $\mathcal{F}$ are found not in a complex way, then the method is rather simple for computer implementation. Note that, in practice, the $\mathcal{F}(y)$ function definition domain is most commonly specified by the form

$$a_{i_k} \leq y_{i_k} \leq b_{i_k}\,.$$

In this case, projection onto the region D reduces to simple calculations.

§**18.4.** Consider the application of the above parametric decomposition approach to the separable programming problem:

$$\min \sum_{j=1}^{n} c_j(x_j)\,,$$

$$\sum_{j=1}^{n} a_{ij}(x_j) \le b_i,\ i = \overline{1,m};\ x_j \ge 0,\ j = \overline{1,n}\,.$$

The subproblem of form (18.2) – (18.4) in this case takes the form

$$\min \sum_{j=1}^{n} c_j(x_j)\,,$$

$$a_{ij}(x_j) \le y_{ij}, \quad i = \overline{1,m}; \quad j = \overline{1,n}\,;$$

$$x_j \ge 0, \quad j = \overline{1,n}\,,$$

where the values y_{ij} are defined as

$$\sum_{j=1}^{n} y_{ij} \le b_i, \quad i = \overline{1,m}\,.$$

In the present case, the subproblem reduces to the simplest problems of nonlinear programming.

For the separable programming problems, the equivalent problem is solved by the subgradient projection method when $c_j(x)$, $a_{ij}(x)$ are convex functions.

Chapter 5

DECOMPOSITION BASED ON SEPARATION OF VARIABLES

§19. CONSTRAINT RELAXATION METHOD FOR THE CONVEX PROGRAMMING PROBLEM

§19.1. Consider the convex programming problem

$$\max f(x),$$
$$g_i(x) \geq 0, \quad i = \overline{1,m}, \tag{19.1}$$
$$x \in s,$$

where $f, g_i\,(i = \overline{1,m})$ are the convex functions, s is the convex subset E^n.

Consider the case with the number of constraints being rather large. Then the reasonable strategy of solution consists in relaxation, i.e. temporary rejection of some of these constraints and solutions of the remaining problem. If the problem is unsolvable, then so is the original problem. If it is solvable, then provided the solution obtained satisfies the relaxed constraints, it is also optimal for original problem (19.1). Otherwise, it is necessary to introduce one or more of the constraints omitted and repeat the procedure. The relaxation method is particularly effective when it is known in advance that a small number of constraints is substantial at the optimal point.

Denote $M = \{1, 2, \ldots, m\}$ and let $R \subset M$. Introduce into consideration the problem

$$\max f(x)\,,$$
$$g_i(x) \geq 0, \quad i \in R\,, \tag{19.2}$$
$$x \in s\,.$$

§19.2. We will set forth the relaxation method algorithm. Assume that it is possible to find such an initial set R that (19.2) has a finite maximum.

Step 0: Suppose $\bar{f} = \infty$ and choose the initial set R such that f is bounded on the constraints of problem (19.2).

Step 1: Solve (19.2). If it is unsolvable, then (19.1) has no solution. Otherwise we get the solution x^R of problem (19.2).

Step 2: If $g_i(x^R)$, $i \in M \setminus R$, x^R is optimal for original problem (19.1). Go to step 6.

Step 3: Otherwise let V be the subset $M \setminus R$ involving the indices of, at least, one of the unfulfilled constraints. Introduce the subset of indices $D \subset R$, for which the constraints $g_i(x) \geq 0$ are fulfilled strictly on the optimal solution x^R:

$$D = \{i | g_i(x^R) > 0,\, i \in R\}\,.$$

Step 4: If $f(x^R) = \bar{f}$, then replace R by $R' = R \cup V$. Go to step 1.

Step 5: If $f(x^R) < \bar{f}$, then replace R by $R' = R \cup V \setminus D$ set $\bar{f} = f(x^R)$. Go to step 1.

Step 6: Terminate.

Thus, if the functional optimal value in (19.2) decreases, then the violated constraints are introduced and the constraints inessential for x^R are removed. If the optimum is maintained, only then is introduction of constraints performed.

§19.3. Show that the above method, after solving a finite number of problems (19.2) either leads to the optimal solution of original problem (19.1) or the unsolvability of the problem is established.

Call problem (19.2) the p^R-problem.

Theorem 19.1: If x^R is a solution to p^R, then x^R is a solution to $p^{R\setminus D}$. Further, if x^R is a unique solution to p^R, then it a unique solution to p^{R-D} as well.

Proof: Suppose we have the vector $\hat{x}$ satisfying the constraints $\{R \setminus D\}$ and such that $f(\hat{x}) > f(x^R)$. Consider the points

$$x_\lambda = \lambda\hat{x} + (1-\lambda)x^R, \quad 0 < \lambda < 1\,.$$

Since the set S is convex and all g_i are concave, $x_\lambda \in S$ and x_λ satisfies the constraints $\{R \setminus D\}$ for all $0 < \lambda < 1$. Consider the constraints in D. Since g_i are concave, then

$$g_i(x_\lambda) \geq \lambda g_i(\hat{x}) + (1-\lambda)g_i(x^R)\,.$$

Since $g_i(x^R) > 0$, $i \in D$, then for the sufficiently small $\lambda g_i(x_\lambda) > 0$, $i \in D$. Therefore, for such λ the point $x_{\bar{\lambda}}$ is feasible in p^R.

However, due to the concavity of f

$$f(x_\lambda) \geq \lambda f(\hat{x}) + (1-\lambda)f(x^R) > f(x^R)\,,$$

which contradicts the optimality of x^R.

Theorem 19.2:

$$f(x^{R'}) \leq f(x^R)\,.$$

Proof: If $R' = R \cup V$, then the theorem evidently holds. Due to theorem (19.1):

$$f(x^{R\setminus D}) = f(x^R)\,,$$

hence

$$f(x^{(R\setminus D)\cup V}) \leq f(x^{R-D})\,.$$

Theorem 19.3: Relaxation procedure terminates after having solved a finite number of the problems P^R and leads either to the optimal solution of the original problem or to the construction of such constraints set that the current problem P^R proves to be unsolvable.

Proof: Since the set M is finite, then it has a finite number of different subsets. With $f(x)$ decreasing from iteration to iteration, none of the subsets can be repeated. Inasmuch as none of the constraints is withdrawn, if $f(x^{R'}) = f(x^R)$, and at least one constraint is added, then f can remain constant only over a finite number of iterations. Thus, for the finite number of steps the procedure must terminate at step 1 or step 2.

§20. THE RITTER METHOD FOR THE BLOCK PROBLEM WITH THE TYING VARIABLES AND CONSTRAINTS

§20.1. Consider the LP block problem with the tying variables and constraints:

$$\max z = \sum_{i=1}^{p} c_i x_i + c_0 y\,, \tag{20.1}$$

$$\sum_{i=1}^{p} A_i x_i + D_0 y = b_0\,, \tag{20.2}$$

$$B_i x_i + D_i y = b_i, \quad i = \overline{1,p} \tag{20.3}$$

$$x_i \geq 0, \quad i = \overline{1,p}, \quad y \geq 0 \tag{20.4}$$

and the problem dual with respect to it:

$$\min \vartheta = \sum_{i=1}^{p} b_i u_i + b_0 u_0\,, \tag{20.5}$$

$$\begin{aligned} &\sum_{i=1}^{p} D_i' u_i + D_0' u_0 \geq c_0\,, \\ &B_i' u_i + A_i u_0 \geq c_i, \quad i = \overline{1,p}\,, \end{aligned} \tag{20.6}$$

where $A_i, B_i, D_i - (m_0 \times n_i), (m_i \times n_i), (m_i \times n_0)$ are matrices, respectively.

Ritter[32] has developed the relaxation method for the problems of form (19.1). At each iteration, variable problems separate into two subsets – the subset S_1, in which non-negativity constraints are relaxed, and the subset S_2, in which they are retained. If the variables in S_1 prove to be non-negative, then the current solution will be optimal for the original problem.

Assume that each matrix $B_i\, (i = \overline{1,p})$ has the rank m_i and, consequently, contains a nonsingular submatrix B_{i_1}. Suppose the initial vector y_0 is known such that the feasible solution of the constraint system exists

$$B_i x_i = b_i - D_i y_0, \quad i = \overline{1,p}\,,$$

then a set of initial bases can be found by solving subproblems

$$\max c_i x_i ,$$
$$B_i x_i = b_i - D_i y_0, \quad x_i \geq 0$$

and using optimal basis matrices as B_{i_1}. After isolating B_{i_1} constraints (20.3) can be rewritten as

$$B_{i_1} x_{i_1} + B_{i_2} x_{i_2} + D_i y = b_i \,. \tag{20.7}$$

Since each matrix B_{i_1} is nonsingular, equation (20.7) can be solved with respect to x_{i_1}:

$$x_{i_1} = B_{i_1}^{-1} b_i - B_{i_1}^{-1} B_{i_2} x_{i_2} - B_{i_1}^{-1} D_i y \,. \tag{20.8}$$

Divide variables in objective function (20.1) and tying constraint (20.2):

$$z = \sum_{i=1}^{p} (c_{i_1} x_{i_1} + c_{i_2} x_{i_2}) + c_0 y \,, \tag{20.9}$$

$$\sum_{i=1}^{p} (A_{i_1} x_{i_1} + A_{i_2} x_{i_2}) + D_0 y = b_0 \,. \tag{20.10}$$

Substituting (20.8) in (20.9), (20.10), obtain the reduced problem

$$\begin{gathered} \max \left\{ z - \alpha = \sum_{i=1}^{p} d_i x_{i_2} + d_0 y \right\} , \\ \sum_{i=1}^{p} M_i x_{i_2} + M_0 y = b \,, \\ x_{i_2} \geq 0, \quad y \geq 0 \,, \end{gathered} \tag{20.11}$$

where

$$\alpha = \sum_{i=1}^{p} c_{i_1} B_{i_1}^{-1} b_i \,,$$
$$d_i = c_{i_2} - c_{i_2} B_{i_1}^{-1} B_{i_2} \,,$$

$$d_0 = c_0 - \sum_{i=1}^{p} c_{i_1} B_{i_1}^{-1} D_i \,,$$
$$M_i = A_{i_2} - A_{i_1} B_{i_1}^{-1} B_{i_2} \,,$$
$$M_0 = D_0 - \sum_{i=1}^{p} A_{i_1} B_{i_1}^{-1} D_i \,,$$
$$b = b_0 - \sum_{i=1}^{p} A_{i_1} B_{i_1}^{-1} b_i \,.$$

The problem has as many constraints as the original problem contains the tying constraints.

The constraints in (20.11) prescribe the same set of feasible solutions as initial constraints (20.2) – (20.3) do. However, since the reduced problem does not explicitly involve the variables x_{i_1}, then it is impossible to fix in it that $x_{i_1} \geq 0$, i.e. the constraints relax. The feasible set of problems (20.11) evidently involves a feasible set of the original problem so that if the reduced problem is unsolvable, then so is the original. It is evident, however, that the reduced problem can have an unconstrained solution in those cases when the original problem has a finite optimal solution. To exclude the possibility, one can add into the reduced problem the constraint of the form

$$\sum_{i=1}^{p} l_i' x_{i_2} + l_0' y \leq t \,, \tag{20.12}$$

where t is a sufficiently large positive number. If, on completion of algorithm operation, this constraint proves to be significant, then the optimum is unbounded. Suppose that an optimal solution to the reduced problem $(\{x_{i_2}^0\}, y_0)$ has been found. Substituting it in the right-hand side of (20.8), we obtain the solution $(\{x_{i_1}^0\}, \{x_{i_2}^0\}, y_0)$ of the original problem without assuming $x_{i_1} \geq 0$. If the constraints are fulfilled, the solution obtained is the solution of the original problem.

Theorem 20.1: (Optimality criterion). The solution $(\{x_{i_1}^0\}, \{x_{i_2}^0\}, y_0)$ constitutes the solution of the original problem if

$$x_{i_1}^0 \geq 0, \quad i = \overline{1,p}\,. \tag{20.13}$$

§**20.2.** Consider the reduced problem transformation in the case of unfulfillment of optimality condition (20.13). In this case it is necessary to construct a new reduced problem, in which some of the violated conditions for non-negativity are introduced as necessary, whereas the requirements for non-negativity of the variables x_{i_2}, for which $x^0_{i_2} > 0$, are omitted.

Consider an arbitrary vector x_{j_1} characterized by the fact that even one of the components $x^0_{j_1}$ in the solution obtained is negative. For definiteness, assume that the first l of the components $x^0_{j_1}$ are negative:

$$(x^0_{j_1})_1 < 0,\ (x^0_{j_1})_2 < 0, \ldots, (x^0_{j_1})_l < 0,\ l \geq 1 \tag{20.14}$$

Suppose $x^0_{j_2}$ has the first q components being positive:

$$(x^0_{j_2})_1 > 0,\ (x^0_{j_2})_2 > 0, \ldots, (x^0_{j_2})_q > 0,\ q \geq 0\,. \tag{20.15}$$

Let us write the relation of (20.8) connecting x_{j_1}, x_{j_2} as:

$$x_{j_1} + B^{-1}_{j_1} B_{j_2} x_{j_2} = b^{-1}_{j_1} b_j - B^{-1}_{j_1} D_j y\,. \tag{20.16}$$

Two cases are possible here.

Case 20.1: There exists at least one nonzero element of the submatrix formed by the first l rows and the first q columns in $B^{-1}_{j_1} B_{j_2}$. In this case, one can perform the leading operation using, as the leading element, an arbitrary nonzero element of the submatrix mentioned. Such an operation transforms (20.16), introducing into the basis the component x_{j_2} positive in $x^0_{j_2}$ and withdrawing the component x_{j_1} negative in $x^0_{j_1}$. This substitution leads to a new basic matrix $\hat{B}_{j_1}$ which is employed for constructing a new reduced problem.

Case 20.2: The submatrix formed by the first l rows and the first q columns $B^{-1}_{j_1} B_{j_2}$ contains only zero elements. Then it is impossible to accomplish substitution of the basis. Choose a negative component $x^0_{j_1}$, say, $(x^0_{j_1})_k, 1 \leq k \leq l$ and use (16) to express explicitly the condition for non-negativity of this variable by x_{j_2} and y:

$$(x_{j_1})_k = \alpha_{j_k} - \bar{b}_{j_k} x_{j_2} - \bar{d}_{j_k} y \geq 0\,, \tag{20.17}$$

where $\alpha_{j_k}, \bar{b}_{j_k}, \bar{d}_{j_k}$ are the kth rows of the matrices

$$B^{-1}_{j_1} b_j, B^{-1}_{j_1} B_{j_2}, B^{-1}_{j_1} D_j$$

respectively. This constraint is added to the cases 1 or 2 for each vector $x_{i_1}^0$ possessing negative components. We come to a new reduced problem in which the matrices B_{i_1} can be changed and the constraints of form (20.17) can be added. Before applying the procedures indicated in cases 1 and 2, these constraints are to be examined according to the rules given below.

Case 20.3: If constraints (20.17) are not significant on the optimal solution, i.e. if

$$\alpha_{j_k} - \bar{b}_{j_k} x_{j_2}^0 - \bar{d}_{j_k} y^0 > 0\,,$$

then the constraints are withdrawn from the reduced problem. By virtue of theorem (9.1), §9 this does not alter the optimal solution of the problem as a whole.

Case 20.4: If constraints (20.17) are significant on the optimal solution, i.e. if

$$\alpha_{j_k} - \bar{b}_{j_k} x_{j_2}^0 - \bar{d}_{j_k} y^0 = 0\,, \tag{20.18}$$

then the constraint is omitted when alteration is possible in the basis related to substituting $(x_{j_1})_k$ by a component x_{j_2} which is positive in $x_{j_2}^0$. Here are visible the elements $\bar{b}_{j_k}$ related to the positive components $x_{j_2}^0$; if there is a nonzero element, then it is used as the leading and, after performing the leading operation, an appropriate constraint of (20.17) is omitted, and the optimal solution again remains unaltered.

Case 20.5: If (20.18) is fulfilled and all the components $\bar{b}_{j_k}$ corresponding to the positive components in $x_{j_2}^0$ are equal to zero, then constraints (20.17) are preserved and conditions (20.18) take the form

$$\bar{d}_{j_k} y^0 = \alpha_{j_k}\,.$$

§**20.3.** Formulate the algorithm for solving problem (20.1) – (20.4).

<u>*Step 1*</u>: Choose the initial basic matrices B_{i_1} $(i = \overline{1,p})$. This can be done in an arbitrary way or by solving the subproblems

$$\max\, c_i x_i\,,$$
$$B_i x_i = b_i - D_i y_0\,, \quad x_i \geq 0$$

and using their optimal bases as B_{i_1}. Reduced subproblem (20.11) is constructed with the use of these matrices.

Step 2: Obtain $(\{x^0_{i_2}\}, y^0)$ as a solution to (20.11). Calculate $x^0_{i_1}$ from formula (20.8). If $x^0_{i_1} \geq 0$, $i = \overline{1,p}$, then $(\{x^0_{i_1}\}, \{x^0_{i_2}\}, y^0)$ is an optimal solution to original problem (20.1) – (20.4). Go to step 6.

Step 3: If $x^0_{i_1} \ngeq 0$, $i \in I$ and the additional constraints of form (20.17) exist in the reduced problem, then we consider them with the use of the procedures indicated in cases 3–5.

Step 4: The leading operation is used for each $i \in I$, as in case 1, or constraints are added, as in case 2.

Step 5: Using the new basic matrices and constraints obtained at stages (steps) 3–4, we construct a new reduced problem and go to step 2.

Step 6: Terminate.

§**20.4.** Consider the issue of convergence.

Since the above algorithm constitutes a relaxation procedure, then theorems 9.1, §9 can be applied. Owing to theorem 9.2, §9, the sequence of objective functions obtained when optimizing the reduced problems, is monotonic decreasing. Theorem 9.3, §9 yields the following conditions for convergence for the finite number of iterations.

Theorem 20.2: If the requirements for non-negativity are omitted only on the iterations when the reduced problem objective function decreases, then the algorithm converges, for the finite number of iterations, either to the optimal solution of original problem (20.1) – (20.4) or to constructing the reduced problem without a feasible solution.

§**20.5.** Show that the Ritter algorithm is dual.

Theorem 20.3: Each solution $(\{x^0_{i_1}\}, \{x^0_{i_2}\}, y^0)$ obtained on the kth iteration corresponds to an extreme point of the feasible set of dual problem (20.5), (20.6) $\{u^0_i\}$ with equal objective functions:

$$\sum_{i=1}^{p} b_i u^0_i = c_0 y^0 + \sum_{i=1}^{p} (c_{i_1} x^0_{i_1} + c_{i_2} x^0_{i_2}) \,. \tag{20.19}$$

Proof: The solution $(\{x^0_{i_1}\}, \{x^0_{i_2}\}, y^0)$ is the solution of the modified original problem prescribed by conditions (20.1) – (20.3) and requirements for non-negativity:

$$x_{i_2} \geq 0, \quad i = \overline{1,p}, \quad y \geq 0\,, \tag{20.20}$$

$$(x_{i_k 1})_{j_k} \geq 0\,, \tag{20.21}$$

where (20.21) corresponds to the kth constraint of form (20.17). The problem dual with respect to the modified original one involves the equality instead of inequalities (20.6) corresponding to the components $\{x_{i_1}\}$ not restricted in sign. By the duality theorem, if the modified original problem is a finite optimal solution, then the problem dual to it also has this solution. Therefore, the modified dual problem achieves an optimum at the extreme point $\{u^0_i\}$ satisfying (20.19). And since any extreme point of the modified dual problem simultaneously represents the extreme point of original dual problem (20.5), (20.6), then the proof of the theorem is completed.

§21. THE ROSEN DIVISION METHOD FOR LINEAR PROGRAMMING PROBLEMS

§21.1. Rosen proposed the division method for solving the pair of dual problems of the form:

Direct Problem:

$$\min\left\{ z = \sum_{i=1}^{p} c_i x_i \right\}, \tag{21.1}$$

$$\sum_{i=1}^{p} A_i x_i = b_0\,, \tag{21.2}$$

$$B_i x_i = b_i, \quad i = \overline{1,p}\,, \tag{21.3}$$

$$x_i \geq 0, \quad i = \overline{1,p}\,, \tag{21.4}$$

Dual Problem:

$$\max\left\{ v = \sum_{i=1}^{p} b_i u_i + b_0 y \right\}, \tag{21.5}$$

$$B'_i u_i + A'_i y \leq c_i, \quad i = \overline{1,p}\,, \tag{21.6}$$

where the matrices A_i, B_i are the $(m_0 \times n_i), (m_i \times n_i)$-dimensional matrices, respectively.

The Rosen method can be regarded as a specialization of the Ritter method described in the preceding paragraph.

We show here that a series of simplifications is possible as opposed to the Ritter method:

(1) it is not required to introduce additional constraints into a reduced problem so that the problem will have no more than m_0 rows;

(2) consecutive reductions of the problem can be solved by the dual simplex method.

Further in this paragraph, we will employ the following assumption.

Assumption 21.1: The direct problem constraint matrix has the full rank.

Theorem 21.1: If assumption 1 is fulfilled, then each matrix B_i contains the nonsingular submatrix B_{i_1} of the m_i dimension.

Proof: If B_i has the rank smaller than m_i, then rows are of the form (21.3), which contradicts assumption (21.1). Therefore, $n_i \geq m_i, B_i$ contains the nonsingular submatrix B_{i_1}. We can also assume that $n_i > m_i$ since, if $n_i = m_i$, then the variables x_i are uniquely defined. The initial set of basic matrices can be obtained by solving the problems:

$$\min c_i x_i \,,$$
$$B_i x_i = b_i, \quad x_i \geq 0\,. \tag{21.7}$$

Assumption 21.2: The initial basic matrices B_{i_1} satisfy the conditions

$$B_{i_1}^{-1} b_i \geq 0, \quad i = \overline{1,p}\,.$$

The basic matrices B_{i_1} are employed for transforming direct and dual problems:

Direct Problem:

$$\min \sum (c_{i_1} x_{i_1} + c_{i_2} x_{i_2})\,, \tag{21.8}$$

$$\sum(A_{i_1}x_{i_1} + A_{i_2}x_{i_2}) = b_0\,,$$
$$B_{i_1}x_{i_1} + B_{i_2}x_{i_2} = b_i\,, \qquad (21.9)$$
$$x_{i_1} \geq 0, \quad x_{i_2} \geq 0, \quad i = \overline{1,p},$$

Dual Problem:

$$\max \sum_i (b_i u_i + b_0 y)\,, \qquad (21.10)$$

$$\begin{aligned} B'_{i_1}u_i + A'_{i_1}y &\leq c_{i_1}\,, \\ B'_{i_2}u_i + A'_{i_2}y &\leq c_{i_2}\,. \end{aligned} \qquad (21.11)$$

As in the Ritter method, relaxing the constraints $x_{i_1} \geq 0$ and using (21.9) for exclusion of x_{i_1} from the coupling constraints of (21.8), we come to a reduced problem. Thus

$$x_{i_1} = B_{i_1}^{-1}b_i - B_{i_1}^{-1}B_{i_2}x_{i_2}\,. \qquad (21.12)$$

The reduced problem takes the form:

$$\begin{aligned} \min \sum d_i x_{i_2}\,, \\ \sum M_i x_{i_2} = \bar{b}_0, \quad x_{i_2} \geq 0\,, \end{aligned} \qquad (21.13)$$

where

$$d_i = c_{i_2} - c_{i_1}B_{i_1}^{-1}B_{i_2}\,,$$
$$M_i = A_{i_2} - A_{i_1}B_{i_1}^{-1}B_{i_2}\,,$$
$$\bar{b}_0 = b_0 - \sum_{i=1}^{p} A_{i_1}B_{i_1}^{-1}b_i\,.$$

Evidently, the feasible set determined by (21.12), (21.13) comprises the original problem feasible set so that if the reduced problem has no feasible solutions, then the original problem is unsolvable. The reduced problem can have an unconstrained solution whereas the original problem has a finite optimal solution, therefore, as above, it is possible to employ the constraint of the form

$$\sum l'_j x_{j_2} \leq t\,,$$

where t is a sufficiently large positive number.

Let $\{x^0_{i_2}\}$ be the solution of (21.13). By formula (21.12), obtain the values of x_{i_1}:

$$x^0_{i_1} = B^{-1}_{i_1} b_i - B^{-1}_{i_1} B_{i_2} x^0_{i_2}, \quad i = \overline{1,p}.$$

Theorem 21.2: (optimality criteria). The vectors $\{x^0_{i_1}\}, \{x^0_{i_2}\}$ constitute a solution to the original problem if and only if

$$x^0_{i_1} \geq 0, \quad i = \overline{1,p}. \tag{21.14}$$

If the optimality criteria is not satisfied, then, as above, it is necessary to take account of at least one of the unviolated constraints of non-negativity and, in doing so, relax the constraints of non-negativity on positive variables. In the Ritter method, for the problem with the tying variables and the tying constraints, we did not always succeed in performing this transformation with the aid of the leading operation and we had to introduce additional constraints. In this case, however, the sequence of the leading operations, which leads to a solution, always exists.

Theorem 21.3[33]**:** If, for an i, $x^0_{i_1}$ has negative components, then we can form a sequence of the leading operations in the ith block introducing into the basis the component x_{i_2}, that has been positive in $x^0_{i_2}$, and the component x_{i_1}, that has been negative in $x^0_{i_1}$.

The proof of the theorem needs the following lemma.

Lemma 21.1[39]**:** Consider the linear system $Ax = b$, where A is the $(m \times n)$ matrix, $n \geq m$, rank $(A) = m$. If the system has a non-negative solution, then it also has a basic non-negative solution.

The proof of theorem (21.3). For notation simplicity, omit the i block index and consider a particular block. There are two solutions of the system

$$B_1 x_1 + B_2 x_2 = b, \quad x_1 \geq 0, \quad x_2 \geq 0, \tag{21.15}$$

such as the feasible $[x_{B_1}, 0]'$, the unfeasible $[x^0_1, x^0_2]$, where $x_{B_1} = B^{-1}_1 b$. The former solution is non-negative (see assumption 21.2) while the latter has negative components in x^0_1. All the vectors of the form

$$(1 - \Theta) \begin{bmatrix} x_{B_1} \\ 0 \end{bmatrix} + \Theta \begin{bmatrix} x^0_1 \\ x^0_2 \end{bmatrix} \tag{21.16}$$

satisfy the conditions

$$B_1x_1 + B_2x_2 = b, \quad x_2 \geq 0\,.$$

Let $I = \{j : x^0_{1_j} < 0\}$. The largest Θ, with which (21.16) remains nonnegative, is bounded by the components with the indices $j \in I$. Under the assumption that these components in (21.16) are equal to null, obtain

$$x_{B_{1_j}} + \Theta(x^0_{1_j} - x_{B_{1_j}}) = 0, \quad j \in I$$

or

$$\Theta_j = \frac{x_{B_{1_j}}}{x_{B_{1_j}} - x^0_{1_j}}, \quad j \in I\,.$$

Note that $0 \leq \Theta_j \leq 1$. The largest Θ, with which (21.16) remains negative, is equal

$$\Theta_s = \min_{j\in I} \Theta_j\,.$$

Let

$$K = \{k : x^0_{2_k} > 0\}$$

must be nonempty since if all $x^0_{2_k} = 0$, then $x^0_1 = B_1^{-1}b \geq 0$, which contradicts our hypothesis that x^0_1 has negative components. When $\Theta = \Theta_s$ the vector (21.16) satisfies the equation

$$B_1(x_{B_1} + \Theta_s(x^0_1 - x_{B_1})) + \sum_{k\in K} B^k_2(\Theta_s x^0_{2_k}) = b\,, \tag{21.17}$$

where B^k_2 is the kth column B_2. By defining Θ_s, the coefficient with the column B^s_1 is equal to null. Since the matrix B_1 is unexpressed and has $(m \times m)$ dimension, the matrix $[B_1, B_2]$ has the rank m. Omit the column B^s_1 in (17) and check the rank of the system obtained. Multiplying (21.17) by B_1^{-1}, obtain

$$\sum_{\substack{i=1\\ i\neq s}}^{m} v^0_i l_i + \sum_{k\in K} \bar{B}^k_2(\Theta_s x^0_{2_k}) = B_1^{-1}b\,, \tag{21.18}$$

where v^0_i is the ith component $(x_{B_1} + \Theta_s(x^0_1 - x_{B_1}))$, $\bar{B}^k_2$ the kth column $B_1^{-1}B_2$. The system has the same rank as system (21.17), without B^s_1.

Moreover, at least one of the vectors $\bar{B}_2^k, k \in K$ say, $\bar{B}_2^{k^*}$ must have a nonzero sth component because for $j \in I$ the vector

$$x_1^0 = x_{B_1} - B_1^{-1} B_2 x_2^0$$

has negative components. Since $s \in I$, if all the components $B_1^{-1} B_2$ in the row s and columns $k \in K$ are equal to zero, then

$$x_{1s}^0 = x_{B_1 s} \geq 0 ,$$

which yields a contradiction. Thus, the vectors l_i, $i = \overline{1,m}$, $i \neq s$, $\bar{B}_2^{k^*}$ are independent, and system (21.17) with the omitted B_1^s has the rank m. Since the system has a non-negative solution, then, due to lemma 21.1, it also has a basic non-negative solution. Any solution of that kind does not comprise B_1^s and can include only the columns $B_2^k, k \in K$.

The proof is completed.

The theorem does not state that the unique leading operation, giving rise to a new basis $\bar{B}_i$, exists. These bases, however, can be obtained from solving the problem:

$$\min x_{1s} ,$$

$$\sum_{k=1}^{m} B_1^k x_{1k} + \sum_{k \in K} B_2^k x_{2k} = b , \tag{21.19}$$

$$x_{1k} \geq 0, \quad x_{2k} \geq 0 ,$$

where s is the index selected from the condition

$$\Theta_s = \min_{j \in I} \Theta_j . \tag{21.20}$$

The initial basis in the problem is B_1. That kind of problem should be solved for each non-negative block.

§21.2. Formulate the algorithm for solving problem (21.1) – (21.4).

<u>*Step 1*</u>: Choose the initial bases B_1 for each block. It can be done by solving subproblem (21.7).

Step 2: With the use of the current set of bases, construct and solve the reduced problem (21.13), obtaining optimal solutions $\{x^0_{i_2}\}$.

Step 3: Calculate $x^0_{i_1}$ from formula (21.12).

Step 4: If $x^0_{i_1} \geq 0$, $i = \overline{1,p}$, then $\{x^0_{i_1}, x^0_{i_2}\}$ is a solution to the original problem (21.1) – (21.4). Go to step 7.

Step 5: Otherwise for each i such that $x^0_{i_1} \ngeq 0$ we calculate the index s according to (21.20). Solve problem (21.19).

Step 6: Using the new basic matrices $\bar{B}_{i_1}$ obtained when solving (21.19), go to step 2.

Step 7: Terminate.

§21.3. A few words must be said about the above algorithm convergence.

Since the algorithm constitutes special implementation of the relaxation procedure, convergence for the finite number of iterations follows from theorem (10.2), §10. In this case, however, we need not take account of the non-negativity condition violations by incorporating additional constraints into the reduced problem. If none of the additional constraints is introduced, then the assumption that each reduced problem has no other optimal solutions satisfying the newly incorporated conditions for non-negativity ensures convergence for the finite number of steps.

§22. THE ROSEN DIVISION METHOD FOR NONLINEAR PROGRAMMING

§22.1. The Rosen method for LP problems, described in the preceding paragraph, was extended by Rosen to particular nonlinear problems of the form:

$$\min \left\{ \sum_{i=1}^{p} c_i(y)x_i + c_0(y) \right\}, \tag{22.1}$$

$$A_i(y)x_i \geq b_i(y), \quad i = \overline{1,p}, \quad p \geq 1\,. \tag{22.2}$$

The distinctive feature of the problem structure is that for the fixed y the problem separates into p independent linear subproblems with respect to the variables x_i. Thus, the natural iterative procedure exists, in which first y changes, and then x_i. The mathematical proof of convergence to the global minimum prevails when in (22.1), (22.2) c_i, A_i $(i = \overline{1,p})$ do not depend on y, i.e. for the problem of the form

$$\min \sum_{i=1}^{p} c_i x_i \,, \tag{22.3}$$

$$A_i x_i \geq b_i(y), \quad i = \overline{1,p} \,. \tag{22.4}$$

Consider first this type of problems and then a more general problem (22.1), (22.2).
We make additional assumptions:

Assumption 22.1: For $i = \overline{1,p}$ each $b_i(y)$ is the differentiable convex function y. Hence it follows that (22.3), (22.4) constitute the convex programming problem for which the local minimum is global.

Assumption 22.2: There exist the feasible vectors (x_i, y), being interior with respect to nonlinear constraints, i.e. such that $A_i x_i \geq b_i(y)$ where the strict inequality for the nonlinear components $b_i(y)$ is fulfilled. This secures existence of the saddle point of the Lagrange function. Assumptions 1, 2 provide that the Kuhn-Tucker conditions are necessary and sufficient for an optimal solution.

Assumption 22.3: None of the matrix rows A_i $(i = \overline{1,p})$ is a zero vector.

Assumption 22.4: The feasible vector y_0 exists and is known, i.e. such y_0 is known for which the x_i satisfying (22.4) exist. Problem (22.3), (22.4) can be solved via minimization by x_i with the fixed y, and then via minimization of a result by y:

$$\min_{y \in S} \left\{ \sum_{i=1}^{p} \min\{c_i x_i | A_i x_i \geq b_i(y)\} \right\}, \tag{22.5}$$

where

$$S = \bigcap_{i=1}^{p} s_i, \quad s_i = \{y | \exists x_i : A_i x_i \geq b_i(y)\} \,.$$

The vectors $y \in S$ are termed feasible.

Define the function

$$\psi_i(y) = \min_{x_i}\{c_i x_i | A_i x_i \geq b_i(y)\}.$$

Then problem (22.5) can be rewritten as:

$$\min\ \psi(y) = \sum_{i=1}^{p} \psi_i(y), \tag{22.6}$$

$$y \in S. \tag{22.7}$$

We will show that under assumption 22.1, problem (22.6), (22.7) is the convex programming problem.

Theorem 22.1: If each function $b_i(y)$ is convex, then the set S is convex and each function ψ_i is convex on S.

Proof:

(1) Show convexity of the set S. To do this, it suffices to say that each $s_i\ (i = \overline{1,p})$ is convex, i.e. from the fact that $y_1 \in s_i$, $y_2 \in S$ it follows that $\lambda y_1 + (1-\lambda)y_2 \in s_i$, $0 < \lambda < 1$. Since $y_1 \in s_i\, \exists\, \bar{x}_i$, which $A_i\bar{x}_i \geq b_i(y_1)$. Similarly, $\exists\ \bar{\bar{x}}_i$ which $A_i\bar{\bar{x}}_i \geq b_i(y_2)$. These inequalities and convexity $b_i(y)$ imply:

$$\begin{aligned} b_i(\lambda y_1 + (1-\lambda)y_2) &\leq \lambda b_i(y_1) + (1-\lambda)b_i(y_2) \\ &\leq \lambda A_i\bar{x}_i + (1-\lambda)A_i\bar{\bar{x}}_i \leq A_i(\lambda\bar{x}_i + (1-\lambda)\bar{\bar{x}}). \end{aligned}$$

The proof of s_i convexity is completed. And since the intersection of convex sets is the convex set, then S is also convex.

(2) We will show that the function $\psi_i\ (i = \overline{1,p})$ is convex on S, i.e.

$$\begin{aligned} &\psi_i(\lambda y_1 + (1-\lambda)y_2) \leq \lambda\psi_i(y_1) + (1-\lambda)\psi_i(y_2), \\ &\forall y_1, y_2 \in S, \quad 0 < \lambda < 1, \\ &\psi_i(y_1) = \min_{x_i}\{c_i x_i | A_i x_i \geq b_i(y_1)\} = c_i x_i(y_1), \\ &\psi_i(y_2) = \min_{x_i}\{c_i x_i | A_i x_i \geq b_i(y_2)\} = c_i x_i(y_2). \end{aligned} \tag{22.8}$$

The convexity of $b_i(y)$ implies

$$\begin{aligned} b_i(\lambda y_1 + (1-\lambda)y_2) &\le \lambda b_i(y_1) + (1-\lambda)b_i(y_2) \\ &\le A_i(\lambda x_i(y_1) + (1-\lambda)x_i(y_2)), \end{aligned}$$

i.e. inequality (22.8) can be rewritten as

$$c_i(\lambda x_i(y_1) + (1-\lambda)x_i y_2) = \lambda c_i x_i(y_1) + (1-\lambda)x_i(y_2).$$

The proof of the theorem is completed.

The original problem, due to the theorem, reduces to solving the convex programming problem with respect to y. The problem, however, is made difficult by the fact that in order to estimate the function $\psi_i(y)$ we need to solve the linear subproblem of the form

$$\min c_i x_i, \tag{22.9}$$

$$A_i x_i \ge b_i(y). \tag{22.10}$$

Let y_0 be a feasible vector. Suppose that each of the problems (22.9), (22.10) can be solved for $y = y_0$. With the aid of this solution we divide the matrices A_i, which have more rows than columns, into the square nonsingular basic matrices A_{i_1}, and the nonbasic matrices A_{i_2}. The matrices A_{i_1} define the solution vectors x_i^0 by the equations

$$A_{i_1} x_1^0 = b_{i_1}.$$

Subproblems (22.9), (22.10) are written as:

$$\min c_i x_i, \tag{22.11}$$

$$A_{i_1} x_i = b_{i_1}(y), \tag{22.12}$$

$$A_{i_2} x_i \ge b_{i_2}(y). \tag{22.13}$$

We handle (22.12) with respect to x_i:

$$x_i = A_{i_1}^{-1} b_{i_1}(y). \tag{22.14}$$

Employ (22.14) to eliminate x_i from (22.11), (22.13):

$$c_i x_i = c_i A_{i_1}^{-1} b_{i_1}(y) = (u_{i_1}^0) b_{i_1}(y), \tag{22.15}$$

$$A_{i_2} A_{i_1}^{-1} b_{i_1}(y) \ge b_{i_2}(y), \quad i = \overline{1,p}, \tag{22.16}$$

where

$$u^0_{i_1} = c_i A^{-1}_{i_1} \geq 0 \tag{22.17}$$

is the vector of the dual variables related to the A_{i_1} optimal basis. Define the functions $\psi_i(y)$ by relation (22.15). Here (22.16) defines the set s_i.

Thus, the coordinating problem in the y space has the objective function

$$f(y) = \sum_{i=1}^{p} (u^0_{i_1}) b_{i_1}(y) \tag{22.18}$$

and constraints (22.16). Objective function (22.18) is convex since b_{i_1} are convex and $u^0_{i_1} \geq 0$, although constraints (22.16) are nonlinear. This means that the global optimal solution cannot be ensured. Any coordinating program, however, serves only to improve the performance, if any, of the y feasible vector. With this in mind we will linearize constraints (22.16) with respect to the current point y_0.

Let

$$Q_i = A_{i_2} A^{-1}_{i_1} . \tag{22.19}$$

Introduce into consideration Jacobian $J_{ij}(y)$ where $J_{ij}(y)$, $(j = 1, 2)$ is the matrix, the rows of which are the b_{ij} component gradients calculated at the point y. Then linearization (22.16) at the point y_0 yields:

$$[Q_i J_{i_1}(y_0) - J_{i_2}(y_0)](y - y_0) \geq b_{i_2}(y_0) - Q_i b_{i_1}(y_0) . \tag{22.20}$$

Consider the coordinating problem which has convex objective function (22.18) and linear constraints (22.20):

$$\min f(y) = \sum_{i=1}^{p} (u^0_{i_1}) b_{i_1}(y) , \tag{22.21}$$

$$[Q_i J_{i_1}(y_0) - J_{i_2}(y_0)](y - y_0) \geq b_{i_2}(y_0) - Q_i b_{i_1}(y_0) , \tag{22.22}$$

$$i = \overline{1, p} .$$

Let y^1 be the solution of (22.21), (22.22). Since y_0 is feasible for it, then

$$f(y^1) \leq f(y_0) . \tag{22.23}$$

It is possible that equality will occur in (22.23). In this case the current solution can be tested for optimality.

Theorem 22.2: (Optimality criterion). Let x_i^0, $i = \overline{1,p}$ be the solution of (22.9), (22.10) for $y = y_0$. Then the necessary and sufficient conditions for $(\{x_i^0\}, y_0)$ to be the solution of the original problem (22.3), (22.4) lies in the fact that y_0 is the solution of the coordinating problem and the following conditions are fulfilled

$$u_{i_1}^* = u_{i_1}^0 - Q_i u_{i_2}^1 \geq 0, \quad i = \overline{1,p}, \tag{22.24}$$

where $u_{i_2}^1$ is a vector of the dual variables corresponding to the ith set of constraints (22.22) in the optimum of the coordinating problem.

Proof. Sufficiency: When carrying out propositions (22.1) – (22.4), the solution is optimal for the original problem if and only if the Kuhn-Tucker conditions are fulfilled. We write the problem in the divided form

$$\min \sum_{i=1}^{p} c_i x_i, \tag{22.25}$$

$$A_{i_1} x_i \geq b_{i_1}(y), \tag{22.26}$$

$$A_{21} x_i \geq b_{i_2}(y), \quad i = \overline{1,p}. \tag{22.27}$$

Let u_{i_1}, u_{i_2} be the dual variables corresponding to constraint (22.26), (22.27).

Write the Lagrange function

$$\mathcal{L} = \sum_{i=1}^{p} (c_i x_i + u_{i_1}(b_{i_1}(y) - A_{i_1} x_i) + u_{i_2}(b_{i_2}(y) - A_{i_2} x_i)). \tag{22.28}$$

Write the Kuhn-Tucker conditions at the point $(\{x_i^0\}, y_0)$ $(i = \overline{1,p})$:

$$A'_{i_1} u_{i_1} + A'_{i_2} u_{i_2} = c_i, \tag{22.29}$$

$$J'_{i_1}(y_0) u_{i_1} + J'_{i_2}(y_0) u_{i_2} = 0, \tag{22.30}$$

$$u'_{i_1}(b_{i_1}(y_0) - A_{i_1} x_i^0) = u'_{i_2}(b_{i_2}(y_0) - A_{i_2} x_i^0) = 0, \tag{22.31}$$

$$A_{i_1} x_i^0 \geq b_{i_1}(y_0), \quad A_{i_2} x_i^0 \geq b_{i_2}(y_0), \tag{22.32}$$

$$u_{i_2} \geq 0, \tag{22.33}$$

$$u_{i_1} \geq 0. \tag{22.34}$$

The optimality conditions for linear subproblems are as follows

$$u_{i_1}^0 = (A_{i_1}^{-1})' c_i \geq 0 \,, \tag{22.35}$$

$$A_{i_1} x_i^0 = b_{i_1}(y_0), \quad A_{i_2} x_i^0 \geq b_{i_2}(y_0) \,. \tag{22.36}$$

The Kuhn-Tucker conditions for the coordinating problem at the point $y = y_0$ are obtained from the Lagrangian

$$\begin{aligned}\hat{\mathcal{L}} = \sum_{i=1}^{p} \{ (u_{i_1}^0) b_{i_1}(y) + (u_{i_2}^1)[b_{i_2}(y_0) - Q_i b_{i_1}(y_0) \\ - (Q_i J_{i_1}(y_0) - J_{i_2}(y_0))(y - y_0)] \}\end{aligned} \tag{22.37}$$

and, for $i = \overline{1,p}$, take the form

$$J_{i_1}'(y_0)[u_{i_1}^0 - Q_i' u_{i_2}^1] + J_{i_2}'(y_0) u_{i_2}^1 \,, \tag{22.38}$$

$$(u_{i_2}^1)(b_{i_2}(y_0) - Q_i b_{i_1}(y_0)) = 0 \,, \tag{22.39}$$

$$u_{i_2}^1 \geq 0 \,. \tag{22.40}$$

Comparing (22.29) – (22.34) with (22.35), (22.36), (22.38) – (22.40) we find that only conditions (22.29), (22.30), (22.34) may not be satisfied. Comparing (22.38) and (22.30) involves our assumption of

$$u_{i_1} = u_{i_1}^* = u_{i_1}^0 - Q_i u_{i_2}^1 \,, \tag{22.41}$$

therefore (22.30) is satisfied, multiplying (22.41), by A_{i_1} obtain

$$A_{i_1}' u_{i_1}^* + A_{i_1}' Q_i u_{i_2}^1 = A_{i_1}' u_{i_1}^0 \,, \tag{22.42}$$

since, due to (22.19), $Q_i' = (A_{i_1}^{-1})' A_{i_2}'$, $u_{i_1}^0 = (A_{i_1}^{-1})' c_i$, (22.42) changes to (22.29). Thus, if

$$u_{i_1}^* \geq 0, \quad i = \overline{1,p} \,, \tag{22.43}$$

the Kuhn-Tucker conditions are satisfied in $(\{x_i^0\}, y_0)$, here $u_{i_1}^*$, $u_{i_2}^1$ are the Lagrange optimal multipliers, i.e. the solution is optimal.

Necessity is proved in the same way.

Consequence: The sufficient condition for $(\{x_i^0\}, y_0)$ to be optimal resides in the fact that y_0 solves the coordinating problem, and none of the linear constraints is fulfilled as equality.

Proof: If y_0 is the interior minimum if, due to (22.39), $u^1_{i_2} = 0$, due to (22.41), $u^*_{i_1} = u^0_{i_1}$. Since $u^0_{i_1} \geq 0$, then $(\{x^0_i\}, y_0)$ is optimal. If the criterion of optimality is violated, then either $f(y^1) < f(y_0)$ or $f(y^1) = f(y_0)$ and some $u^*_{i_1} \not\geq 0$. Consider first the case $f(y^1) < f(y_0)$ since the coordinating problem has been formed under the assumption of $A_{i_1} x_i = b_{i_1}(y)$ and linearization of constraints (22.16), then the objective function $f(y)$ of the coordinating problem is equal to $\psi(y)$ only for such y that A_{i_1} is an optimal basis of a linear subproblem (22.9), (22.10). Actually we want to reduce ψ rather than f. Further, due to linearization y^1 may not satisfy (22.16), and, thus, it may not be in S. The following theorem deals with these difficulties. It shows that either $y^1 \in S$ and satisfies $\psi(y^1) < \psi(y_0)$ or, if the nonzero step Θ_m can be accomplished along the line from y_0 to y^1, then it is possible to construct a new vector $y_0 + \Theta_m\ (y^1 - y_0)$ that exhibits the same property.

Theorem 22.3: (Substantiation of linearization). Assume that an optimal solution of the coordinating problem satisfies the condition

$$f(y^1) < f(y_0)\,. \tag{22.44}$$

If y^1 satisfies the conditions

$$h_i(y^1) = Q_i b_{i_1}(y^1) - b_{i_2}(y^1) \geq 0, \quad i = \overline{1,p}\,, \tag{22.45}$$

then

$$\psi(y^1) \leq f(y^1) < f(y_0) = \psi(y_0)\,. \tag{22.46}$$

If y^1 does not satisfy (22.45) we will introduce Θ_m as the largest step from y_0 to y^1, where the feasibility is retained:

$$\Theta_m = \max\{\Theta | h_i(y_0 + \Theta(y^1 - y_0)) \geq 0,\, i = \overline{1,p},\, 0 \leq \Theta \leq 1\}\,. \tag{22.47}$$

Although if $\Theta_m > 0$, then

$$\psi(y_0 + \Theta_m(y^1 - y_0)) < \psi(y_0)\,. \tag{22.48}$$

Proof: Inequalities (22.45) constitute, due to (22.12) – (22.16) the necessary and sufficient conditions for the basis A_{i_1} being feasible for the ith linear subproblem with $y = y^1$. Since A_{i_1} may not be an optimal basis when $y = y^1$, then

$$\psi(y^1) \le f(y^1)\,. \tag{22.49}$$

Since A_{i_1} is the optimal basis with $y = y_0$, then

$$f(y_0) = \psi(y_0)\,. \tag{22.50}$$

Combining (22.49), (22.50), (22.44), we obtain (22.46). To prove (22.48), we set

$$\hat{y} = y_0 + \Theta_m(y^1 - y_0)\,.$$

Since f is convex, then

$$f(\hat{y}) \le f(y_0) + \Theta_m(f(y^1) - f(y_0))\,.$$

The basic matrices A_{i_1} may not be optimal in (22.9), (22.10) when $y = \hat{y}$ so that $\psi(\hat{y}) \le f(\hat{y})$.

Using (22.50) obtain

$$\psi(\hat{y}) \le \psi(y_0) + \Theta_m(f(y^1) - f(y_0))\,.$$

Under the theorem conditions

$$f(y^1) < f(y_0), \quad \Theta_m > 0\,,$$

so that

$$\psi(\hat{y}) < \psi(y_0)$$

and this is exactly (22.48).

Consider the situation when $f(y^1) = f(y_0)$ and a particular $u^*_{i_1} \not\geq 0$. Here the result is similar to that described in the preceding paragraph.

Theorem 22.4: (Alternative optimal basis). If $f(y^1) = f(y^0)$ and for a particular i, $u^*_{i_1}$ there are some negative components, the optimal solution exists for the ith linear problem ($y = y_0$) such that at least one row of the original basis A_{i_1} corresponding to the negative component $u^*_{i_1}$ is absent.

§**22.2.** Formulate the algorithm of the above method for solving problem (22.3) – (22.4).

Step 1: Choose the vector $y_0 \in S$.

Step 2: Solve subproblems (22.9) – (22.10) with $y = y_0$, obtain the optimal solution x^0_i, the basic matrices A_{i_1} and the dual variables $u^0_{i_1}$.

Step 3: Construct the coordinating problem (22.21) – (22.22), solve it, obtain the optimal solution y_1.

Step 4: If $f(y^1) = f(y_0)$ and all $u^*_{i_1} \geq 0$ (22.24), the solution $(\{x^0_i\}, y_0)$ is optimal. Go to step 8.

Step 5: If $f(y^1) = f(y_0)$ and $u^*_{i_1} \not\geq 0$ for $i \in I$, then by theorem (22.4) there exists an alternative optimal basis for the ith linear subproblem in which at least one row A_{i_1} corresponding to the negative component $u^*_{i_1}$ is absent. Such a basis can be found by solving the linear problem (11.19), §11. We will carry out at least one alteration in the basis and go to step 3 with a new basic matrix and dual variables.

Step 6: If $f(y^1) < f(y_0)$, all $h_i(y^1) \geq 0$, then due to theorem 3, $\psi(y^1) < \psi(y_0)$. Suppose that $y_0 = y^1$ and go to step 2.

Step 7: If $f(y^1) < f(y_0)$ some $h_i(y^1) \not\geq 0$, calculate

$$\Theta_m = \max\{\Theta | h_i(y_0 + \Theta(y^1 - y_0)) \geq 0, \, i = \overline{1,p}, \, 0 \leq \Theta \leq 1\}.$$

The following cases are possible here:

(1) $\Theta_m > 0$. Due to theorem (22.3), $\psi(y_0 + \Theta_m(y^1 - y_0)) < \psi(y_0)$. Return to step 2, set $y_0 = y_0 + \Theta_m(y^1 - y_0)$.

(2) $\Theta_m = 0$. For some i, j there exists $h_{ij}(y_0) = 0$, $h_{ij}(y^1) < 0$. It can be shown that the row j of the matrix A_{i_2} can be replaced by a row A_{i_1}, which leads to a new optimal basis in the ith subproblem. Using the new basis go to step 3.

Step 8: Terminate.

Prove the statement made in step 7, in case 2. From $h_{ij}(y_0) = 0$ it follows that

$$A_{i_1} x_i^0 = b_{i_1}(y_0),$$
$$A_{i_2}^j x_i^0 = b_{i_2}^j(y_0),$$

where $A_{i_2}^j, b_{i_2}^j$ are the jth rows A_{i_2}, b_{i_2}. By virtue of assumption 3 $A_{i_2}^j \neq 0$, so that $A_{i_2}^j \ A_{i_1}^{-1} \neq 0$, i.e. $A_{i_2}^j$ can be replaced by a row A_{i_1}, which leads to $\tilde{A}_{i_1}$ a nonsingular matrix.

The relation

$$\tilde{A}_{i_1} x_i^0 = b_{i_1}(y_0) \tag{22.51}$$

must hold and, since $\tilde{A}_{i_1}$ is nonsingular, x_i^0 is a unique solution (22.51). The dual problem objective function $c_i x_i$ does not change, so that $\tilde{A}_{i_1}$ is a new optimal basis.

§**22.3.** The optimal algorithm finite convergence is proved by the theorem.

Theorem 22.5: If optimal bases do not recur at the algorithm steps 5, 7, then an optimal solution is obtained for the finite number of iterations.

Proof: At each iteration, where the optimality criterion is not satisfied, the coordinating problem objective function $f(y)$ decreases or the optimal basis is found in one or more subproblems. Whenever $f(y)$ decreases we have to turn to a new set of optimal bases $\{A_{i_1}\}$ differing from the employed in any preceding cycle because iteration of the set will bring about iteration of the coordinating problem and, therefore, that of the minimal value of f, as well. Consider the cycle at which $f(y)$ remains the same.

We have a finite number of optimal bases for every y value. If none of these bases recurs at steps 5 and 7 of the algorithm, then at each iteration the current set of optimal bases cannot coincide with the set existing at the preceding iteration. Since the complete number of permissible sets $\{A_{i_1}\}$ is finite, then an optimal set can be obtained for the finite number of iterations.

§23. BENDERS METHOD FOR A SPECIAL MATHEMATICAL PROGRAMMING PROBLEM

§**23.1.** The Benders method based on variable division deals with the mathematical programming problem of the form

$$\min\{cx + f(y)\}\,, \tag{23.1}$$

$$Ax + g(y) \geq b\,, \tag{23.2}$$

$$x \geq 0, \quad y \in S\,, \tag{23.3}$$

where c, x are n-dimensional vectors, y is a p-dimensional vector, $f(y)$ a scalar function, b an m-dimensional vector, $g(y)$ an m-dimensional vector function, A is an $(m \times n)$-dimensional matrix, S a particular subset of E^p, e.g. a set of integer vectors. Since problem (23.1) – (23.3) is linear by x with the fixed y, it is natural to attempt solving it via fixing y, solving the linear problem with respect to x, obtaining a "better" value of y, etc. Let

$$R = \{y | \exists\, x \geq 0 : Ax \geq b - g(y), \quad y \in S\}\,. \tag{23.4}$$

The vectors $y \in R$ will be called feasible. The set R can be specified explicitly, for which we will employ the Farkash lemma.

The Farkash lemma. There exists the vector $x \geq 0$ satisfying the conditions $Bx = a$ if and only if $a'u \geq 0$ for all u satisfying $B'u \geq 0$. Fix y and apply the lemma to the linear problem.

$$Ax - s = b - g(y)\,,$$
$$x \geq 0, \quad s \geq 0\,.$$

Hence it follows that y is feasible if and only if the following condition is fulfilled

$$(b - g(y))'u \leq 0 \tag{23.5}$$

for all u satisfying

$$A'u \leq 0, \quad u \geq 0\,.$$

Since the cone

$$C = \{u | A'u \leq 0, \quad u \geq 0\} \tag{23.5'}$$

is polyhedral, then it is determined by a finite number of generatrices, i.e. the vectors u_i^r $(i = \overline{1, n_r})$ exist such that any element $u \in C$ can be presented in the form

$$u = \sum_{i=1}^{n_r} \lambda_i u_i^r, \quad \lambda_i \geq 0\,. \tag{23.6}$$

Substituting (23.6) in (23.5), obtain

$$\sum_{i=1}^{n_r} \lambda_i (b - g(y))' u_i^r \le 0 \,,$$

which holds for all $\lambda_i \ge 0$ if and only if

$$(b - g(y))' u_i^r \le 0, \quad i = \overline{1, n_r} \,. \tag{23.7}$$

Thus, the set R in (23.4) can be represented in the form

$$R = \{y | (b - g(y))' u_i^r \le 0, \, i = \overline{1, n_r}, \, y \in S\} \,.$$

If R is empty, then the original problem (23.1) – (23.3) has no (feasible) solution. In case R is nonempty, (23.1) – (23.3) will be rewritten as

$$\min_{y \in R} \{f(y) + \min\{c'x | Ax \ge b - g(y), \, x \ge 0\}\} \,. \tag{23.8}$$

For the fixed y, the interior minimization of (23.8) is the LP problem.

We will write this problem and its dual problem. The direct problem

$$\begin{aligned} &\min cx \,, \\ &Ax \ge b - g(y), \quad x \ge 0 \,. \end{aligned} \tag{23.9}$$

The dual problem

$$\begin{aligned} &\max (b - g(y)) u \,, \\ &A'u \le c, \quad u \ge 0 \,. \end{aligned} \tag{23.10}$$

According to the duality theorem the following equality of the optimal functionals of problems (23.9), (23.10) exists:

$$\min\{cx | Ax \ge b - g(y), \, x \ge 0\} = \max\{(b - g(y))u, \, A'u \le c, \, u \ge 0\} \,. \tag{23.11}$$

Substituting (23.11) in (23.8), we come to a new form of the original problem

$$\min_{y \in R} \{f(y) + \max\{(b - g(y))'u | A'u \le c, \, u \ge 0\} . \tag{23.12}$$

Consider the constraint set of the dual problem (23.10)

$$P = \{u | A'u \le c, \, u \ge 0\} \,. \tag{23.13}$$

The set does not depend on y, and the magnitudes u_i^r in (23.5′) are the P set extreme rays. If P is empty, then the criterion value for problem (23.9) and, therefore, that for problem (23.1) – (23.3) are not bounded below. If P is nonempty, then the interior maximum in (23.12) is achieved at one of the extreme points of P or approaches ∞ when moving along the extreme ray of P. In the latter case, direct problem (23.9) is unacceptable, which contradicts the initial assumptions so that we confine ourselves to considering only the extreme points of P. And such points constitute a finite number. Denote them by u_i^p, $i = \overline{1, n_p}$. Problem (23.12) is now rewritten as

$$\min_{y \in R}\{f(y) + \max_{i=\overline{1,n_p}} (b - g(y))' u_i^p\}\,. \tag{23.14}$$

And the problem is equivalent to the following

$$\begin{gathered} \min z\,, \\ z \geq f(y) + (b - g(y))' u_i^p, \quad i = \overline{1, n_p}\,, \\ y \in R\,. \end{gathered} \tag{23.15}$$

Using the definition of the set R, (23.15) is written as

$$\begin{gathered} \min z\,, \\ z \geq f(y) + (b - g(y))' u_i^p, \quad i = \overline{1, n_p}\,, \\ (b - g(y))' u_i^r, \quad i = \overline{1, n_r}\,, \\ y \in S\,. \end{gathered} \tag{23.16}$$

Thus, the relation between the original problem and problem (23.16) is clear. These results are summarized as the theorem.

Theorem 23.1:

(1) Problem (23.16) has a feasible solution $\Leftrightarrow$ (23.1) – (23.3);

(2) if (z^0, y^0) is a solution to (23.16), then (x^0, y^0) is the solution of original problem (23.1) – (23.3), where x^0 is a solution to (23.9) with $y = y^0$;

(3) if (x^0, y^0) is a solution to (23.1) – (23.3) and $z^0 = c'x^0 + f(y^0)$, then (z^0, y^0) is a solution to (23.16).

Unfortunately, the process of solving problem (23.16) is made difficult due to an enormous number of constraints. Therefore it is natural to apply the relaxation procedure described in §9. We make the following assumptions:

(1) S is closed and bounded;

(2) $f(y)$ and the components $g(y)$ are continuous on S.

Lemma 23.1: If assumptions (1), (2) are fulfilled and (23.16) is feasible, then z has no finite lower bound $\Leftrightarrow$ P is an empty set.

Proof: See Ref. 19.

Consider problem (23.16) on a constraint set

$$\begin{gathered} \min z\,, \\ z \geq f(y) + (b - g(y))' u_i^p, \quad i \in I_1\,, \\ (b - g(y))' u_i^r \leq 0, \quad i \in I_2\,, \\ y \in S\,, \end{gathered} \tag{23.17}$$

where $I_1 \subset \{1, \ldots, n_p\}$; $I_2 \subset \{1, \ldots, n_r\}$. When satisfying the assumptions (1), (2), if (23.17) is feasible, then it has the finite optimal solution (z^0, y^0). The solution is optimal for (23.16) $\Leftrightarrow$:

$$(b - g(y^0))' u_i^p \leq z^0 - f(y^0), \quad i = \overline{1, n_p}\,, \tag{23.18}$$

$$(b - g(y^0))' u_i^r \leq 0, \quad i = \overline{1, n_r}\,. \tag{23.19}$$

In order to get rid of a large excess, we will introduce the auxiliary linear problem

$$\begin{gathered} \max(b - g(y^0))u\,, \\ u \in P\,. \end{gathered} \tag{23.20}$$

The index of the extreme point, at which solution (23.20) is achieved, coincides with the index of the most unsatisfied constraint (23.18). Problem (23.20) coincides with dual problem (23.10) with $y = y^0$. Since z^0 is assumed to be finite, then, due to lemma 23.1, P is nonempty. Thus, (23.20) has a finite optimal solution or an unbounded solution. In case it is not bounded, the objective function tends to $+\infty$ along the half-line $u_i^p + \lambda u_i^r$, $\lambda \geq 0$. In this case

$$(b - g(y^0))' u_i^r > 0$$

for a certain i. Hence it follows that one of constraints (23.19) is violated. Thus, (23.18), (23.19) is satisfied by $\Leftrightarrow$:

$$\max\{(b - g(y^0))'u | A'u \leq c,\ u \geq 0\} \leq z^0 - f(y^0)\,. \tag{23.21}$$

z^0 cannot diminish, however, without violating some constraint. Therefore, if (23.21) is fulfilled, then it is also fulfilled as an equality.

Theorem 23.2: (Optimality criteria). If (z^0, y^0) is optimal for (23.17), then it is optimal for (23.16) $\Leftrightarrow$:

$$\max\{(b - g(y^0))'u | A'u \leq c,\ u \geq 0\} = z^0 - f(y^0)\,. \tag{23.22}$$

If the optimality criterion does not work, then some of constraints (23.18), (23.19) are not satisfied so that

$$\max\{(b - g(y^0))'u | A'u \leq c,\ u \geq 0\} > z^0 - f(y^0)\,. \tag{23.23}$$

If the maximum in (23.23) is achieved at the extreme point u^0 then

$$(b - g(y^0))'u^0 > z^0 - f(y^0)\,. \tag{23.24}$$

(23.24) establishes the new constraints that should be introduced into a new problem (17).

If linear subproblem (23.20) is unconstrained, then the simplex method leads to the extreme point $P - u^0$ and the extreme ray P denoted by v^0 so that $(b - g|y^0))'$ approaches $+\infty$ along the half-line

$$u = u^0 + \lambda v^0, \quad \lambda \geq 0 \tag{23.25}$$

and, in order it might occur, v^0 has to satisfy the conditions

$$(b - g(y^0))'v^0 > 0\,, \tag{23.26}$$

which prescribes a new constraint in form (23.19) to be added to (23.17) for its new formation.

§**23.2.** Consider in greater detail the solution of problem (23.17).

Although the above procedure can be formally applied to a very wide class of problems, in practice it is efficient when (23.17) can be efficiently solved. Consider the case when it is possible.

(1) S is a set of vectors in E^p with the non-negative integer components $g(y) = By$, $f(y) = d'y$. In this instance, (23.17) is an integer linear problem. And since sequential problems (23.17) differ only in adding one or two constraints, then, for their solution, the cut-off methods of the Gomory method type are appropriate.

(2) S is determined by a set of linear and nonlinear inequalities

$$S = \{y | \bar{g}_i(y) \leq 0,\ i = \overline{1,r},\ g_j' y \leq b_j,\ j = \overline{1,s}\},$$

where g_i are nonlinear continuous differentiable functions. The functions $f(y)$, $\bar{g}(y)$ can be nonlinear, and also continuous and differentiable. Then (23.17) has a linear objective function but nonlinear constraints.

Whatever the applications may be, the Benders procedure displays the ability to maintain any special structure of the matrix A. For instance, if the matrix A is of a transportation type, then (23.19) also is a transportation problem in which right-hand parts change at each step. And since at each iteration of (23.19) only right-hand sides change, the optimal dual solution at iteration i remains feasible for a new dual problem at iteration $i+1$, as well. Thus, the dual simplex method is supposed to be the most suitable for its solution.

§23.3. Formulate the Benders method algorithm for solving problem (23.1) – (23.3).

Step 1: We start the procedure with permutation of problem (23.17) in which there are several solutions.

Step 2: Solve problem (23.17). If it is unfeasible, then so is the original problem. Otherwise, we will obtain the finite optimal solution (z^0, y^0) or data on solution unboundedness. If $z^0 = -\infty$, then the y^0 derivative element is taken as S. Go to step 3.

Step 3: Solve dual problem (23.10). If it is unfeasible, then, due to lemma 1, the original problem has the unconstrained solution. If the dual problem is unbounded, then go to step 6.

Step 4: If the objective function optimal value in step 3 is equal to $z^0 - f(y^0)$, then the solution of $(z^0,\ y^0)$ constitutes the solution of (23.16). If x^0 is the solution of (23.9), then $(x^0,\ y^0)$ is the solution of (23.1) – (23.3). Go to step 7.

Step 5: If in step 3 the criterion of optimality is not satisfied and dual problem (23.10) has the finite optimal solution u^0, then

$$z^0 < f(y^0) + (b - g(y^0))'u^0 ,$$

so that the current solution of (23.17) does not obey the constraint

$$z \geq (b - g(y))'u^0 + f(y) . \tag{23.27}$$

Add this constraint to (23.17) and return to step 2.

Step 6: If (23.10) has the unconstrained solution, then the simplex method enables one to find the ray v^0 and the point u^0 such that the objective function (10) tends to $+\infty$ along the ray:

$$u = u^0 + \lambda v^0, \quad \lambda \geq 0 . \tag{23.28}$$

The inequality

$$(b - g(y^0))'v^0 > 0$$

is satisfied so that y^0 does not obey the constraint

$$(b - g(y))'v^0 \leq 0 .$$

Add this constraint to (23.17). If

$$z^0 < f(y^0) + (b - g(y^0))'u^0$$

also exists for the extreme point u^0 indicated in (23.28), then add the constraint of form (23.27) to (23.17). Return to step 2.

Step 7: Terminate.

§**23.4.** Consider the problem of the above algorithm convergence.

The algorithm convergence follows immediately from the finite number of constraints (23.16).

Theorem 23.3: The iterative procedure described terminates for the finite number of iterations after an optimal solution of the original problem has been obtained or with information on unfeasibility or unboundedness of this problem.

Proof: Problem (23.16) has only a finite number of constraints. If the optimality criterion is not satisfied, then one or more new constraints are added to (23.17). Therefore, for the finite number of iterations the optimality criterion is satisfied or the full set of constraints is employed. (23.16) is unfeasible if and only if (23.17) is unfeasible at a particular step. (23.16) is unbounded $\Leftrightarrow$ dual problem (23.10) is unfeasible.

The algorithm features the possibility of calculating the upper and lower bounds of the optimal objective function value. The upper estimate is generated by the sequence of feasible solutions (23.1) – (23.3) so that the better of these estimates can be taken as a solution if the procedure terminates before an optimum is achieved.

Assume that (z^i, y^i) are the solutions of (23.17) at the step i and linear problem (23.9) has a feasible solution with $y = y^i$. If x^i is the solution of (23.9), then (x^i, y^i) is feasible for (23.1) – (23.3) and the following inequality exists

$$cx^i + f(y^i) \geq \min\{z|(z, y) \in G\}, \tag{23.29}$$

where $G = \{(z, y)|(z, y)$ obeys the constraints of problem $(23.16)\}$ obtaining the lower bound follows from the fact that constraints are added to (23.17) so that if z^i is the objective function optimal value for (23.17) at step i, then

$$z^i \leq z^{i+1} \leq \min\{z|(z, y) \in G\}. \tag{23.30}$$

Conditions (23.29), (23.30) at the step i give the estimates

$$z^i \leq \min\{z|(z, y) \in G\} \leq \min_{j=1,i}(cx^j + f(y^j)). \tag{23.31}$$

With the iterative procedure converging, the right- and left-hand sides of (23.31) converge to the value confined between them. And the optimality criterion represents the condition for the upper and lower bounds being equal.

§**23.5.** We will study the relation between the decomposition methods of Benders and Dantzig-Wulf.

Consider instead of original problem (23.1) – (23.3) the LP problem:

$$\begin{aligned} &\min(cx + dy)\,, \\ &Ax + By \geq b\,, \\ &x \geq 0, \quad y \geq 0 \end{aligned} \tag{23.32}$$

and the problem dual with respect to it:

$$\begin{gathered} \max bu\,, \\ B'u \leq d, \quad A'u \leq c, \quad u \geq 0\,. \end{gathered} \tag{23.33}$$

Assume that the set

$$P = \{u | A'u \leq c,\, u \geq 0\}$$

is bounded. Then problem (23.16) for (23.32) takes the form

$$\begin{gathered} \min z\,, \\ z + (By - b)'u_j \geq dy,\, \forall j\,, \\ y \geq 0\,, \end{gathered} \tag{23.34}$$

where u_j is the jth extreme point of P. Problem (23.17) for (23.32) has the same form, though only a certain subset of constraints with j numbers are taken into consideration. The optimality criterion of the point (z^0, y^0) constitutes fulfillment of the inequality

$$\max\{(b - By^0)u | A'u \leq c,\, u \geq 0\} = z^0 - dy^0\,, \tag{23.35}$$

where (z^0, y^0) is the solution of problem (23.34) for a certain set of constraints with j numbers. To test the criterion, we need to solve the linear problem:

$$\begin{aligned} &\max(b - By^0)u\,, \\ &A'u \leq c\,, \\ &u \geq 0\,. \end{aligned} \tag{23.36}$$

We apply now the Dantzig-Wulf decomposition principle to problem (23.33). Write any $u \in P$ as a convex combination of the extreme points of P:

$$u = \sum_j \lambda_j u_j, \quad \sum_j \lambda_j = 1, \quad \lambda_j \geq 0\,, \tag{23.37}$$

substitute the result in the remaining constraints of (23.33) and its objective function, which leads to the coordinating problem:

$$\begin{aligned} \max \sum_j (bu_j)\lambda_j \,, \\ \sum_j (B'u_j)\lambda_j \le d \,, \\ \sum_j \lambda_j = 1 \,, \\ \lambda_j \ge 0 \,. \end{aligned} \tag{23.38}$$

The problem dual with respect to (23.38) has the form:

$$\begin{aligned} \min(dy + v) \,, \\ v + (By)'u_j \ge bu_j \,, \\ y \ge 0 \end{aligned} \tag{23.39}$$

or, setting

$$z = dy + v \tag{23.40}$$

we get

$$\begin{aligned} \min z \,, \\ z + (By - b)'u_j \ge dy, \quad \forall j \\ y \ge 0 \,. \end{aligned} \tag{23.41}$$

Problem (23.41) coincides with problem (23.34) in the Benders method. Let (y^0, v^0) be the dual variables corresponding to the optimal solution of the reduced coordinating problem. Consider the local subproblem of the decomposition principle

$$\begin{aligned} \max(b - By^0)'u \,, \\ A'u \le c \,, \\ u \ge 0 \,. \end{aligned} \tag{23.42}$$

It is clear that (23.42) coincides with subproblem (23.36) in the Benders algorithm.

The optimality criterion of the coordinating problem is as follows

$$\max\{(b - By^0)u | A'u \le c, \quad u \ge 0\} \le v^0 \,. \tag{23.43}$$

It has been shown above that the inequality is actually fulfilled as equality.

(23.40) implies

$$z^0 = dy^0 + v^0$$

and (23.43) transforms to the form

$$\max\{(b - By^0)u | A'u \leq c, u \geq 0\} = z^0 - dy^0 ,$$

which coincides with the optimality criterion in the Benders method. Thus the methods of Benders and Dantzig-Wulf are dual to one another. The Benders algorithm is applied, however, to a wider class of problems, e.g. the partially integer problems of mathematical programming.

§**23.6.** And as conclusion of the paragraph we will consider one search algorithm of approximate solutions for the following partially integer problem[18]:

$$\begin{aligned} &\max\{f_1(x) + f_2(y)\} , \\ &\varphi_1(x) + \varphi_2(y) \leq b , \\ &y \in Y , \end{aligned} \tag{23.44}$$

where x is the n vector, y is the integer m vector, b the l vector, Y the bounded set, $f_2(y)$, $\varphi_2(y)$ the arbitrary functions, and $f_1(x)$ the twice continuous differentiable concave function. Assume that the feasible set of problem (23.44) is bounded and nonempty. The maximum of the Euclidean norm of the vector x on the set is denoted by M.

Formulate the search algorithm of the approximate solution of problem (23.1) based on the Benders decomposition concept. In the algorithm description we will meet two sets of vectors $\{p_i\}, i = 1, 2, \ldots, r_k$, $\{\lambda_j\}$, $j = 1, 2, \ldots, q_k$, three sets of scalars $\{p_i\}, i = 1, 2, \ldots, r_k$, $\{\mu_j\}$, $\{\Lambda_j\}$, $j = 1, 2, \ldots, q_k$, where k is the number of iterations.

<u>*Step 1*</u>: Choose the vector $y_1 \in Y$ and the number d_1 being larger than the objective function optimum value of problem (23.1). Set $k = 1$, $r_0 = 0$, $q_0 = 0$. Introduce the accuracy parameters of the scheme ε_i, $i = \overline{1,5}$ the meaning of which will be clear from the subsequent steps of the algorithm.

Step 2: Solve the problem

$$\min_x \psi(x, y_k, d_k),$$

$$\begin{aligned}\psi(x, y_k, d_k) &= (\max\{0, d_k - f_1(x) - f_2(y_k)\})^2 \\ &\quad + \sum_{i=1}^{l} (\max\{0, (\varphi_1(x) + \varphi_2(y_k) - b)_i\})^2 \\ &= [d_k - f_1(x) - f_2(y_k)]_+^2 \\ &\quad + \|(\varphi_1(x) + \varphi_2(y_k) - b)_+\|_E^2 .\end{aligned}$$

In this case, it is supposed that the desired minimum is found with sufficient accuracy, when we obtain the point x_k in which even one of the following three situations is realized.

Case 23.1: $\psi(x_k, y_k, d_k) \le \varepsilon_1$. The pair x_k, y_k is taken as the approximate solution of the original problem. Go to step 5.

Case 23.2: $\psi(x_k, y_k, d_k) > \varepsilon_1$, $1/2|(\psi'_x(x_k, y_k, d_k), x_k)| \le \varepsilon_3 \psi(x_k, y_k, d_k)$, $1/2\|\psi'_x(x_k, y_k, d_k)\|_E < \varepsilon_2[d_k - f_1(x) - f_2(y_k)]$.

In this case we set $r_k = r_{k-1} + 1$, $q_k = q_{k-1}$, the vector $p_{r_k} = [\varphi_1(x_k) + \varphi_2(y_k) - b]^T + [d_k - f_1(x_k) - f_2(y_k)]^{-1}$ is introduced into the set $\{p_i\}$, and the value

$$p_{r_k} = (p_{r_k}, b - \varphi_1(x_k) + \varphi_1'(x_k)x_k) + f_1(x_k) - (f_1'(x_k), x_k) - b$$

into the set $\{p_i\}$. Go to step 3.

Case 23.3:

$$\begin{aligned}&\psi(x_k, y_k, d_k) > \varepsilon_1, \\ &\|[\varphi_1(x_k) + \varphi_2(y_k) - b]^T + \varphi_1'(x_k)\|_E < -\varepsilon_4 \\ &\quad \times ([\varphi_1(x_k) + \varphi_2(y_k) - b]^T{}_+, b - \varphi_1(x_k) - \varphi_2(y_k) + \varphi_1'(x_k)x_k) .\end{aligned}$$

In this case we set $q_k = q_{k-1} + 1, r_k = r_{k-1}$, the vector $\lambda_{q_k} = [\varphi_1(x) + \varphi_2(y_k) - b]_+^T$ is introduced into the set $\{\lambda_j\}$, the value $\Lambda_{q_k} = (\lambda_{q_k}, b - \varphi_1(x_k) + \varphi_1'(x_k)x_k)$ into the set $\{\Lambda_j\}$, and the value $\mu_{q_k} = \|\lambda_{q_k}\varphi_1'(x_k)\|_E$ into the set $\{\mu_j\}$.

<u>*Step 3*</u>: Solve the problem of the form

$$\begin{aligned} &\max_y \{ \min_{i=\overline{1,r_k}} \{p_i - (p_i, \varphi_2(y))\} + f_2(y)\} \\ &\varepsilon_5[\Lambda_j - (\lambda_j, \varphi_2(y))] \geq -\mu_j, \quad j = \overline{1, q_k}, \quad y \in Y. \end{aligned} \tag{23.45}$$

In the case of unsuccessful choice of the parameters $\varepsilon_4, \varepsilon_5$, the feasible set of problem (23.45) can prove to be empty and calculations terminate, though the approximate solution of problem (23.44) has not been found.

<u>*Step 4*</u>: Let y_{k+1} be a solution to problem (23.45). Define the value

$$d_{k+1} = \min\{d_k, f_2(y_{k+1}) + \min_{i=\overline{1,r_k}} \{p_i - (p_i, \varphi_2(y_{k+1}))\}\}$$

and form the function $\psi(x, y_{k+1}, d_{k+1})$. Go to step 2.

<u>*Step 5*</u>: Terminate.

The algorithm convergence to the approximate solution of problem (23.44) for the finite number of steps is shown by the following theory.[18]

Theorem 23.4: Proceeding from any initial approximation and with $0 < \varepsilon_4 \leq \varepsilon_5 \leq 1/M$, $0 < \varepsilon_3 < 1$, $\varepsilon_1 > 0$, $\varepsilon_2 > 0$ the algorithm described will find the approximate solution of problem (23.44) for the finite number of steps. Here directing $\varepsilon_1, \varepsilon_2$ towards zero we can provide any degree of closeness to the true solution.

If we assume that the problem $\min_x \psi(x, y_k, d_k)$ is accurately solved, i.e. ε_1, $\varepsilon_2 = 0$, then the algorithm yields an exact solution to problem (23.44) either in the limit or for the finite number of steps (here the finite convergence occurs in the linear case).

Chapter 6

DECOMPOSITION BASED ON OPTIMIZATION TECHNIQUE

The present chapter deals with application of numerous well-known methods for finding the dynamic programming extrema, the conditional gradient, the componentwise descent, etc., to the problems with a specific structure, which enables one to carry out, to a certain extent, their decomposition.

§24. APPLICATION OF THE COMPONENTWISE DESCENT METHOD FOR SOLVING THE PROBLEMS OF MATHEMATICAL PROGRAMMING AND OPTIMAL MANAGEMENT

§24.1. We set forth first the componentwise descent method. Consider the problem

$$\min f(x) = f(x_1, \ldots, x_n), \tag{24.1}$$

$$x_i \in M_x^i, \ \ i = \overline{1,n}, \tag{24.2}$$

where M_x^i are the bounded line segments.

The idea of the method resides in the following: if the approximation $x^k = (x_1^k, \ldots, x_n^k)$ is formed, then the following approximation is also determined by solving the minimization problems n $(i = \overline{1,n})$ with one variable

$$\min_{x_i \in M_x^i} f(x_1^{k+1}, \ldots, x_{i-1}^{k+1}, x_i, x_{i+1}^k, \ldots, x_n^k), \tag{24.3}$$

$k = 0, 1, \ldots; x_1^0, \ldots, x_n^0$ are initial approximations.

The method convergence has been proved (see, e.g. Ref. 4) under assumption that $f(x)$ is the convex differentiable function in the region involving the closed, bounded convex set $M_x = M_x^1 \times \cdots \times M_x^n$. In this case there exists the equality

$$\lim_{k \to \infty} f(x^k) = \min_{x \in M_x} f(x). \tag{24.4}$$

§**24.2.** With a particular form of functions (1), the componentwise descent method generates the hierarchic optimization scheme. In Ref. 35 we consider the problem

$$\min \sum_{i=1}^{m} f_i(x_i, u), \tag{24.5}$$

$$x_i \in M_x^i, \quad i = \overline{1, m}, \quad u \in v. \tag{24.6}$$

By virtue of the specific structure of the objective function in (24.5), minimization by x_i $(i = \overline{1, m})$ with the use of the method (24.3) can be carried out independently and simultaneously. Thus we obtain the following scheme for solving problem (24.5) – (24.6) at two levels:

(1) Select the initial approximation $u^0 \in v$.

(2) If the approximation $x^k = (x_1^k, \ldots, x_m^k)$, u^k is constructed, then x^{k+1} is constructed according to the rules:

   (a) solve the problems

$$\min f_i(x_i, u^k), \quad i = \overline{1, m} \tag{24.7}$$

   the solutions of which are taken as x_i^{k+1};

   (b) at the second level, we solve the problem

$$\min_{u \in v} \sum_{i=1}^{m} f_i(x_i^{k+1}, u) \tag{24.8}$$

   the solution of which gives u^{k+1}.

The present approach can be generalized for constructing the optimization schemes at several levels. For instance, we need to solve the problem

$$\min \sum_{i=1}^{m} \sum_{j=1}^{m_i} f_{ij}(x_{ij}, x_i, u)\,, \tag{24.9}$$

$$x_{ij} \in M_x^{ij}, \quad x_i \in M_x^i, \quad u \in v\,. \tag{24.10}$$

In this case a new approach $x^{k+1} = (x_{11}^{k+1}, \dots, x_{mm}^{k+1}, x_1^{k+1}, \dots, x_m^{k+1}, u^{k+1})$ is constructed as:

(1) the first level subsystems solve the problems

$$\min_{x_{ij} \in M_x^{ij}} f_{ij}(x_{ij}, x_i^k, u^k), \quad i = \overline{1,m}, \quad j = \overline{1,m_i}$$

the solutions of which are taken as x_{ij}^{k+1};

(2) the second level subsystems solve the problems

$$\min_{x_i \in M_x^i} \sum_{j=1}^{m_i} f_{ij}(x_{ij}^{k+1}, x_i, u^k), \quad i = \overline{1,m}$$

from which we obtain x_i^{k+1};

(3) the third level solves the problem

$$\min_{u \in v} \sum_{i=1}^{m} \sum_{j=1}^{m_i} f_{ij}(x_{ij}^{k+1}, x_i^{k+1}, u)\,.$$

The process is iterated with $k = k + 1$.

Evidently, the solutions of the problem

$$\min \sum_{i_1=1}^{m} \sum_{i_2=1}^{m_{i_1}} \cdots \sum_{i_{n-1}=1}^{m_{i_1 \dots i_{n-2}}} f_{i_1 \dots i_{n-1}}(x_{i_1 \dots i_{n-1}}, x_{i_1 i_2}, x_{i_1}, u)\,,$$

$$x_{i_1 \dots i_l} \in M_x^{i_1, \dots, i_l}, \dots, u \in v, l = \overline{1, n-1}$$

with the aid of (24.3) we obtain the scheme of optimization at n levels. In conclusion it should be noted that problem (24.5) – (24.6) has an economic application.

Assume that we have m objects (subsystems) utilizing various (their own) resources $x_i \in M_x^i\,(i = \overline{1,m})$ and the common (central) resource $u = u_1 + \cdots + u_m \in v$. The profit of the ith subsystem is $f_i(x_i, u_i)$, the profit of the whole system is $\sum_{i=1}^{m} f_i(x_i, u_i)$.

The solution of the problem at two levels:

$$(1) \qquad \max_{x_i \in M_x^i} f_i(x_i, u_i^k) \to x_i^{k+1}\,;$$

$$(2) \qquad \max_{\substack{u \in v \\ u_i \in v_i}} \sum_{i=1}^{m} f_i(x_i^{k+1}, u_i) \to u^{k+1}\,.$$

Similarly, it is possible to formulate the problems leading to the schemes of optimization at many levels.

§**24.3.** Consider application of the componentwise descent method for optimization at two levels of a discrete dynamic system involving m centre controlled independent subsystems:

$$\min \sum_{i=1}^{m} \sum_{k=0}^{N-1} f_{ik}(x_k^i, u_k^i, x_k, u_k) + \sum_{i=1}^{m} f_{iN}(x_N^i) + f_N(x_N)\,, \tag{24.11}$$

$$x_{k+1}^i = c_k^i x_k^i + D_k^i u_k^i + G_k^i u_k, \quad x_0^i \tag{24.12}$$

$$x_{k+1} = c_k x_k + D_k u_k, \quad x_0\,(k = \overline{0, N-1}), \quad u_k^i \in v^i; u_k \in v\,, \tag{24.13}$$

where x_k^i are the phase variables of the ith subsystem, x_k are the phase variables of the centre, u_k^i the ith subsystem "own" controls, u_k "central" controls.

Denote by $\phi^i(t_k, t_l)$, $\phi(t_k, t_l)$ fundamental matrices for the ith subsystem and the center, respectively. Using the Cauchy formula, (24.12), (24.13) are rewritten as:

$$x_k^i = \phi^i(t_k, t_0) x_0^i + \sum_{t_0 \le t_j < t_k} \phi^i(t_k, t_{j+1})(D_j^i u_j^i + G_j^i u_j)\,, \tag{24.14}$$

$$x_k = \phi(t_k, t_0) x_0 + \sum_{t_0 < t_j < t_k} \phi(t_k, t_{j+1}) D_j u_j\,. \tag{24.15}$$

Substituting (24.14), (24.15) in (24.11), we obtain the minimization problem of form (24.5) – (24.6):

$$\min \left\{ \sum_{i=1}^{m} \sum_{k=0}^{N-1} f_{ik} \left(\phi^i(t_k, t_0) x_0^i + \sum_{t_0 \le t_j < t_k} \phi^i(t_k, t_{j+1})(D_j^i u_j^i + G_j^i u_j)\,, \right. \right.$$

$$\left. u_k^i, \phi(t_k, t_0) x_0 + \sum_{t_0 \le t_j < t_k} \phi(t_k, t_{j+1}) D_j u_j, u_k \right)$$

$$+ \sum_{i=1}^{m} f_{iN} \left(\phi^i(t_N, t_0) x_0^i + \sum_{t_0 \le t_j < t_k} \phi^i(t_N, t_{j+1})(D_j^i u_j^i + G_j^i u_j) \right)$$

$$\left. + f_N \left(\phi(t_N, t_0) x_0 + \sum_{t_0 \le t_j < t_N} \phi(t_N, t_{j+1}) D_j u_j \right) \right\}.$$

Here, in (24.5) – (24.6), x_i correspond to u_k^i, and u are u_k equations.

§25. CONDITIONAL GRADIENT METHOD AND DECOMPOSITION OF PROBLEMS OF MATHEMATICAL PROGRAMMING AND OPTIMAL CONTROL

§25.1. Consider the convex programming problem

$$\min f(x), \qquad x \in X\,,$$

where $f(x)$ is the convex functional, X is the closed, constrained convex set.

We set forth briefly, for this problem, the conditional gradient method in which the following iteration process is formed:

(1) select the initial approximation $x^0 \in X$;

(2) if the approximation x^k is found, then the following approximation x^{k+1} is constructed according to the rules:

(a) solve the minimization problem

$$\min_{x \in X} (\operatorname{grad} f(x^k), x)\,; \tag{25.1}$$

(b) $x^{k+1} = x^k + \alpha_k(\bar{x}^k - x^k)$, $0 \le \alpha_k \le 1$, where $\bar{x}^k$ is the solution of (25.1) ($k = 0, 1, \dots$). There exist various possibilities for selecting the step α_k according to which different versions of the above method are obtained. The method convergence has been proved under the assumption that $f(x)$ is differentiable, and the gradient $\operatorname{grad} f(x)$ satisfies the Lifshitz condition (see, e.g. Ref. 4).

Then it is proved that with the appropriate choice α_k

(1) $\lim_{k\to\infty} f(x^k) = f^* = \inf_{x\in X} f(x)$;

(2) $f(x^k) - f^* \le \frac{c}{k}$, where $c = \text{const} > 0$;

(3) there exists the sequence x^k converging to the minimum point x^*.

§**25.2.** Consider application of the conditional gradient method for decomposition of large nonlinear programming problems (26).

Assume that we have the problem

$$\min_{x_i \in X_i} f(x_1, \dots, x_n), \tag{25.2}$$

where $f(x)$ is convex, X_i ($i = \overline{1,n}$) are the closed bounded convex sets.

The very form of problem (25.2) suggests that in the system under consideration one can isolate n subsystems each of which is in agreement with the vector x_i. We will construct the iteration process for solving (25.2) based on minimizing some objective functions for each system separately (by every vector x_i).

Denote

$$\operatorname{grad} f(x) = (\mathcal{F}_1(x), \dots, \mathcal{F}_n(x)).$$

Since in the present case

$$(\operatorname{grad} f(x^k), x) = \sum_{i=1}^{n} (\mathcal{F}_i(x^k), x^i),$$

then problem (25.1) falls into n independent minimization problems

$$\min_{x_i \in X_i} (\mathcal{F}_i(x^k), x^i), \quad i = \overline{1,n}.$$

The given circumstance enables one to organize the following solution procedure for problem (25.2) at two levels:

(1) select $x^0 = (x_1^0, \dots, x_n^0)$, $x_i^0 \in X_i$;

(2) if x^k is found, then

(a) the second level calculates $\mathcal{F}_i(x^k)$, transfers it to the ith subsystem $(i = \overline{1,n})$;

(b) the ith subsystem solves the minimization problem

$$\min_{x_i \in X_i} (\mathcal{F}_i(x^k), x_i)\,, \tag{25.3}$$

the solution of which is $\bar{x}_i^k$;

(c) by $\bar{x}_i^k$, calculate a new approximation from the formula

$$x_i^{k+1} = x_i^k + \alpha_k(\bar{x}_i^k - x^k), \quad (i = \overline{1,n}), \quad k = 0, 1, \dots . \tag{25.4}$$

It has been noted above that different possibilities exist for selecting the step α_k (see, e.g. Ref. 4). The two-level procedure described provides an example of decentralized planning.

Every subsystem $i = \overline{1,n}$ reports its plan x_i^k to a central unit at the kth moment of time. Using the plans obtained as the base, the central unit formulates and reports to subsystems $\mathcal{F}_i(x^k)$, i.e. general directions of changing the plans formed by them. Thereupon every subsystem i derives its new plan x_i^{k+1}, reports it to the central unit and so on.

§**25.3.** We apply the conditional gradient method to decomposition of large problems of optimal control when the equations of moving with respect to phase coordinates lead to the block-triangular form[36]:

$$\min f(u) = \int_{t_0}^{T} \mathcal{F}(x(t), u(t), t)dt + \phi(x(t))\,, \tag{25.5}$$

$$u_i(t) \in v_i \tag{25.6}$$

the phase variables $x(t) = (x_1(t), \dots, x_n(t))$, the controls $u(t) = (u_1(t), \dots, u_n(t))$ are related by a system of linear equations

$$\dot{x}_i(t) = A_{ii}(t)x_i + \sum_{j=i+1}^{n} A_{ij}(t)x_j + \sum_{j=1}^{n} B_{ij}(t)u_j\,, \tag{25.7}$$

$$x_i(t_0) = x_{i_0}, \quad i = \overline{1,n} \tag{25.8}$$

where $x_i(t)$, $u_i(t)$ are the vectors of the n_i dimension and the matrices A_{ij}, B_{ij} have the $(n_i \times n_j)$ dimension. Assume that v_i are the closed, bounded convex sets. The formula

$$\text{grad}\, f = \mathcal{F}_u - B^T \psi \,, \tag{25.9}$$

where $\psi(t)$ is a solution of the adjoint system

$$\frac{d\psi_i}{dt} = \sum_{j=1}^{i} A_{ji}^T(t)\psi_j + \mathcal{F}_{x_i} \,, \tag{25.10}$$

$$\psi_i(t) = -\phi_{x_i}(x(T)), \;\; B^T(t) = (B_{ji}^T(t)) \,, \tag{25.11}$$

is valid.

To implement the conditional gradient method, we have to calculate the grad f with the fixed $u(t) = u^k(t)$, and to do this requires the appropriate $x(t)$, $\psi(t)$. Since systems (25.7), (25.10) are block-triangular, then in the present case the linear inhomogeneous systems are successively integrated for finding $x_i(t)$, $\psi_i(t)$

$$\begin{aligned} \dot{x}_i &= A_{ii}x_i + f_i(t), \;\; x_i(t_0) = x_{i0} \quad (i = n, \dots, 1) \,, \\ \dot{\psi}_i &= A_{ii}^T \psi_i + g_i(t), \;\; \psi_i(T) = -\phi'_{x_i}(x(T)), \;\; i = \overline{1, n} \,, \end{aligned} \tag{25.12}$$

where

$$f_i(t) = \sum_{j=i+1}^{n} A_{ij}x_j(t) + \sum_{j=1}^{n} B_{ij}u_j(t) \,,$$

$$g_i(t) = -\sum_{j=1}^{i-1} A_{ji}^T \psi_j(t) + \mathcal{F}_{x_i}(x(t), u(t), t) \,.$$

It is known that

$$\begin{aligned} x_i(t) &= T_i(t)x_{i0} + T_i(t) \int_{t_0}^{t} T_i^{-1}(\tau) f_i(\tau) d\tau, \;\; (i = n, \dots, 1) \,, \\ \psi_i(t) &= -[T_i^T(t)]^{-1} T_i^T(T) \phi_{x_i}(x(T)) \\ &\quad + [T_i^T(t)]^{-1} \int_{T}^{t} T_i^T(\tau) g_i(\tau) d\tau, \;\; (i = 1, \dots, n) \end{aligned} \tag{25.13}$$

where T_i $(i = \overline{1,n})$ are fundamental matrices for the system

$$\dot{x}_i = A_{ii}x_i(t), \quad i = \overline{1,n}\,.$$

From formulae (25.13) it is possible, given $T_i(t)$, $T_i^T(t)$, $T_i^{-1}(t)$, $[T_i^T(t)]^{-1}$, to calculate successively $x_n, \ldots, x_1, \psi_1, \ldots, \psi_n$ with the fixed $u(t)$.

We set forth a general scheme of solving (25.5) – (25.8) at two levels:

(1) select the initial approximation

$$u^0(t) = (u_1^0(t), \ldots, u_n^0(t)), \quad u_i^0(t) \in v_i, \quad i = \overline{1,n}\,,$$

(2) if the approximation u^k is found, then u^{k+1} is calculated according to the rules:

(a) from formulae (25.13) calculate $x_i^k(t), \psi_i^k(t)$ (with the fixed $u^k = u^k(t)$) in the following order

$$\begin{array}{lcc}
\text{the } n\text{th subsystem} & x_n^k(t) & \psi_n^k(t) \\
 & \downarrow & \uparrow \\
\text{the } (n-1) \text{ subsystem} & x_{n-1}^k(t) & \psi_{n-1}^k(t) \\
 & \downarrow & \uparrow \\
\cdots & \cdots & \cdots \\
 & \vdots & \vdots \\
 & \downarrow & \uparrow \\
\text{the 1st subsystem} & x_1^k(t) & \psi_1^k(t)
\end{array}$$

(b) by these functions, calculate from formula (25.9)

$$\operatorname{grad} f(u^k) = (G_1(u^k), \ldots, G_n(u^k))\,,$$

where $G_i(u^k) = \mathcal{F}_{u_i} - (B^T\psi)_i$;

(c) the subsystems solve minimization problems (and their solutions exist since a linear functional achieves a minimum on the closed, bounded, convex set)

$$\min_{u_i \in v_i} (G_i(u^k), u_i)\,, \tag{25.14}$$

where $G_i = (G_{i_1}, \dots, G_{i_{n_i}})$; $u_i = (u_{i_1}, \dots, u_{i_{n_i}})$;

$$(G_i(u^k), u_i) = \int_{t_0}^{T} \sum_{j=1}^{n_i} G_{ij}(u^k(t))u_{ij}(t)dt\,;$$

denote the solutions of subproblems (25.14) by

$$\bar{u}_i^k(t), \quad i = \overline{1,n}\,;$$

(d) by the $\bar{u}_i^k(t)$ derived, calculate a new approximation

$$u_i^{k+1}(t) = u_i^k(t) + \alpha_k(\bar{u}_i^k(t) - u_i^k(t)), \quad i = \overline{1,n}\,.$$

Subsequently, the process is iterated ($k = 0, 1, \dots$).

§26. UTILIZATION OF A PENALTY CONSTANT IN DECOMPOSITION OF THE MATHEMATICAL PROGRAMMING PROBLEM

§26.1. Consider the block programming problem

$$\begin{gathered} \min F(x) = \sum_{i=1}^{N} f_i(x_i)\,, \\ \sum_{i=1}^{N} g_k(x_k) \leq b, \quad x_k \in R_k, \quad k = \overline{1,N}\,, \end{gathered} \tag{26.1}$$

here $x_k = (x_{k_1}, \dots, x_{k_{s_k}}) \in E^{s_k}$ $(k = \overline{1,N})$,

$$x = (x_1, \dots, x_N) \in E^n \quad \left(n = \sum_{k=1}^{N} s_k\right),$$

$\sum_{k=1}^{N} g_k(x_k)$ is an m-dimensional vector function,

$$b = (b_1, \dots, b_m), \quad R_k \subset E^{s_k} \ (k = \overline{1,N})\,,$$

assume that R_k $(k = \overline{1,N})$ are convex, closed and bounded, $F(x)$ is the convex function differentiable on $R_1 \times \cdots \times R_n$, $\sum_{i=1}^{N} g_k(x_k)$ is the convex vector function which is continuous on $R_1 \times \cdots \times R_n$.

Introduce the variables $u = (u_1, \ldots, u_N)$, $v = (v_1, \ldots, v_N)$, where $u_k = (u_{k_1}, \ldots, u_{k_m})$, $v_k = (v_{k_1}, \ldots, v_{k_m})$, $(k = \overline{1, N})$, $(u, v) \in E^{2mN}$ and consider the problem equivalent to problem (26.1):

$$\begin{gathered} \min F(x)\,, \\ g_k(x_k) - u_k \leq 0, \;\; k = \overline{1, N}\,, \\ x_k \in R_k, \;\; k = \overline{1, N}\,, \\ \|u - v\|^2_{E^{mN}} = 0, \;\; \sum_{k=1}^{N} v_k \leq b, \;\; \|v\|^2 \leq R\,, \end{gathered} \tag{26.2}$$

where R satisfies the condition

$$R \geq \sum_{k=1}^{N} \max_{x_k \in R_k} \|g_k(x_k)\|^2_{E^m}\,.$$

Denote

$$G_k = \{(x_k, u_k) | g_k(x_k) - u_k \leq 0, \;\; x_k \in R_k, \;\; \|u_k\|^2 \leq R\} \;\; (k = \overline{1, N})\,,$$

$$G_{N+1} = \left\{ v \,\middle|\, \sum_{k=1}^{N} v_k \leq b, \;\; v^2_{r_k} \leq R \;\; (r = \overline{1, m}; \;\; k = \overline{1, N}) \right\},$$

$$G = G_1 \times \cdots \times G_N \times G_{N+1}\,.$$

Select a strictly increasing sequence of positive numbers $\{M_l\}$, for which $M_l \underset{l \to \infty}{\longrightarrow} \infty$ and reduce solving problem (26.2) to solving a sequence of the problems $(2l)$, $(l = 1, 2, \ldots)$:

$$\begin{gathered} \min\{F(x) + M_l \|u - v\|^2\}\,, \\ (x, u, v) \in G\,. \end{gathered}$$

Describe the iterative process of solving problem (26.2):

(1) Fix v^{l-1} (with $l = 1, v^0$ is chosen arbitrarily) and select $(x^{1l}, u^{1l}) \in G_1 \times \cdots \times G_N$ on which

$$\min\{\mathcal{F}(x) + M_l \|u - v^{l-1}\|^2\} \tag{26.3}$$

is achieved. And to do this requires solving N convex programming problems of the form:

$$\begin{gathered} \min\{f_k(x_k) + M_l \|u_k - v_k^{l-1}\|^2\}\,, \\ (x_k^{1l}, u_k^{1l}) \in G_k\,. \end{gathered}$$

(2) Fix the solution of (26.3) (x^{1l}, u^{1l}) and define the point $v^{1l} \in G_{N+1}$ at which a minimum is achieved

$$\mathcal{F}(x^{1l}) + M_l \|u^{1l} - v\|^2 . \tag{26.4}$$

And the problem is expanded into m subproblems: derive $(v_{r_1}^{(1l)}, \ldots, v_{r_N}^{(1l)})$, at which

$$\min \sum_{k=1}^{N} (u_{r_k}^{1l} - v_{r_k})^2 ,$$
$$\sum_{k=1}^{N} v_{r_k} \le b_r, \quad v_{r_k}^2 \le R \ \ (k = \overline{1, N}) , \tag{26.5}$$

is achieved, here $r = \overline{1, m}$. Problems (26.5) constitute strictly convex square programming problems.

Fix v^{1l} and go to point (1).

In this way one forms the sequence $\{x^{il}, u^{il}, v^{il}\}$ involving a finite or infinite number of convergent subsequences. Isolate among these sequences the one converging to $(\bar{x}^l, \bar{u}^l, \bar{v}^l)$. Let M^{2l} be a set of all solutions $(2l)$.

Denote $\Delta^l = \|\bar{u}^l - \bar{v}^l\|^2$, $\bar{x}$, $\bar{u}^l$, $\bar{v}^l$ is a solution to $(2l)$ and show that for all $(x, u, v) \in M^{2l} \|u - v\|^2 = \Delta^l$. Introduce the function

$$\mu(M, \Delta^l) = \min_{(x,u,v) \subset G} [\mathcal{F}(x) + M(\|u - v\|^2 - \Delta^l)] \tag{26.6}$$

and show that each $M_l > 0$ is in agreement with a uniquely defined Δ^l.

From the form of function (26.6) we see that M_l minimizes $\mu(M, \Delta^l)$ when there exists the point $(x, u, v) \in M^{2l}$ such that $\|u - v\|^2 - \Delta^l = 0$, and such a point in M^{2l} represents $(\bar{x}^l, \bar{u}^l, \bar{v}^l)$. If $\Delta^l > 0$ then there exists $(x, u, v) \in G$ for which $\|u - v\|^2 < \Delta^l$ and by the Kuhn-Tucker theorem for each point $(\bar{x}, \bar{u}, \bar{v}) \in M^{2l}$, $(\bar{x}, \bar{u}, \bar{v}, M_l)$ is the saddle point of the function $\mathcal{F}(x) + M(\|u - v\|^2 - \Delta^l)$ in the region $G, M \ge 0$. But for all points $(x, u, v) \in M^{2l}$ however, $M_l(\|u - v\|^2 - \Delta^l) = 0$ must hold, and since $M_l > 0$, then for all these points $\|u - v\|^2 = \Delta^l$. If $\Delta^l = 0$, then $\|u - v\|^2 = \Delta^l$ also holds for all $(x, u, v) \in M^{2l}$. Indeed if we assume that the point $(x, u, v) \in M^{2l}$ will be obtained for which $\|u - v\|^2 = \Delta^l > 0$ then, iterating the preceding reasoning, we get that $\Delta^l = \Delta > 0$. It is by this that we have demonstrated that a uniquely defined Δ^l corresponds to $M_l > 0$.

Lemma: The chosen $\{M_l\}$ corresponds to the sequence $\{\Delta^l\}$ for which $\Delta^l \underset{l\to\infty}{\longrightarrow} 0$.

Proof: See Ref. 21. Denote by M^2 the set of all solutions of problem (26.2).

Theorem 26.1: The chosen $\{M_l\}$ is in agreement with the sequence $\{\bar{x}^l\}$ for which

$$\rho(\bar{x}^l, M^2) \underset{l\to\infty}{\longrightarrow} 0 .$$

Proof: Denote

$$X(\Delta^l) = \{x | \mathcal{F}(x) \le \mathcal{F}(\bar{x}), \ \ \|u - v\|^2 \le \Delta^l, \ \ (x, u, v) \in G\}, \ \ (l = 1, 2, \dots) .$$

The set $X(\Delta^l)$ involves a set of all $\bar{x}^l$ minimizing $F(x)$ with the constraints $\|u - v\|^2 \le \Delta^l$ and $(x, u, v) \in G$. Indeed, from

$$\bar{x}^l \in \{x | F(x) \le F(\bar{x}^l), \|u - v\|^2 \le \Delta^l, (x, u, v) \in G\}$$

it follows that

$$\bar{x}^l \in X(\Delta^l) , \tag{26.7}$$

since $F(\bar{x}^l) \le F(\bar{x})$ for $l = 1, 2, \dots$. From $\Delta^l \underset{l\to\infty}{\longrightarrow} 0$ (by the lemma) it follows that $X(\Delta^l) \underset{l\to\infty}{\longrightarrow} M^2$ and the latter, considering (7), implies

$$\rho(\bar{x}^l, M^2) \underset{l\to\infty}{\longrightarrow} 0 .$$

The proof is completed.

§**26.2.** We consider here the penalty method for the mathematical programming problem which generates an iterative process at N levels. Moreover, calculations can be performed at $N - 1$ levels, and this gives the reason to call the process two-level as well.

Consider the following problem of mathematical programming

$$\min\{F(x) | G_i(x) \le 0, \quad i = \overline{1, N}\} , \tag{26.8}$$

here $x \in E^n$, $F(x)$ is a function, $G_i(x)$, $(i = \overline{1, N})$ a vector function. Introduce the point variables $y_i \in E^n$ $(i = \overline{2, N})$.

Denote

$$G_1 = \{x | G_1(x) \le 0\}, \quad G_i = \{y_i | G_i(y_i) \le 0\}, \quad i = \overline{2, N},$$
$$G = G_1 \times \cdots \times G_N .$$

Consider the problem equivalent to problem (26.8)

$$\min \left\{ F(x) \middle| \sum_{i=2}^{N} \|x - y_i\|^2 = 0, \quad (x, y_2, \dots, y_N) \in G \right\}. \tag{26.9}$$

Assume that G_1 is a bounded set, G is convex and closed, the function $F(x)$ on G_1 is convex and differentiable. We will set forth the penalty method for solving (26.8), using for this purpose problem (26.9).

Let $\{M_l\}$ be a strictly increasing sequence of positive numbers (penalties) such that $M_l \underset{l\to\infty}{\longrightarrow} \infty$. Define the problem

$$\min \left\{ F(x) + M_l \sum_{i=2}^{N} \|x - y_i\|^2 \middle| (x, y_2, \dots, y_N) \in G \right\} \tag{26.9l}$$

and denote its solution by $(\bar{x}^l, \bar{y}_2^l, \dots, \bar{y}_N^l)$. Thus, we have reduced solving problem (26.8) to solving a sequence of problems (26.9l) ($l = 1, 2, \dots$). We will solve problems (26.9l) by the componentwise descent method, the iterative process of which in the present case resides in the following

(1) Fix $y_i^{0l} = y_i^{l-1}$ $(i = \overline{2, N})$ (with $l = 1$ y_i^{0l} $(i = \overline{2, N})$ are chosen arbitrarily) and define x^{1l} as a solution of the problem

$$\min \left\{ F(x) + M_l \sum_{i=2}^{N} \|x - y_i^{0l}\|^2 \middle| x \in G_1 \right\}.$$

(2) Fix x^{1l} and define y_i^{1l} $(i = \overline{2, N})$ as a solution of the problems

$$\min\{\|x^{1l} - y_i\|^2 | y_i \in G_i\}, \quad (i = \overline{2, N}).$$

(3) Fix y_i^{1l} $(i = \overline{2, N})$, define x^{2l}, etc. The process of applying the method forms the sequence $\{x^{kl}, y_2^{kl}, \dots, y_N^{kl}\}$, any convergent sequence of which,

under the above assumptions, converges to one of the solutions of problem (26.21). Following the analysis carried out in §26.1, denote

$$\Delta^l = \sum_{i=2}^{N} \|\bar{x}^l - \bar{y}_i^l\|^2 \tag{26.10}$$

and replace everywhere $\|u-v\|^2_{E^{mN}}$ by $\sum_{i=2}^{N} \|x-y_i\|^2$ and (u,v) by the point $(y_2,\dots,y_N)$. Following theorem §26.1, we write

$$\rho(\bar{x}^l, M^1) \underset{l\to\infty}{\longrightarrow} 0\,, \tag{26.11}$$

where M^1 is a set of all solutions of problem (26.8). According to the lemma from §26.1 $\Delta^l \underset{l\to\infty}{\longrightarrow} 0$. Therefore, considering (26.10), (26.11):

$$\rho(\bar{y}_i^l, M^1) \underset{l\to\infty}{\longrightarrow} 0 \quad (i = \overline{2,N})\,.$$

Note that the method described when seeking the solution of the system $G_i(x) \le 0\,(i = \overline{1,N})$ provides the solution of the problem

$$\min\left\{ \sum_{i=2}^{N} \|x-y_i\|^2 \,\middle|\, (x, y_2, \dots, y_N) \in G \right\}.$$

§27. DECOMPOSITION BASED ON SIMPLEX METHOD MODIFICATIONS

§27.1. Consider some decomposition approaches for LP problems when implementing the simplex method which, adjusting itself to specific character of the problem, carries out partitioning into subproblems and thus enables one to reduce calculations. To begin with, we introduce into consideration a new designation of matrices indicating a set of their rows and columns. Thus, the symbol $A[M,N]$ denotes a matrix, the row indices of which run over the set M and the column indices stand for the set N. In this case $A[I,J]$ designates part of the matrix $A[M,N]$ composed of its elements standing at the intersections of the rows with the index from I and the columns with the indices from J. The present paragraph deals with LP problems the matrices $A[M,N]$ of which involve a special submatrix $A[M_0,N_0]$ of a sufficiently large dimension. Let $M_1 = M \setminus M_0$, $N_1 = N \setminus N_0$ and in the general case we obtain the following square partition of the matrix $A[M,N]$:

$$\left[\begin{array}{c|c} A[M_1,N_1] & A[M_1,N_0] \\ \hline A[M_0,N_1] & A[M_0,N_0] \end{array}\right].$$

If the purely special classes ($M_1 = N_1 = \varnothing$) are eliminated from consideration, then we get three classes of problems: problems with horizontal bordering ($N_1 = \varnothing$), problems with vertical bordering ($M_1 = \varnothing$), problems with two-side bordering ($M_1 \neq \varnothing, N_1 \neq \varnothing$). We shall not make any suppositions about the number of elements of the sets M_1, N_1, though application of the following decomposition methods to LP problems with bordering is worthwhile only when M_1, N_1 is substantially less than the rank of the matrix $A[M,N]$.

Note that the Dantzig-Wulf decomposition method described in chapter 3, §9 constitutes a simplex method modification for solving LP problems with horizontal bordering, and those with vertical bordering when the original problem is replaced by the dual.

§**27.2.** Consider general description and substantiation of one decomposition method based on the simplex method – the method for decomposing system.[3]

Assume that a system of constraints for the LP problem is divided into two parts so that the problem takes the form

$$\min c[N]x[N]\,, \tag{27.1}$$

$$A[M_1,N]x[N] = b[M_1], \quad x[N] \geq 0 \tag{27.2}$$

$$A[M_2,N]x[N] = b[M_2]\,. \tag{27.3}$$

Denote by $X(I)$ for any $I \subset M_1 \cup M_2$ a set of solutions to a system of constraints with indices from I.

Suppose there is $x^0[N] \in X(M_2)$ and the set $I_0 \subset M_2$, which contains indices of all the constraints active on $x^0[N]$. Solve the following problem

$$\min\{c[N]x[N],\ x[N] \in X(I_0 \cup M_1)\}\,. \tag{27.4}$$

If $X(I_0 \cup M_1) = \varnothing$, then also $X(M_1 \cup M_2) = \varnothing$ with the objective function being unconstrained in (27.4) the set I_0 expands (if $c[N]x[N]$ is constrained in general on $X(M_1 \cup M_2)$). Denote by $\bar{x}^0[N]$ the minimum obtained, and by $\bar{I}_0$ a set of the indices active for it so that $\bar{I}_0 \subset I_0 \cup M_1$. If $\bar{x}^0[N] \in X(M_2 \setminus I_0)$ then this is the solution of the original LP problem. Otherwise, consider the points $x_\lambda[N] = \lambda\bar{x}^0[N]+(1-\lambda)x^0[N]$, $\lambda \in [0,1]$ and find among them the point $x_{\lambda_0}[N] \in X(M_2)$ corresponding to the largest possible λ_0. Since the point $\bar{x}^0[N] \in X(I_0)$, then $\lambda_0 > 0$ and if only one such index $i_0 \in M_2 - I_0$ is found that a constraint with this index is active in $x_{\lambda_0}[N]$. Denote by I_1 any subset in M_2 containing the indices active on $x_{\lambda_0}[N]$ constraints, and take as $x^1[N]$ any point in $X(M_2)$ the active constraints of which are in I_1. For instance, one can take $x^1[N] = x_{\lambda_0}[N]$, and collect in I_1 the indices of its active constraints. The choice, however, can also be made otherwise. In any case, the constraints with indices from $\bar{I}_0 \cap I_0$ are active at the point $x_{\lambda_0}[N]$ and, therefore, $\bar{I}_0 \subset M_1 \cup I_1$. Moreover, the point $\bar{x}^0[N]$ does not obey the constraint with the index $i_0 \in I_1$. Again solve problem (27.4), this time on the set $X(I_1 \cup M_1)$. If the set is empty, then so is the set $X(M_1 \cup M_2)$. Otherwise, the form $c[N]x[N]$ achieves a minimum on $X(I_1 \cup M_1)$ since it achieves the minimum at the point $\bar{x}^0[N]$ on the set $X(\bar{I}_0)$, and $\bar{I}_0 \subset I_1 \cup M_1$. Denote by $\bar{x}^1[N]$ the point of the minimum obtained.

Since $\bar{x}^1[N]$ obeys the constraint with the i_0 index, and $\bar{x}[N]$ does not, then it implies that $\bar{x}^0[N] \neq \bar{x}^1[N]$. And since the constraint set, with which the point $\bar{x}^1[N]$ is obtained, involves all the active constraints of the point $\bar{x}^0[N]$ then $c[N]\bar{x}^1[N] \geq c[N]\bar{x}^0[N]$. For the pair of points $(x^1[N], \bar{x}^1[N])$ it is possible to iterate all constructions obtaining the points $x^2[N], \bar{x}^2[N]$, etc. Thus, we get the sequence of sets $I_k \subset M_2$, $k = 0, 1, \ldots$ and the sequence $\bar{x}^k[N]$, $k = 0, 1, \ldots$ of minimum points $c[N]x[N]$ on the set $X(I_k \cup M_1)$ for which

$$c[N]\bar{x}^{k+1}[N] > c[N]\bar{x}^k[N], \quad k = 0, 1, \ldots$$

And since the number of $c[N]\bar{x}^k[N]$ is uniquely determined by the set I_k, then the sequence $\bar{x}^k[N], k = 0, 1, \ldots$ is finite. Therefore, for the finite number of steps we will either obtain the solution of the original problem or establish its unsolvability.

§**27.3.** We consider application of the method described to the problem in canonic form.

Now, assume that problem (27.1) – (27.3) is specified. We take system (27.2) as the second group of constraints. The issue of selecting the initial point $x^0[N]$ can be solved in many ways. If form (27.1) is bounded below on a set of solutions of system (27.2), then one can take its minimum point as $x^0[N]$. In a more general case one can proceed from the fact that there is a known row $y_1[M_2]$ with which

$$(c[N] - y_1[M_2]A[M_2, N])x[N] \tag{27.5}$$

is bounded below on the solutions $x[N]$ of system (27.2). Let $x^0[N]$ be a minimum point of (27.5), and J_0 its associated basic set ($J_0 \subset N$). If we denote by $y_1[M_1]$ the solution of the problem dual with respect to (27.5), (27.2), then the row $y_1[M_1 \cup M_2]$ satisfies the conditions

$$\begin{aligned} y_1[M_1 \cup M_2]A[M_2 \cup M_1, N \setminus J_0] \leq c[N \setminus J_0]\,, \\ y_1[M_1 \cup M_2]A[M_1 \cup M_2, J_0] = c[J_0]\,. \end{aligned} \tag{27.6}$$

Because of this, either the system

$$x[N \setminus J_0] \geq 0, \quad A[M_1, N]x[N] = b[M_1]\,, \tag{27.7}$$

$$A[M_2, N]x[N] = b[M_2] \tag{27.8}$$

is inconsistent and, therefore, so is system (27.2) – (27.3) or form (27.1) attains the minimum on the set of solutions (27.7) – (27.8).

Now, obtain the minimum point of (27.1) with the constraints (27.7) – (27.8). Since the matrix $A[M_1, J_0]$ is nondegenerate, and the unknown $x[J_0]$ are not related by the non-negativity condition, then, using (27.7), we write

$$x[J_0] = D[J_0, M_1](b[M_1] - A[M_1, N \setminus J_0]x[N \setminus J_0])\,, \tag{27.9}$$

where $D[J_0, M_1] = (A[M_1, J_0])^{-1}$.

Substituting (27.9) in (27.1), (27.8), we get the problem

$$\begin{aligned} \min \tilde{c}[N \setminus J_0]x[N - J_0]\,, \\ x[N \setminus J_0] \geq 0\,, \end{aligned} \tag{27.10}$$

$$\tilde{A}[M_2, N \setminus J_0]x[N \setminus J_0] = b[M_2] - A[M_2, J_0]x^0[J_0]\,, \tag{27.11}$$

where the constant summand is omitted in (27.10) and

$$\begin{aligned}\tilde{c}[N\setminus J_0]x[N\setminus J_0]&=(c[N\setminus J_0]-y_1[M_1]A[M_1,N\setminus J_0])x[N\setminus J_0],\\ \tilde{A}[M_2,N\setminus J_0]&=A[M_2,N\setminus J_0]-A[M_2,J_0]D[J_0,M_1]A[M_1,N\setminus J_0].\end{aligned} \tag{27.12}$$

Evidently, the virtual calculation of all the matrix columns (27.12) can be justified only with the block diagonal matrix $A[M_1, N]$ of a small dimension, though.

Problem (27.10) – (27.11) can also be solved in the following way. Suppose for this problem there exists the next basic set $K \subset N \setminus J_0$, the columns of the basic matrix $\tilde{A}[M_2, K]$ are obtained or the matrix $\tilde{D}[K, M_2]$ inverse to it is known. Obtain the row $y[M_2] = \tilde{c}[K]\tilde{D}[K, M_2]$ and test the optimality condition

$$y[M_2]\tilde{A}[M_2, N \setminus J_0] \leq \tilde{c}[N \setminus J_0]\,. \tag{27.13}$$

If we first calculate the row

$$y[M_1] = y_1[M_1] - y[M_2]A[M_2, J_0]D[J_0, M_1]\,, \tag{27.14}$$

then (27.13) will be written as

$$y[M_2 \cup M_1]A[M_2 \cup M_1, N \setminus J_0] \leq c[N \setminus J_0]\,. \tag{27.15}$$

We need not know the matrix $D[J_0, M_1]$ since its utilization can be replaced by solving the system

$$w[M_1]A[M_1, J_0] = y[M_2]A[M_2, J_0]\,.$$

Then $y[M_1] = y_1[M_0] - w[M_0]$. Obtain the number $j' \in N \setminus J_0$, on which (27.15) is violated, and calculate $\tilde{c}[j'] = c[j'] - y_1[M_1]A[M_1, j']$, and then solve the system

$$A[M_1, J_0]g[J_0] = A[M_1, j']$$

and obtain the column $\tilde{A}[M_2, j'] = A[M_2, j'] - A[M_1, J_0]g[J_0]$, to be introduced into the basis.

Suppose the optimal solution $\bar{x}^0[N \setminus J_0]$ of problem (27.10), (27.11) has been obtained. Substituting it in the right-hand side of (27.9) or solving the system

$$A[M_1, J_0]x[J_0] = b[M_1] - A[M_1, N \setminus J_0]\bar{x}^0[N \setminus J_0]$$

we get the remaining components of the minimum point $\bar{x}^0[N]$ of form (27.1) on a set of solutions to (27.7) – (27.8).

If $\bar{x}^0[J_0] \geq 0$, then the column of $\bar{x}^0[N]$ constitutes the optimal solution of problem (27.1) – (27.3). Otherwise, we obtain

$$\lambda_0 = \min\left\{\frac{x^0[j]}{x^0[j] - \bar{x}^0[j]},\ j \in J_0,\ \bar{x}^0[j] < 0\right\} \tag{27.16}$$

and set $x_{\lambda_0}[N] = \lambda_0 \bar{x}^0[N] + (1-\lambda_0)x^0[N]$. If $\bar{J}_0$ is the basic set with which solving problem (27.10) – (27.11) terminates, j_0 is the number transferring a minimum to (27.16), then the numbers of all positive components are in the set $\bar{J}_0 \cup (J_0 \setminus j_0)$. They can be taken as the initial for the next step, though the set may not be basic for (27.2). It can be remedied in the following way.

We solve the following LP problem

$$\begin{gathered} \min \sum_{j\in k} x[j]\,, \\ x[\bar{J}_0 \cup J_0] \geq 0\,, \\ A[M_1, \bar{J}_0 \cup J_0]x[\bar{J}_0 \cup J_0] = b[M_0]\,. \end{gathered} \tag{27.17}$$

As an initial feasible set for (27.17), we can take J_0 and, after solving it, we will obtain a basic set J_1 and a feasible solution of $x^1[\bar{J}_0 \cup J_0]$. Since $x_{\lambda_0}[J_0 \cup J_0]$ satisfies the constraints of (27.17) and the functional for it is equal to zero, then $x^1[k_0] = 0$. For the set J_1 obtained and $x^1[N]$ we can carry out the following step of the method. Evidently, the methods for solving LP problems with horizontal bordering can also be applied to the problems with vertical bordering. For this purpose, it suffices to turn to the dual problem. Moreover, these methods can be applied for problems with two-side bordering, cutting off first a horizontal bordering part, as described, and considering the remaining problem with vertical bordering as the special. In turn it is possible to use the decomposition method each time when solving the special problem.

§27.4. Consider one more decomposition scheme based on the simplex method for the general LP problem[15]

$$\max\{cx | Ax = b,\ x \geq 0\}\,, \tag{27.18}$$

where $A = A[M, N]$, $M = \{1, \ldots, m\}$, $N = \{1, \ldots, n\}$, $m \leq n$. Partition the set M into the subset M_k so that $M = \bigcup_{k=1}^{t} M_k$, $M_k \cap M_j = \varnothing$, $k \neq j$ and isolate the submatrices $A[M_k, N]$ of the matrix $A[M, N]$. Assume that the blocks $A[M_k, N]$ are loosely bound, i.e. when solving (27.18) by the simplex method all the basic matrices involve a considerable number of the columns $A[M, j\,]$ among which only one from $A[M_k, j\,]$ is different from the zero one. At each iteration of the simplex method we need to solve two systems of the equations

$$\begin{aligned} y'[M]A[M, R] &= c'[R]\,, \\ A[M, R]q[R] &= A[M, j\,], \quad j \in N \setminus R\,, \end{aligned} \tag{27.19}$$

where $A[M, R]$ is a basic matrix, $R = \{j_1, \ldots, j_m\} \subset N$ is a subset of basic numbers.

We will employ the above assumption to transform $A[M, R]$. By rearranging the columns (if it is necessary) we can transform it to the form

$$A[M, R] = \begin{pmatrix} A[M_1, P] & A[M_1, Q_1] & \ldots & 0 \\ \cdot & \cdot \;\; \cdot & & \\ A[M_t, P] & 0 & \ldots & A[M_t, Q_t] \end{pmatrix} \tag{27.20}$$

where the matrices $A[M_s, P]$, $A[M_s, Q_s]$, $s = \overline{1, t}$ can also be nonzero, though not both at once. Since the matrix $A[M, R]$ is not expressed, then the Binet-Cauchy formula implies the existence of the partition $P = \bigcup_{s=1}^{t} P_s$ such that the matrices

$$A[M_s, k_s] = (A[M_s, P_s], A[M_s, Q_s]), \quad s = \overline{1, t}$$

are square nonsingular ($k_s = P_s \cup Q_s$).

We form the block diagonal matrix

$$D[M, R] = \begin{pmatrix} A[M_1, k_1] & \ldots & 0 \\ 0 & \ldots & A[M_t, k_t] \end{pmatrix}$$

and consider the product $D^{-1}[M, R]A[M, R]$.

It is evident that

$$A^{-1}[k_s, M_s]A[M_s, P_s] = \begin{pmatrix} I(P_s, P_s) \\ O(Q_s, P_s) \end{pmatrix},$$

$$A^{-1}[k_s, M_s]A[M_s, Q_s] = \begin{pmatrix} O(P_s, Q_s) \\ I(Q_s, Q_s) \end{pmatrix}.$$

Isolating in each block of the matrix the upper P_s rows and the first P columns, we get the square nonsingular matrix

$$G[P,P] = \begin{pmatrix} A^{-1}[P_1, M_1]A[M_1, P] \\ \dots \\ A^{-1}[P_t, M_t]A[M_t, P] \end{pmatrix}$$

denote

$$\begin{aligned} \bar{A}[k_s, j\,] &= A^{-1}[k_s, M_s]A[M_s, j\,], \\ \bar{y}'[k_s] &= y'[M_s]A[M_s, k_s]. \end{aligned} \tag{27.21}$$

With the aid of $\bar{A}[P_s, j\,]$, $q[P_s]$, $\bar{y}[P_s]$, $c[P_s]$, $s = \overline{1,t}$ form the vectors

$$\bar{A}[P, j\,],\ q[P],\ \bar{y}[P],\ c[P], \tag{27.22}$$

where $P = \bigcup\limits_{s=1}^{t} P_s$. We employ now the matrix $D[M, R]$ to transform equations (27.19)

$$\begin{aligned} y'[M]D[M,R]D^{-1}[R,M]A[M,R] &= c'[R], \\ D^{-1}[R,M]A[M,R]q[R] &= D^{-1}[R,M]A[M,j\,]. \end{aligned} \tag{27.23}$$

Considering (27.21), (27.22), the definition of the matrix $G[P,P]$ and the structure $D[M,R]$, $A[M,R]$, obtain from (27.23)

$$\begin{aligned} \bar{\sigma}'_s[P] &= c'[Q_s]A^{-1}[Q_s, M_s]A[M_s, P], \\ \bar{y}'[P] &= (c'[P] - \sum_{s=1}^{t} \bar{\sigma}'_s[P])G^{-1}[P,P], \\ \bar{y}'[K_s] &= (\bar{y}'[P_s], c'[Q_s]), \\ y'[M_s] &= y'[k_s]A^{-1}[k_s, M_s], \\ q[P] &= G^{-1}[P,P]\bar{A}[P,j\,], \\ q[Q_s] &= A^{-1}[Q_s, M_s](A[M_s, j\,] - A[M_s, P]q[P]). \end{aligned} \tag{27.24}$$

Relations (27.24) show the possibility of decomposition when solving (27.19). At each iteration of the simplex method a change is made in the matrices $A^{-1}[k_s, M_s]$, $G^{-1}[P, P]$.

Thus, the main part of the simplex method calculations is carried out, using (27.24) as the base, while all the other operations are performed in the ordinary way.

Remark: If a particular block has a special structure, then it can be completely included in the algorithm.

§27.5. Consider the algorithm for solving the large problems of linear dynamic programming which represents a further development of the above method.[16]
The linear dynamic programming problem in its general statement takes the form

$$\max_{x_k, u_k} \left(\sum_{k=0}^{N+1} c_k x_k + \sum_{k=0}^{N} d_k u_k \right), \tag{27.25}$$

$$x_{k+1} + A_k x_k + B_k u_k = f_k \,, \tag{27.26}$$

$$A_k^1 x_k + B_k^1 u_k = f_k^1, \quad k = \overline{0, N}\,. \tag{27.27}$$

We suppose, without restricting generality, that all the vectors of the states x_k, $k = \overline{1, N}$ are free variables (otherwise it will require only introducing auxiliary variables and equations into (27.27). Next we assume that a non-negativity condition is imposed on the vectors x_0, x_{N+1}, u_k, $k = \overline{0, N}$.

Problem (27.25) – (27.27) can be regarded as an LP problem and solved by the simplex method, at each iteration of which we need to solve two adjoint systems of equations of form (27.19). Evidently, the constraint matrix of LP problem takes the form

$$\Delta = \begin{pmatrix} E & B_N & A_N & & & & & & 0 \\ 0 & B_N^1 & A_N^1 & & & & & & \\ & & E & B_{N-1} & A_{N-1} & & & & \\ & & & B_{N-1}^1 & A_{N-1}^1 & & & & \\ & & & & & \ddots & & & \\ & & & & & & E & B_0 & A_0 \\ 0 & & & & & & & B_0^1 & A_0^1 \end{pmatrix},$$

which implies that the basic matrix of the problem at each iteration is as follows

$$\Delta_0 = \begin{pmatrix} \tilde{E} & \tilde{B}_N & A_N & & & & & & \\ & \tilde{B}_N^1 & A_N^1 & & & & & & \\ & & E & \tilde{B}_{N-1} & A_{N-1} & & & & \\ & & & \tilde{B}_{N-1}^1 & A_{N-1}^1 & & & & \\ & & & & & \ddots & & & \\ & & & & & & E & \tilde{B}_0 & \tilde{A}_0 \\ & & & & & & & \tilde{B}_0^1 & \tilde{A}_0^1 \end{pmatrix},$$

where $\tilde{B}_k\,(k=\overline{0,N})$, $\tilde{A}_0$, $\tilde{B}_k^1\,(k=\overline{0,N})$, $\tilde{A}_0^1$, $\tilde{E}$ are the submatrices composed of the columns of appropriate matrices. Since the variables x_k, $k=\overline{1,N}$ are free, then they are simultaneously basic at each iteration.

We rewrite the principal equations of the simplex method (27.19) as

$$\Delta_0[M,k]q[k] = b^j[M], \quad y'[M]\Delta_0[M,k] = c'[k]\,, \tag{27.28}$$

where M, k are the appropriate sets of the basic matrix rows and columns, and the vector $b^j[M]$ has zero components everywhere with the exception of the rows pertaining to the jth moment of time. If we introduce decomposition of the set $M = \bigcup_{k=0}^{N} M_k$ according to the moments of time, then

$$b^j[M] = \begin{pmatrix} 0 \\ b^j[M_j] \\ 0 \end{pmatrix}$$

and, moreover, we partition the sets M_k into M_{k_0}, M_{k_1} according to the constraints (27.26), (27.27).

At first we will consider the principal equations of the method described in §27.4 for the case of two-strip partition of major equations. The ordered series of such partitions underlies the method suggested for solving linear dynamic programming problems. Note the first strip corresponding to zero reading, i.e. isolate the rows of the set M_0 in the matrix Δ_0, and transfer the rows of the set $M^1 = \bigcup_{k=1}^{N} M_k$ to the second strip. Take as connecting the components of the vector x_1 which correspond, in the first strip, to the coefficient matrix

$$J^0 = \begin{pmatrix} E \\ 0 \end{pmatrix}$$

and, in the second, to

$$J^1 = \begin{pmatrix} 0 \\ \vdots \\ 0 \\ A_1 \\ A_1^1 \end{pmatrix}.$$

Following the method described in §27.4, we will introduce the square nonsingular matrices D_0, Δ_1 characterized by the Δ_0 columns multiplication coefficients in appropriate strips (i.e. products of D_0^{-1} and Δ_1^{-1} by corresponding local columns) form unit vectors. Naturally, the matrices D_0, Δ_1 are not uniquely determined by these conditions, but later on it will be shown how to make a better choice.

Replace equations (27.28) with the equivalent

$$H_0^{-1}\Delta_0 q = H_0^{-1} b^j, \quad y' H_0 H_0^{-1}\Delta_0 = c', \tag{27.29}$$

where

$$H_0 = \begin{pmatrix} \Delta_1 & 0 \\ 0 & D_0 \end{pmatrix}.$$

Considering the definitions of Δ_1, D_0 we have the columns of the matrix $H_0^{-1}\Delta_0$, with the exception of those pertaining to the coefficients of the vector x_1, will become unit vectors. Isolate the submatrix G_1 of the matrix $H_0^{-1}\Delta_0$ formed by the elements standing at the intersection of the above columns and rows which do not contain other nonzero elements. Let k^1 be the set of columns of the matrix Δ_1, k_0 – the set of columns of the matrix D_0 and let $k = k^1 \cup k_0$. Introduce the matrices $E[X_1, k^1]$, $E[X_1, k_0]$ composed of rows-unit vectors with units being at the places corresponding to the rows $\Delta_1^1 J^1$, $D_0^{-1} J_0$, which form the matrix G_1, i.e.

$$G_1 = \begin{pmatrix} E[X_1, k^1]\Delta_1^{-1} J^1 \\ E[X_1, k_0] D_0^{-1} J_0 \end{pmatrix}.$$

Somewhat changing the notations, we employ formulae (27.24)

$$
\begin{gathered}
q[X_1] = G_1^{-1} \begin{pmatrix} E[X_1, k^1]\Delta_1^{-1} & b^j[M^1] \\ E[X_1, k_0]D_0^{-1} & b^j[M_0] \end{pmatrix}, \\
q[k_0 \setminus X_1] = E[k_0 \setminus X_1, k_0]D_0^{-1}\left(b^j[M_0] - J_0 q[X_1]\right), \\
q[k^1 \setminus X_1] = E[k^1 \setminus X_1, k^1]\Delta_1^{-1}\left(b^j[M^1] - J' q[X_1]\right), \\
\sigma_0'[X_1] = E[X_1, k_0]c'[k_0]D_0^{-1}J_0, \qquad (27.30) \\
\sigma_1'[X_1] = E[X_1, k^1]c'[k_1]\Delta_1^{-1}J^1, \\
\bar{y}_1'[X_1] = (c'[X_1] - \sigma_0'[X_1] - \sigma_1'[X_1])\,G_1^{-1}, \\
y'[M_0] = \{(\bar{y}_1'[X_1] - c'[X_1])\,E[X_1, k_0] + c'[k_0]\}D_0^{-1}, \\
y'[M^1] = \{(y_1'[X_1] - c'[X_1])\,E[X_1, k^1] + c'[k^1]\}\Delta_1^{-1},
\end{gathered}
$$

where $E[k_0 \setminus X_1, k_0]$ are submatrices of the unit matrix $E[k_0, k_0]$, as supplementary (possibly after rearranging rows) to $E[X_1, k_0]$, the remaining matrices and vectors are formed in the same way.

Formulae (27.30) show the possibility of decomposition at one step. But the main difficulties of solution are associated with recurrence of the matrix Δ_1 which has the same structure as the entire basic matrix Δ_0. At the same time, the analysis of equations indicates that the same operations are carried out with Δ_1 as with the matrix Δ_0. This enables us to construct the recurrent process of solution, the construction of which is characterized by the fact that we can keep the structure of the remaining part unchanged by separating successively the bordering associated with the moments of time. With the basis changing, however, this feature causes us to make efforts to maintain the structure.

We describe the recurrent process of solving (27.29). Similarly to the definition of the matrix G_1, D_0, $\Delta_1, \ldots$ of the vectors $q[X_1]$, $q[k_0 \setminus X_1]$, $q[k^1 \setminus X_1]$, $\sigma_0'[X_1]$, $\sigma_1'[X_1]$ for the basic matrix Δ_0 we introduce for the matrix Δ_i the corresponding matrices and vectors with the index i. In this case the set of columns of the matrices D_i is denoted by k_i and $k^{i+1} = \bigcup_{i+1}^{N} k_j$. Also denote

$$\bar{b}[k_i] = D_i^{-1} b[M_i]\,. \tag{27.31}$$

Due to the basic matrix definition and the fact that the basis can involve only the vectors in $b[M]$ possessing nonzero elements only at one of the moments of time, we get for a particular s

$$\bar{b}[k_i] = 0, \quad i \neq s\,.$$

The direct move of the algorithm for solving equations (27.29) has the form

$$q^N[X_{N+1}] = E[X_{N+1}, k_N]\bar{b}[k_N]\,,$$

$$\bar{\bar{b}}[k_i] = \bar{b}[k_i] - J_0^i q^i[X_{i+1}]\,,$$

$$q^{i-1}[X_i] = G_i^{-1}\begin{pmatrix} E[X_i, k_i] & \bar{\bar{b}}[k_i] \\ E[X_i, k_{i-1}] & \bar{b}[k_{i-1}] \end{pmatrix},$$

$$\begin{aligned} y'_N[M^N] &= y'_N[M_N] = c'[k_N]D_N^{-1}\,, \\ \sigma'_i[X_i] &= y'_i[M_i]\begin{pmatrix} A_i \\ A_i^1 \end{pmatrix}, \\ \sigma'_{i-1}[X_i] &= c'[k_{i-1}]D_{i-1}^{-1}J_0^{i-1}\,, \\ \bar{y}'_i[X_i] &= \left(c'[X_i] - \sigma'_i[X_i] - \sigma'_{i-1}[X_i]\right)G_i^{-1}\,, \\ y'_{i-1}[M_{i-1}] &= \{(\bar{y}'_i[X_i] - c'[X_i])E[X_i, k_{i-1}] + c'[k_{i-1}]\}D_{i-1}^{-1}\,, \\ i &= N-1, \ldots, 1\,. \end{aligned} \tag{27.32}$$

From this we get $q_0[X_1] = q[X_1]$, $y'_0[M_0] = y'[M_0]$, i.e. the subvectors of the solution vectors of (27.29). Besides, if $\bar{b}[k_s] \neq 0$, then

$$q^i[X_{i+1}] = 0, \quad i = N-1, \ldots, s+1\,.$$

The inverse move of the algorithm is performed with the formulae

$$q[X_{i+1}] = q^i[X_{i+1}] - G_{i+1}^{-1} \begin{pmatrix} 0 \\ E[X_{i+1}, k_i]D_i^{-1} \begin{pmatrix} A_i \\ A_i^1 \end{pmatrix} q[X_i] \end{pmatrix},$$

$$q[v_i] = \bar{b}[v_i] - E[v_i, k_i]D_i^{-1} \begin{pmatrix} A_i \\ A_i^1 \end{pmatrix} q[X_i] - E[v_i, X_{i+1}]q[X_{i+1}], \quad (27.33)$$

$$i = \overline{0, N}.$$

$$\gamma'[k_i] = c'[k_i] + (\bar{y}'[X_i] - c'[X_i])\, E[X_i, k_i],$$

$$\sigma_i'[X_{i+1}] = \gamma'[k_i]D_i^{-1} J_0^i,$$

$$\bar{y}'[X_{i+1}] = \left(c'[X_{i+1}] - \bar{\sigma}_i'[X_{i+1}] - \sigma_{i+1}'[X_{i+1}]\right) G_{i+1}^{-1},$$

$$y'[M_i] = \{(\bar{y}'[X_{i+1}] - c'[X_{i+1}])\, E[X_{i+1}, k_i] + \gamma'[k_i]\}D_i^{-1},$$

$$i = \overline{1, N}.$$

Thus, (27.31), (27.32) prescribe a sequence of matrix operations to solve the basic equations of the simplex method. The remaining calculations are carried out in the regular way. From formulae (27.31), (27.32) it is seen that in the process of solving the basic equations one should maintain a comparatively small amount of information apart from going from one iteration to another (D_i^{-1}, G_i^{-1}), viz. the vectors $q^i[X_{i+1}]$, $\sigma_i'[X_i]$, $i = \overline{1, N}$.

Chapter 7

DECOMPOSITION AND AGGREGATION

The present chapter deals with the decomposition methods based on introducing aggregative variables (macrovariables). Aggregative variables are frequently introduced as sums of some components of initial unknown values. In other cases the weighted sums are taken. Thus, each time we have an intermediate problem in the aggregative variable case, which involves a smaller number of unknowns as compared to their quantity in the original problem. The aggregative problem is either solved in a finite form or acts as the coordination in the iterative process. An exact or approximate solution of the original problem is conclusively constructed according to some rules.

§28. AGGREGATION METHOD FOR SOLVING A SYSTEM OF LINEAR EQUATIONS

§28.1. Consider the following system of linear equations

$$x = Ax + b, \tag{28.1}$$

where x, b are n-dimensional vectors, A is $(n \times n)$ matrix possessing the properties:

$$0 \le a_{ij} \le 1, \quad i, j \in \{1, \ldots, n\}, \quad \sum_{i=1}^{n} a_{ij} \le 1, \quad j \in \{1, \ldots, n\},$$

$$b_i \ge 0, \quad i \in \{1, \ldots, n\}.$$

Let $x = (x_1, \ldots, x_n)$ be the solution of system (28.1). By adding all the equations of system (28.1), obtain the scalar equation

$$X = aX + \bar{b}\,,$$

where

$$X = x_1 + \cdots + x_n, \quad a = \sum_{i=1}^{n} a_i p_i; \quad p_i = \frac{x_i}{X}, \qquad \bar{b} = b_1 + \cdots + b_n, \quad a_i = \sum_{j=1}^{n} a_{ji}\,. \tag{28.2}$$

And conversely, if the X solution of equation (28.2) is known, then

$$x_i = \left(\sum_{j=1}^{n} a_{ij} p_j\right) X + b_i\,. \tag{28.3}$$

It is evident that problem (28.1) will be solved if we succeed in defining the p_i numbers so that

$$p_i = \frac{x_i}{X}, \qquad \sum_{i=1}^{n} p_i = 1\,.$$

We will set forth the iterative method for solving the system of equations (28.1) described in Refs. 43, 44. Let $x^0 = (x_1^0, \ldots, x_n^0)$ be specified as an initial approximation of solving (28.1). Let

$$X^{(0)} = \sum_{i=1}^{n} x_i^0, \qquad p_i^{(0)} = \frac{x_i^{(0)}}{X^{(0)}}\,.$$

Consider X^1 according to the following rule:

$$X^{(1)} = (a_1 p_1^{(0)} + \cdots + a_n p_n^{(0)}) X^{(1)} + \bar{b}\,,$$

or, denoting

$$a^{(0)} = a_1 p_1^{(0)} + \cdots + a_n p_n^{(0)}\,,$$

obtain

$$X^{(1)} = a^0 X^{(1)} + \bar{b} \tag{28.4}$$

and further

$$X^{(N+1)} = \left(\sum_{j=1}^{n} a_j p_j^{(N)} \right) X^{(N+1)} + \bar{b} \,, \tag{28.5}$$

$$x_i^{(N+1)} = \left(\sum_{j=1}^{n} a_{ij} p_j^{(N)} \right) X^{(N+1)} + b_i , \quad p_j^{(N)} = \frac{x_j^{(N)}}{X^{(N)}} \,. \tag{28.6}$$

From (28.5), (28.6) we get

$$\begin{aligned} p_i^{(N+1)} &= \frac{x_i^{(N+1)}}{X^{(N+1)}} = \sum_{j=1}^{n} a_{ij} p_j^{(N)} + r_i \sum_{j=1}^{n} \beta_j p_j^{(N)} \\ &= \sum_{j=1}^{n} (a_{ij} + r_i \beta_j) p_j^{(N)} \,; \end{aligned} \tag{28.7}$$

$$r_i = \frac{b_i}{\bar{b}} ; \quad \beta_j = 1 - a_j \,. \tag{28.8}$$

Introduce the matrix

$$S = \{s_{ij}\} \,,$$
$$s_{ij} = a_{ij} + r_i \beta_j , \quad i,j \in \{1, \ldots, n\}$$

and rewrite (8) in the form

$$p^{(N+1)} = S p^{(N)} = S^{(N+1)} p^{(0)} \,. \tag{28.9}$$

The matrix S has the properties:

(1) $s_{ij} \geq 0, \quad i,j = \overline{1,n},$

(2) $\sum_{i=1}^{n} s_{ij} = \sum_{i=1}^{n} (a_{ij} + r_i \beta_j) = a_i + \beta_j = 1, \quad j = \overline{1,n}.$

Assume that the matrix S is undecomposable, i.e. there is no such permutation matrix P^* that

$$P^* S (P^*)^{-1} = \left[\begin{array}{c|c} S_1 & S_2 \\ \hline 0 & S_3 \end{array} \right]$$

where S_1 is a square submatrix. A sufficient condition for this is non-decomposability of the matrix A. It is known[1] that

(1) the sequence $\{S^N\}$ converges to a matrix T;
(2) the column vectors of the matrix T are alike and equal to the vector $t > 0$;
(3) for any vector p ($p_i > 0, \sum_i p_i = 1$)

$$\lim S^N p = t, \quad t = ST .$$

From (28.9) and the above properties of the matrix S it follows that

$$p^{(N+1)} \xrightarrow[N\to\infty]{} p, \qquad X^{(N+1)} = \frac{b}{\sum_j \beta_j p_j^{(N)}} \xrightarrow[N\to\infty]{} X$$

and

$$x_i^{(N+1)} \xrightarrow[N\to\infty]{} x_i .$$

Thus the convergence of the iteration process has been shown.

From the calculation standpoint, using the obtained results for constructing the iteration process is more handy.

Definition: X_k is called J_k aggregative variable if

$$X_k = \sum_{i\in J_k} x_i, \quad k = \overline{1,l};$$
$$J_i \cap J_j = 0, \quad i \neq j;$$
$$\bigcup_{i=1}^{l} J_i = \{1, 2, \dots, n\}.$$

The matrix $A_{(n\times n)}$, with such aggregation, is convoluted into the matrix $\Lambda_{(l\times l)}$ the entries of which are obtained by transforming the A matrix elements as follows:

$$\bar{a}_{ij} = \sum_{s,t} a_{st} + \bar{p}_t; \quad s \in J_i, \quad t \in J_j; \quad i, j = 1, 2, \dots, l,$$
$$\bar{p}_t = \frac{x_t}{X_j}, \quad \sum_{t\in J_j} \bar{p}_t = 1 . \qquad (28.10)$$

Denote

$$\bar{b}_i = \sum_{s \in J_i} b_s, \qquad \bar{B} = (\bar{b}_1, \dots, \bar{b}_l)^T .$$

Then system (28.1) transforms into the system

$$X = \bar{A}X + \bar{B} . \tag{28.11}$$

When constructing the iteration process we specify, as before, the initial value of $x_s^{(0)}$, $s = \overline{1,n}$. At the first step we get the values of $X_1, \dots, X_l$ by solving the system

$$X^{(1)} = \bar{A}^{(0)} X^{(1)} + \bar{B} . \tag{28.12}$$

The entries of the matrix $\bar{A}^{(0)}$ are derived from (28.10) if we set

$$\bar{p}_t = \frac{x_t^{(0)}}{X_j^{(0)}}; \quad t \in J_j ;$$

$$X_j^{(0)} = \sum_{t \in J_j} x_t^{(0)}, \quad j = \overline{1,l} .$$

At the same time, the knowledge of the exact solution of (28.12) is basically not necessary.

Solve (28.1) as follows:

Define

$$y^{(0)} = \sum_{k=1}^{l} X_k^{(0)} = Y^{(0)} ,$$

$$q_k^{(0)} = \frac{X_k^{(0)}}{Y^{(0)}}; \qquad \bar{a}_j^{(0)} = \sum_{i=1}^{l} \bar{a}_{ij}^{(0)} ; \tag{28.13}$$

$$\bar{a}^{(0)} = \sum_{j=1}^{l} \bar{a}_j^{(0)} q_j^{(0)}; \quad \bar{b} = \sum_{i=1}^{l} \bar{b}_i .$$

Let $Y^{(1)}$ be obtained from solving the scalar equation

$$Y^{(1)} = \bar{a}^{(0)} Y^{(1)} + \bar{b} . \tag{28.14}$$

Instead of defining $X_k^{(1)}$ from (28.12), we set

$$X_k^{(1)} = Y^{(1)} q_k^{(0)}, \quad k = \overline{1,l} \tag{28.15}$$

and next obtain

$$x_s^{(1)} = \sum_{j=1}^{l} \sum_{t\in J_j} a_{st}\bar{p}_t^{(0)} X_j^{(1)} + b_s, \quad s = \overline{1,n}\,. \tag{28.16}$$

Similarly, the value of x_s at the $(N+1)$ step of iteration is obtained from the formula:

$$\begin{gathered} x_s^{(N+1)} = \sum_{j=1}^{l} \sum_{t\in J_j} a_{st}\bar{p}_t^{(N)} X_j^{(N+1)} + b_s\,, \\ p_t^{(N)} = x_t^{(N)} \Big/ X_j^{(N)}, \quad t \in J_j\,, \\ X^{(N+1)} = Y^{(N+1)} q^{(N)}, \quad j = \overline{1,l}\,, \\ Y^{(N+1)} = \bar{a}^{(N)} Y^{(N+1)} + b, \quad q_j^{(N)} = \frac{X_j^{(N)}}{Y^{(N)}}\,. \end{gathered} \tag{28.17}$$

We will call the process of solving system (28.1) in the first case the single-step iteration process, and the process described just now the two-step iteration process since in the former case all the variables of system (28.1) are aggregated into one variable X, whereas in the latter a similar aggregation process was accomplished for two steps.

As a result of iteration with the number N of the two-step process we obtain the values of the variables $x_s = \bar{x}_s^{(N)}$. Let $x_s = x_s^{(N)}$ be the value of x_s after the Nth iteration in the single-step process. We will demonstrate convergence of the two-step process.

The initial values for both processes are the same, i.e.

$$p_t^0 = \bar{p}_t^0 q_j^0, \quad t \in J_j, \quad j = \overline{1,l} \tag{28.18}$$

Compare (28.3) with (28.14), obtain

$$\begin{aligned} \bar{b} &= \sum_{i=1}^{l} \bar{b}_i = \sum_{j=1}^{n} b_j = b\,, \\ \bar{a}^{(0)} &= \sum_{j=1}^{l} \bar{a}_j^{(0)} q_j^{(0)} = \sum_{i=1}^{l} \sum_{j=1}^{l} \bar{a}_{ij}^{(0)} q_j^{(0)} \\ &= \sum_{\substack{s\in U J_i \\ t\in U J_j}} a_{st} p_t^{(0)} = \sum_{i,j=1}^{n} a_{ij} p_j^{(0)} = a^0\,. \end{aligned} \tag{28.19}$$

From (28.3), (28.14), (28.18), (28.19) we get

$$X^{(1)} = Y^{(1)}\,. \tag{28.20}$$

By using (28.6), (28.13), (28.16), (28.18) and the fact that J_j do not intersect, we obtain

$$\begin{aligned}\bar{x}_s^{(1)} - b_s &= \sum_{j=1}\sum_{t\in J_j} a_{st}\bar{p}_t^{(0)} q_j^{(0)} Y^{(1)} \\ &= \sum_{t\in U J_j} a_{st}p_t^{(0)} X^{(1)} = x_s^{(1)} - b_s\,.\end{aligned}$$

Similarly, by induction we prove that $\bar{x}_s^{(N)} = x_s^{(N)}$. Naturally, the process of solving (28.1) can be represented not only as the two-step but also as the multi-step procedure. The method described enables one to solve (28.1) numerically at the cost of solving the problems each of which accommodates a little more than part of the matrix A or a certain convolution $\bar{A}$ of the matrix A.

§**20.2.** The iterative aggregation method described is particularly efficient if the matrix A in (28.1) has the block diagonal structure

$$A = \begin{pmatrix} a_{11} & 0 & \dots & 0 \\ a_{21} & a_{22} & \dots & 0 \\ \cdot & \cdot & \cdot & \cdot \\ a_{m1} & a_{m2} & \dots & a_{mm} \end{pmatrix}.$$

All the variables of x_i, $(k = \overline{1,m})$, $i \in J_k$ are assume to be aggregated into one magnitude of X_k $(k = \overline{1,m})$. Thus (28.1) is written in the form:

$$\begin{aligned}X_1 &= \bar{a}_{11}X_1 + b_1\,, \\ X_2 &= \bar{a}_{21}X_1 + \bar{a}_{22}X_2 + b_2\,, \\ X_m &= \bar{a}_{m1}X_1 + \bar{a}_{m2}X_2 + \cdots + \bar{a}_{mm}X_m + b_m\,.\end{aligned}$$

We set that at each step of the iterative process (28.1) is uncoupled into m equations to be solved separately.

§29. AGGREGATION METHOD FOR THE BLOCK PROBLEM OF LINEAR PROGRAMMING

§29.1. Consider the following LP problem (23)

$$\min\left\{c_0x_0+\sum_{i=1}^{k}c_ix_i+\sum_{i=1}^{k}d_iy_i\right\},$$

$$A_0x^0+\sum_{i=1}^{k}A_ix_i=b_0\,, \tag{29.1}$$

$$B_ix_i+D_iy_i=b_i,\quad i=\overline{1,k}\,,$$

$$x_0\ge 0;\quad x_i\ge 0,\quad y_i\ge 0,\quad i=\overline{1,k}\,.$$

Assume that the conditions

$$B_ix_i+D_iy_i=b_i,\quad x_i\ge 0,\quad y_i\ge 0$$

prescribe for all $i\in\{1,\dots,N\}$ the constrained polyhedrons.

Aggregation will be taken as the following change of variables

$$x_i=\alpha_iX_i,\quad i=\overline{1,k}\,, \tag{29.2}$$

where $\alpha=(\alpha_1,\dots,\alpha_k)$ is the vector of aggregation weights, α_i $(i=\overline{1,k})$ a particular fixed vector.

We will say that aggregation is feasible if $\alpha_i\ge 0$ for all $i=\overline{1,N}$.

By substituting (29.2) in (29.1), we get the problem with aggregative variables

$$\min\left\{c_0x_0+\sum_{i=1}^{k}\bar c_iX_i+\sum_{i=1}^{k}d_iy_i\right\},$$

$$A_0x_0+\sum_{i=1}^{k}\bar a_iX_i=b_0\,, \tag{29.3}$$

$$b_iX_i+D_iy_i=b_i,\quad i=\overline{1,k}\,,$$

$$x_0\ge 0;\quad X_i\ge 0,\quad y_i\ge 0,\quad i=\overline{1,k}\,,$$

where

$$\bar a_i=A_i\alpha_i,\quad \bar b_i=B_i\alpha_i,\quad \bar c_i=c_i\alpha_i\,. \tag{29.4}$$

Since the aggregation weights can be selected in various ways, then one original problem corresponds to an infinite number of different aggregative problems. We then establish some solvability properties of these problems.

Theorem 29.1: If the original problem is solvable, the aggregative problem has a feasible solution and aggregation is feasible, then the linear form of the aggregative problem is bounded below on the entire set of its programs.

Proof: Let $\{\bar{x}_0,\ X_i,\ \bar{y}_i\}$ be a feasible solution of aggregative problem (29.13). Set $\bar{x}_i = \alpha_i \bar{X}_i$. Evidently, $\{\bar{x}_0, \bar{x}_i, \bar{y}_i\}$ is feasible in the original problem, and the value of its functional on it $\bar{\varphi}$ coincides with the functional value of problem (29.3) on the set $\{\bar{x}_0, \bar{X}_i, \bar{y}_i\}$. From this we get $\bar{\varphi} \geq \varphi^*$, where φ^* is the optimal value of the functional in (29.1) restricted owing to its solvability.

Introduce into consideration the added problem

$$\min\left\{c_0 x_0 + \sum_{i=1}^{k} d_i y_i\right\}, \tag{29.5}$$

$$A_0 x_0 = b_0, \quad D_i y_i = b_i, \quad x_0 \geq 0, \quad y_i \geq 0, \quad i = \overline{1,k}\,.$$

Theorem 29.2: So that any one of the aggregative problems will have a feasible solution, it is necessary and sufficient that the added problem might have optimal solutions.

Proof: Let $\{\bar{x}_0, \bar{y}_i\}$ be the optimal solution of (29.5). By adding to it X_i, $i = \overline{1,k}$, we get the feasible solution of any aggregative problem. If the added problem is unsolvable then so is the aggregative problem obtained with $\alpha = 0$. Thus, if the original and the added problem are solvable and aggregation is feasible, then each of the aggregative problems is solvable, which will be further kept in mind. Let $\{\bar{x}_0, \bar{X}_i, \bar{y}_i\}$ be the optimal solution of (29.5) with a particular vector α, the functional value being $\bar{\varphi}$. Assume that $\bar{x}_i = \alpha_i X_i$ and call the set $\{\bar{x}_0, \bar{x}_i, y_i\}$ a disaggregated solution. It is feasible in the original problem. Thus, if $\{\varphi_\alpha\}$ is a set of optimal values of the functions of aggregative problems, then we have

$$\varphi_\alpha \geq \varphi^*\,. \tag{29.6}$$

Theorem 29.3: Let α_1, α_2 be the aggregation weight vectors of two aggregative problems, here $\alpha_1 = k\alpha_2$ and $k > 0$, then the sets of their disaggregated solutions coincide.

Proof: See Ref. 23.

We will establish now the existence of aggregative problems the functional values of which coincide with the functional optimal value of original problem (29.1). Let $\{x_0^*, x_i^*, y_i^*\}$ be the optimal solution of (29.1). Suppose that $\alpha_i = x_i^*$, $X_i = 1$, $i = \overline{1,k}$, then we have the set $\{x_0^*, 1, y_i^*\}$ being feasible in the corresponding aggregative problem which gives the functional value φ^*. Thus we have the equality

$$\min_{\alpha} \varphi_\alpha = \varphi^* . \tag{29.7}$$

The aggregation vectors α, with which equality (29.7) is achieved, will be termed the optimal aggregation vectors, and their associated aggregative problems the problems with the optimal value of parameters. Aggregation, which leads to the problem with optimal parameters, will be called optimal. The next purpose is to construct the iterative process which, starting with a feasible aggregation, leads to optimal aggregation.

To formulate the basic theorem we have to introduce additional definitions. Let $\{\bar{x}_0, \bar{X}_i, \bar{y}_i\}$ be the optimal solution of the aggregative problem with the aggregation weight vector α. Setting $\bar{b}_0 = b_0 - \sum_{i=1}^{k} \bar{a}_i X_i$ we can construct the LP problem

$$\min c_0 x_0 ,$$
$$A_0 x_0 = \bar{b}_0, \quad x_0 \geq 0 .$$

It is evident that x_0 is one of the optimal solutions of the problem. We will identify the problem as connected. Let δb be an arbitrary vector of the same dimensionality as $\bar{b}_0$. Introduce the so-called connected parametric problem

$$\begin{gathered} \min c_0 x_0 , \\ A_0 x_0 = \bar{b}_0 + \beta \delta b, \quad x_0 \geq 0 . \end{gathered} \tag{29.8}$$

With $\beta = 0$ the solution of (29.8), as we see, exists, and with $\beta > 0$ it may not already exist. We need to fulfill the condition.

Condition 29.1: For any vector δb and a particular $\beta > 0$ the solution of (29.8) exists if it exists for $\beta = 0$. For problem (29.8) we formulate the dual with $\beta = 0$.

$$\max \bar{b}_0 u\,,$$
$$u' A_0 \geq c^0\,.$$

Denote by v_0 the constrained part of the optimal program estimate vector set of this problem, i.e.

$$v^0 = \{u_0 | u_0' A_0 \leq c_0, \quad (c_0, \bar{x}_0) = (\bar{b}_0, u_0), \quad \|u_0\| < \infty\}\,.$$

For any $u_0 \in v^0$ we can form the vectors

$$\sigma_i = u_0 A_i - c_i\,. \tag{29.9}$$

Consider the local problems with each $i \in \{1, \ldots, k\}$:

$$\begin{aligned} &\max\{\sigma_i x_i + d_i y_i\}\,,\\ &B_i x_i + D_i y_i = b_i\,,\\ &x_i \geq 0, \quad y_i \geq 0\,. \end{aligned} \tag{29.10}$$

Let $\{\hat{x}_i, \hat{y}_i\}$ $(i = \overline{1,k})$ be some optimal solutions of problems (29.10), and $\hat{z}_i = \sigma_i \hat{x}_i + d_i \hat{y}_i$ be the functional optimal value of the ith problem.

Establish solvability of the problem. The elements of the disaggregated solution $\{\bar{x}_i, \bar{y}_i\}$ are feasible in the corresponding local problems. Therefore, if $\bar{z}_i = \sigma_i \bar{x}_i + d_i \bar{y}_i$, then the inequality $\bar{z}_i \leq \hat{z}_i$ is fulfilled or, by introducing the designations $\hat{z} = \sum_{i=1}^{k} \hat{z}_i$, $\bar{z} = \sum_{i=1}^{k} \bar{z}_i$, the inequality $\bar{z} \leq \hat{z}$ is completed.

Theorem 29.4: (Aggregation optimality criteria).

If condition 29.1 is met, then for aggregation optimality it is necessary and sufficient that for a particular $u_0 \in v_0$ the following equality might be fulfilled

$$\bar{z} = \hat{z}\,. \tag{29.11}$$

Proof:

Sufficiency:

Consider the problem dual to the aggregative

$$\begin{aligned} &\max\left\{u_0 b_0 + \sum_{i=1}^{k} u_i b_i\right\},\\ &u_0 A_0 \leq c_0\,,\\ &u_0 \bar{a}_i - u_i \bar{b}_i \leq \bar{c}_i\,,\\ &u_i D_i \leq d_i, \quad i = \overline{1,k}\,. \end{aligned} \tag{29.12}$$

Let $\{\bar{x}_0, \bar{X}_i, \bar{y}_i\}$, $\{\bar{u}_0, \bar{u}_i\}$ be the optimal solutions of (29.3), (29.12), respectively. By the second duality theorem we have

$$\bar{X}_i(\bar{u}_0, \bar{a}_i - \bar{u}_i \bar{b}_i - \bar{c}_i) = 0 .$$

By utilizing (29.9), (29.4), we rewrite the above equality as

$$(\sigma_i, \bar{x}_i) - \bar{X}_i(\bar{u}_i, \bar{b}_i) = 0 ,$$

where $\bar{x}_i$ is the element of the disaggregated solution. Moreover, $\bar{b}_i \bar{X}_i = b_i - D_i \bar{y}_i$ and, therefore,

$$\bar{u}_i b_i = \sigma_i \bar{x}_i + \bar{u}_i D_i \bar{y}_i = \bar{z}_i . \tag{29.13}$$

If $\bar{\varphi}$ is the functional optimal value of the aggregative problem, then it is simultaneously the functional value of the original problem on the disaggregated solution which is feasible in this problem. From the optimality in the aggregative problem and (29.13) we get

$$\bar{\varphi} = \bar{u}_0 b_0 + \sum_{i=1}^{k} \bar{u}_i b_i = \bar{u}_0 b_0 + \bar{z} . \tag{29.14}$$

Consider now the problems dual to the local (29.10) with $i = 1, \ldots, k$

$$\begin{gathered} \min v_i b_i , \\ v_i B_i \leq \sigma_i, \quad v_i D_i \leq d_i . \end{gathered} \tag{29.15}$$

Let v_i, $i = \overline{1,k}$ be the optimal solutions of those problems, i.e. $\hat{z}_i = v_i b_i$. Then the set $\{\bar{u}_0, v_i\}$, $(i = \overline{1,k})$ is admissible to the problem being dual to the original. This can be managed if we develop the first inequality in (29.15) with regard to σ_i definition, the functional value of the problem dual to the original being equal

$$\theta = \bar{u}_0 b_0 + \sum_{i=1}^{k} v_i b_i = \bar{u}_0 b_0 + \hat{z} . \tag{29.16}$$

If $\bar{z} = \hat{z}$, then (29.14), (29.16) imply

$$\bar{\varphi} = \theta ,$$

i.e. the functional values of the original and the dual problem coincide on two feasible solutions $\{\bar{x}_0, \bar{x}_i, \bar{y}_i\}$, $\{\bar{u}_0, v_i\}$. This signifies optimality of the latter in appropriate problems. Thus, we conclude about the optimality of integration when (29.11) is valid.

Necessity: Consider formation of local problems with the optimal solution of the dual problem for the connected (29.18).

Let $\bar{u}_0$ be the optimal program estimates corresponding to the basis $\bar{x}_0$. Form the local problems and let $\bar{z} < \hat{z}$. This means that for a particular $k' \le k$ the inequalities $\bar{z}_i < \hat{z}_i$, $i = \overline{1, k'}$ are fulfilled. Furthermore, let $\{\hat{x}_i, \hat{y}_i\}$ be the optimal programs of the local problems with $i = \overline{1, k'}$, and $\{\bar{x}_i\}$ be the disaggregated solution components with $i = \overline{1, k}$. Form a new vector of aggregation weights

$$\alpha_i = \begin{cases} (1-\beta)\bar{x}_i + \beta\hat{x}_i, & i = \overline{1, k'}, \\ \bar{x}_i, & i = \overline{k'+1, k}. \end{cases} \tag{29.17}$$

Aggregation is feasible with $\beta \in [0, 1]$, since $\bar{x}_i \ge 0$, $\hat{x}_i \ge 0$ for all i. The local constraints, which must be satisfied with the variables of the new aggregative problem, reduce to the following inequalities:

$$B_i[(1-\beta)\bar{x}_i + \beta\hat{x}_i]X_i + D_i y_i = b_i, \quad i = \overline{1, k'},$$
$$B_i\bar{x}_i X_i + D_i y_i = f_i, \quad i = \overline{k'+1, k}.$$

Hence it follows that the set

$$X_i = 1, \quad i = \overline{1, k},$$
$$y_i(\beta) = \begin{cases} (1-\beta)\bar{y}_i + \beta\hat{y}_i, & i = \overline{1, k'}, \\ \bar{y}_i, & i = \overline{k'+1, k}, \end{cases}$$

satisfies all the block constraints of the new aggregative problem ($y_i(\beta) \ge 0$, since $\bar{y}_i \ge 0, \hat{y}_i \ge 0$). Its remaining constraints reduce to the condition $A_0 x_0 = \bar{b}_0 + \beta\sigma b$, where

$$\bar{b}_0 = b_0 - \sum_{i=1}^{k} A_i \bar{x}_i = b_0 - \sum_{i=1}^{k} \bar{a}_i X_i,$$

$$\delta b = \sum_{i=1}^{k'} A_i(\bar{x}_i - \hat{x}_i).$$

Let $\bar{x}_0(\beta)$ be the solution of the parametric programming problem

$$\min c_0 x_0 \,, \qquad A_0 x_0 = \bar{b}_0 + \beta \delta b, \quad x_0 \geq 0 \,. \tag{29.18}$$

Then the set

$$\{\bar{x}_0(\beta), X_i, y_i(\beta)\} \tag{29.19}$$

is feasible in the new aggregative problem at least for those β for which solution (29.18) exists. However, with $\beta = 0$, this is the connected problem, therefore, by the condition of (29.1) there exists $\bar{\beta} > 0$ such that (29.18) has a solution at least in the segment $0 \leq \beta \leq \bar{\beta}$. The problem resides in estimating the solution of the functional of the new aggregative problem on its solution (29.19) with $\beta \neq 0$. Assume that $\bar{x}_0 > 0$, i.e. the estimates $\bar{u}_0$ apply to a nondegenerate basic solution of the connected problem. In this case there exists $\beta' > 0$, $\beta' \in [0, \beta]$, with which the basis of the vector $\bar{x}_0$ remains the basis of the optimal solution of (29.18) and with $\beta = \beta'$, as well. Let $\bar{x}_0(\beta')$ be the optimal solution of (29.18) with $\beta = \beta'$ being in the vector basis $\bar{x}_0$. Then we get $c_0\bar{x}_0(\beta') = \bar{u}_0 b_0 + \beta'(\bar{u}_0, \delta b)$. And since $\bar{u}_0$ is the optimal solution of the problem dual to the connected, then

$$c_0\bar{x}_0(\beta') = c_0 x_0 + \beta'(\bar{u}_0, \delta b) \,.$$

The functional of the new aggregative problem on (29.19) is equal

$$\begin{aligned}
\tilde{\varphi} &= c_0 x_0(\beta) + \sum_{i=1}^{k} c_i + \sum_{i=1}^{k} d_i y_i(\beta) \\
&= \bar{\varphi} + \beta \left[\bar{u}_0 \delta b + \sum_{i=1}^{k'} (c_i, \hat{x}_i - \bar{x}_i) + \sum_{i=1}^{k'} (d_i, -\bar{y}_i + \hat{y}_i) \right] \\
&= \bar{\varphi} + \beta \sum_{i=1}^{k'} \{(\bar{u}_0 A_i - c_i, \bar{x}_i - \hat{x}_i) + (d_i, \hat{y}_i - \bar{y}_i)\} \\
&= \bar{\varphi} + \beta(-\hat{z}' + \bar{z}') \,.
\end{aligned} \tag{29.19}$$

And since $\beta > 0$ and $\bar{z}' < \hat{z}'$, then, denoting the functional optimal value of the new aggregative problem by $\bar{\varphi}'$, we get $\varphi^* \leq \bar{\varphi}' \leq \tilde{\varphi} \leq \bar{\varphi}$, $\bar{\varphi} > \varphi^*$, i.e. aggregation is not optimal. The lemma is valid. If the strict inequality $\hat{z} > \bar{z}$ is fulfilled, then aggregation with the vector α is not optimal. This completes theorem (29.4).

§**29.2.** Let us formulate the algorithm for solving problem (29.1). The process of "improving" the aggregating vector with the functional decreasing for the disaggregated solutions admissible to (29.1) is based on formula (29.17) and relation (29.19).

Step 0: Choose $\alpha^{(0)}$ the vector of aggregation weights and the parameter $\beta \in [0,1]$. Let $l = 0$, $\bar{\varphi}^0 = \infty$.

Step 1: Solve aggregative problem (29.3) for the weight vector $\alpha^{(l)}$. Let $\{\bar{x}_0, \bar{X}_i, \bar{y}_i\}$, $(i = \overline{1,k})$ be a solution and $\bar{\varphi}^l$ be the functional value of (29.3). If $\bar{\varphi}^l > \bar{\varphi}^{l-1}$, diminish β.

Step 2: By the optimal solution of (29.3), setting $\bar{b}_0 = b^0 - \sum_{i=1}^{k} \bar{a}_i X_i$, where $\bar{a}_i$ is defined by formula (29.4), we construct the connected problem

$$\min c_0 x_0 \,,$$
$$A_0 x_0 = \bar{b}_0 , \quad x_0 \geq 0$$

and the dual to it

$$\max \bar{b}_0 u \,,$$
$$u' A_0 \geq c_0 \,. \tag{29.20}$$

Let u be the optimal solution of (29.20).

Step 3: Define the vectors $\sigma_i = u_i A_i - c_i$ and formulate local problems for each $i = \overline{1,k}$:

$$\max\,(\sigma_i x_i + d_i y_i) \,,$$
$$B_i x_i + D_i y_i = b_i \,, \tag{29.21}$$
$$x_i \geq 0, \quad y_i \geq 0 \,.$$

Let $(\hat{x}_i, \hat{y}_i)$, $(i = \overline{1,k})$ be the optimal solutions of problems (29.21).

<u>*Step 4*</u>: Define the values

$$\bar{z} = \sum_{i=1}^{k} \bar{z}_i, \quad \hat{z} = \sum_{i=1}^{k} \hat{z}_i,$$

where $\hat{z}_i = \sigma_i \hat{x}_i + d_i \hat{y}_i$, $\bar{z}_i = \sigma_i \bar{x}_i + d_i \bar{y}_i$.

<u>*Step 5*</u>: Test the optimality criterion. If $\bar{z} = \hat{z}$ with a specified accuracy, then go to step 7.

<u>*Step 6*</u>: Construct the set $k' = \{i : \bar{z}_i < \hat{z}_i, i = \overline{1,k}\}$ and define $\alpha^{(l+1)}$ – a new vector of aggregation weights with the formula

$$\alpha_i = \begin{cases} (1-\beta)\bar{x}_i + \beta \hat{x}_i, & i = \overline{1,k'}, \\ \bar{x}_i, & i = \overline{k'+1,k}. \end{cases}$$

Let $l = l + 1$. Go to step 1.

<u>*Step 7*</u>: Define the optimal solution of problem (29.1)

$$\{x_0^*, x_i^*, y_i^*\}, \quad i = \overline{1,k},$$

where $x_i^* = \alpha_i X_i$, $i = \overline{1,k}$.

Terminate.

§30. DECOMPOSITION AND AGGREGATION BASED ON PERTURBATIONS METHOD

The perturbations method is one of the major instruments of applied mathematics. Its basic idea lies in dividing a structure. It is this method that enables one to overcome a barrier existing between simple idealized models and complex real systems. Note that application of the perturbations method calls for deep penetration into the essence of the problem in hand to isolate in it the main and secondary factors.

§30.1. Consider the mathematical programming problem

$$\begin{aligned} &\max f_0(x), \\ &g_0(x) \le 0, \end{aligned} \tag{30.1}$$

where x is the n-dimensional vector, $g_0(x)$ the m-dimensional vector function. Concurrent with (30.1), consider the family of problems

$$\begin{aligned} &\max\left(f_0(x)+\varepsilon f_1(x)\right), \\ &g_0(x)+\varepsilon g_1(x)\le 0, \end{aligned} \tag{30.2}$$

where ε is a nonnegative scalar parameter. Problem (30.1) is identified as generative with respect to the family of problems (30.2) which will be termed the perturbated problem. The successful solution of the problem by the perturbations method is basically determined by two circumstances: firstly, the simplicity of properties of a generative problem, secondly, the proximity of the functions determining the original and the generative problem.

Consider some types of the problems admitting immediate simplification.

(1) $$\max\{\mathcal{F}(z_1(x^1),\dots,z_k(x^k))/x^s\in X_s,\quad s=\overline{1,k}\}, \tag{30.3}$$

where $\mathcal{F}(z_1,\dots,z_k)$ is a monotone increasing function, x^s are the n_s vectors, X_s are the subsets of the n_s-dimensional Euclidean spaces. The solution of the problem reduces to the independent solution of the problem of the form

$$\max\{z_s(x^s)/x^s\in X_s\},\quad s=\overline{1,k}. \tag{30.4}$$

If the generative problem takes form (30.3), it will be called the immediately decomposable.

(2) $$\max\{\mathcal{F}(z_1(x^1),\dots,z_k(x^k))\mid g(z_1(x^1),\dots,z_k(x^k))\le 0\}. \tag{30.5}$$

The solution of (30.5) is equivalent to the solution of the optimization problem in the aggregative variables z_i, $(i=\overline{1,k})$.

$$\max\{\mathcal{F}(z_1,\dots,z_k)|g(z_1,\dots,z_k)\le 0\}, \tag{30.6}$$

here the solutions of problems (30.5) in the initial variables are the roots of the equations

$$z_s(x^s)=z_s^*,\quad s=\overline{1,k}, \tag{30.7}$$

where $\{z_s^*\}$ is the solution of (30.6). We will call the problem of form (30.5) the immediately aggregatable.

(3)
$$\max\{\mathcal{F}(z_1(x^1),\ldots,z_k(x^k))|g(z_1(x^1),\ldots,z_k(x^k))\le 0,\; x_s\in X_s,\quad s=\overline{1,k}\}. \tag{30.8}$$

The set of its solutions constitutes the roots of equation (30.7). Here $\{z_s^*\}$ is the solution of the problem

$$\max\{\mathcal{F}(z_1,\ldots,z_k)|g(z_1,\ldots,z_k)\le 0,\; \check{z}_s\le z_s\le \hat{z}_s,\; s=\overline{1,k}\},$$

where $\check{z}_s = \min\{z_s(x^s)|x^s\in X_s\}$, $\hat{z}_s = \max\{z_s(x^s)|x^s\in X_s\}$. Here, as in (30.5), the monotonicity of $\mathcal{F}$ is not presupposed. It is evident that (30.3), (30.5) constitute generalization. The generative problem of such a structure will be referred to as that admitting immediate aggregation and decomposition.

Suppose that we aim at constructing a solution of the perturbated problem (30.2) with a particular concrete value of the parameter $\varepsilon=\varepsilon_0$. Assume that the immediate solution of the problem presents difficulties while the solution of the generative problem proves to be comparatively simple. Then it is wise to substitute the solution of the original perturbated problem by the sequential procedure where the first stage constitutes construction of the generative problem, and the second is construction of a correction to it.

§**30.2.** The study of connection between the solutions of the perturbative and the generative problem starts with considering the simplest situation where only the objective functional is subject to perturbation, i.e. we consider the following problems:

$$\max\,(f_0(x)+\varepsilon f_1(x)),\quad x\in X; \tag{30.9}$$

$$\max\, f_0(x),\quad x\in X. \tag{30.10}$$

Denote by $X^*, X^*(\varepsilon)$ the optimal solution sets of (30.9) (30.10), and by $\mathcal{F}^*$, $\mathcal{F}^*(\varepsilon)$ the functional optimal values of these problems. The apparently evident statement about asymptomatic (with $\varepsilon \to 0$) nearness of solutions to (30.9), (30.10) can be refuted by the example. In the problem $\min\{x+\varepsilon x^2 \,|x \le -1\}$ the solution $x^*(\varepsilon) = \min\{-1/2\varepsilon, -1\}$ exists with any $\varepsilon > 0$. And in the generative problem there is no solution.

$$\inf\{x|x \le -1\} = -\infty\,.$$

We will set forth the theorem in which the conditions excluding the possibility of similar inconsistency between the solutions of (30.9), (30.10) are isolated.

Theorem 30.1: Let the function $f_0(x)$ be continuous on X and the solution set of problem (30.10) be nonempty. With any continuous function $f_1(x)$ satisfying the constraint $\sup\limits_{x \in X} f_1(x) < \infty$ for the solution set of problem (30.9) to be nonempty and bounded (with all sufficiently small ε), it is necessary and sufficient that one could not isolate the point sequences $\{x_i\}$ such that $x_i \in X$, $\|x_i\| \xrightarrow[i \to \infty]{} \infty$, $\lim\limits_{i \to \infty} f_0(x_i) = \mathcal{F}^*$.

The proof of the theorem as well as that of the following theorems are given in Ref. 30.

Theorem 30.2: Let the functions $f_0(x)$, $f_1(x)$ be continuous on the set X, $X^*(\varepsilon)$ be bounded for all sufficiently small ε, i.e. there exists $\bar{\varepsilon} > 0$ such that with $0 \le \varepsilon \le \bar{\varepsilon}$

$$X^*(\varepsilon) \subset \bar{X}\,,$$

where $\bar{X}$ is the bounded, closed set in R^n. Let the solution x^* uniquely generate problems. Then for any solution $x^*(\varepsilon) \in X^*(\varepsilon)$ of the perturbative problem there exists the limit

$$\lim_{\varepsilon \to 0} x^*(\varepsilon) = x^*\,.$$

As it follows from theorem (30.2), in the solution uniqueness case of the generative problem the range of searching the solutions of the perturbative problem can be narrowed to a certain neighborhood of the point x^*. In this case, after solving the generative problem, the perturbations method reduces, first, to establishing the dimensions of the above neighborhood and obtaining by this technique the accuracy estimates for x^* regarded as an approximate solution to (30.9), and, second, if it is possible, to constructing corrections to this solution. The simplest estimates can be obtained on a basis of the following relations

$$f_0(x^*(\varepsilon)) + \varepsilon f_1(x^*(\varepsilon)) \geq f_0(x^*) + \varepsilon f_1(x^*)\,, \tag{30.11}$$

$$f_0(x^*(\varepsilon)) \leq f_0(x^*)\,, \tag{30.12}$$

$$f_0(x^*) - f_0(x^*(\varepsilon)) \leq \varepsilon\{\max\{f_1(x)|x \in X\} - f_1(x^*)\}\,. \tag{30.13}$$

Relation (30.13) according to the function properties $f_0(x)$ and the set X may serve as a source of estimates for the value $\|x^* - x^*(\varepsilon)\|$. We abandon now the assumption of uniqueness of the generative problem solution and consider the problem

$$\begin{gathered} \max f_1(x)\,, \\ x \in X^*\,, \end{gathered} \tag{30.14}$$

the functional optimal value of which is F_1^*. We will call (30.14) the auxiliary problem. Denote

$$\text{Arg max}\,\{f_1(x)|x \in X\} = \{y \in X, f_1(y) = \max_{x \in X} f_1(x)\}\,.$$

Theorem 30.3: Suppose that all the conditions of theorem (30.2) have been fulfilled except for the assumption of the solution uniqueness of the generative problem. Then

(1) the set X^{**} of limit points of the elements $X^*(\varepsilon)$ obtained with $\varepsilon \to 0$, is contained in the solution set of the auxiliary problem

$$X^{**} \subset \text{Arg max}\,\{f_1(x)|x \in X^*\}\,;$$

(2) the objective functional optimal value $\mathcal{F}^*(\varepsilon)$ of the perturbative problem can be presented in the form

$$\mathcal{F}^*(\varepsilon) = \mathcal{F}^* + \varepsilon \mathcal{F}_1^* + O(\varepsilon)\,;$$

(3) on any solution of the auxiliary problem the objective functional value of the perturbative problem differs from the optimal on $O(\epsilon)$.

The consequence of theorem (30.3) represents the following two statements.

Theorem 30.4: Suppose that the conditions of theorem (30.3) have been fulfilled and, moreover, the solution x^{**} of (30.14) is unique. Then for any solution of $x^*(\varepsilon)$ (30.9) there exists the limit

$$\lim_{\varepsilon \to 0} x^*(\varepsilon) = x^{**}.$$

Theorem 30.5: Suppose that the conditions of theorem (30.3) have been fulfilled. Further assume that there exists $\hat{\varepsilon}_1 > 0$ such that with $0 < \varepsilon \leq \hat{\varepsilon}_1$ the solution $x^*(\varepsilon)$ of the perturbative problem is unique. Then $x^*(\varepsilon)$ is a continuous function on $0 < \varepsilon \leq \hat{\varepsilon}_1$ and there exists the limit

$$\lim_{\varepsilon \to 0} x^*(\varepsilon) = x^{**} \in \operatorname{Arg\,max}\{f_1(x) | x \in X^*\}.$$

§**30.3.** Consider a general formulation when not only the objective functional is subject to perturbations but so are the constraints of the generative problem, i.e. consider problem (30.2). Assume that the generative problem (30.1) satisfies the regularity condition, i.e. at each point of the solution the constraint gradients fulfilled on the set X^* as strict equalities are linearly independent. It is known that, when fulfilling regularity conditions, the Kuhn-Tucker unique vector corresponds to each solution of the mathematical programming problem. Now, the solution of (30.1) corresponds to a unique nonnegative vector of the Lagrange multipliers $\Lambda(x) = \{\lambda_1(x), \ldots, \lambda_m(x)\}$. The auxiliary problem here takes the form

$$\begin{gathered} \max\,(f_1(x) - \Lambda'(x) g_1(x)), \\ x \in X^*. \end{gathered} \tag{30.15}$$

We will set forth several theorems,[30] the statements of which form the basis for applying the technique of the perturbations theory.

Theorem 30.6: Assume that the set X of the generative problem feasible solutions is closed and there exists $\bar{\varepsilon} > 0$ such that with $0 < \varepsilon \leq \bar{\varepsilon}$

$$X^*(\varepsilon) \subset \bar{X}, \quad X^* \subset \bar{X},$$

where $\bar{X}$ is the bounded closed subset in R^n. Suppose that the functions $f_0(x)$, $f_1(x)$, $g_0(x)$, $g_1(x)$ are continuous and have continuous partial derivatives of the first order in a certain neighborhood of $\bar{X}$ and let the generative problem (30.1) satisfy the regularity condition. Then

(1) the set X^{**} of limit points of the elements of $X^*(\varepsilon)$ obtained with $\varepsilon \to 0$ is contained in the solution set of the auxiliary problem, i.e.

$$X^{**} \subset \operatorname{Arg\,max}\{f_1(x) - \Lambda'(x) g_1(x) | x \in X^*\};$$

(2) the optimal value of $\mathcal{F}^*(\varepsilon)$ of the perturbative problem objective functional can be presented in the form

$$\mathcal{F}^*(\varepsilon) = \mathcal{F}^* + \varepsilon \mathcal{F}_1^* + O(\varepsilon),$$

where $\mathcal{F}_1^*$ is the optimal value of the functional in (30.1)

(3) any solution $\bar{x}^*$ of the auxiliary problem can be set in agreement with the vector $\bar{y}(\varepsilon)$ such that

$$\bar{x}(\varepsilon) = \bar{x}^* + \varepsilon \bar{y}(\varepsilon) \in X(\varepsilon), \ \|y(\varepsilon)\| \leq M = \text{const.},$$

and on the solution $\bar{x}(\varepsilon)$ the objective functional value of the perturbative problem differs from the optimal on $O(\varepsilon)$. The consequences of theorem 30.6 constitute the following statements.

Theorem 30.7: Suppose that the conditions of theorem (30.6) have been fulfilled and there exists $\varepsilon_1 > 0$ such that with $0 < \varepsilon \leq \varepsilon_1$ the solution $x^*(\varepsilon)$ of problem (30.2) is unique. Then $\varepsilon_2 > 0$ exists such that with $0 < \varepsilon \leq \varepsilon_2$, $x^*(\varepsilon)$ is the continuous function ε, here

$$\lim_{\varepsilon \to 0} x^*(\varepsilon) = x^* \in \text{Arg max}\,\{f_1(x) - \Lambda'(x) g_1(x) | x \in X^*\}.$$

Theorem 30.8: Suppose that the conditions of theorem (30.6) have been fulfilled and, moreover, the solution x^* of the auxiliary problem is unique. Then for any solution $x^*(\varepsilon)$ of the perturbative problem there exists the limit

$$\lim_{\varepsilon \to 0} x^*(\varepsilon) = x^* = \text{Arg max}\,\{f_1(x) - \Lambda'(x) g_1(x) | x \in X^*\}.$$

Thus, to obtain zero approximation of the perturbative problem in conditions when the generative problem admits not one, but a set of solutions we need to isolate from this set the program which is the solution of the auxiliary problem. Consider the possibility of searching the perturbative problem solution as the series

$$\begin{aligned} x^*(\varepsilon) &= x^* + \varepsilon x_1^* + \cdots, \\ \Lambda^*(\varepsilon) &= \Lambda^* + \varepsilon \Lambda_1^* + \cdots, \end{aligned} \tag{30.16}$$

where x^* is the solution of the auxiliary problem, $\Lambda^* = \Lambda(x^*)$ and calculating x_i^*, Λ_i^* will be described below.

Theorem 30.9: Suppose that the conditions of theorems (7), (8) have been fulfilled and, moreover,

(a) the functions $f_0(x)$, $g_0(x)$, $f_1(x)$, $g_1(x)$ are $(m+1)$-time differentiable in the neighborhood of the point x^*;

(b) the generative problem satisfies at the point x^* complementary nonrigidity conditions;

(c) it is nonzero, define det $\{\nabla\nabla\mathcal{L}(x^*, \Lambda^*)$, where $\mathcal{L}(x,\Lambda) = f_0(x) - \Lambda' g_0(x)$, $\{\nabla\nabla\mathcal{L}(x^*, \Lambda^*)\}$ is the matrix of the function $\mathcal{L}(x,\Lambda)$ second derivatives by the x vector components and the positive components of the vector Λ.

Under these conditions expansion (30.16) exists.

Theorem 30.10: Suppose that the conditions of theorem (30.9) have been fulfilled, but instead of the strict complementary nonrigidity conditions and the nondegeneracy condition $\nabla\nabla\mathcal{L}(x^*, \Lambda^*)$ we presuppose sign definiteness of the matrix $\nabla_x\nabla_x\mathcal{L}(x^*, \Lambda^*)$. Then expansion (30.16) exists. We will describe the procedures of obtaining x_i^*, Λ_i^*.

In conditions of theorem (30.9), x_i^*, Λ_i^* are calculated from the formula

$$\begin{pmatrix} x_i^* \\ \Lambda_i^* \end{pmatrix} = \{\nabla\nabla\mathcal{L}(x^*, \Lambda^*)\}^{-1} \begin{pmatrix} \varphi_i(x^*, x_1^*, \dots, x_{i-1}^*, \Lambda^*, \Lambda_1^*, \dots, \Lambda_{i-1}^*) \\ \psi_i(x^*, x_1^*, \dots, x_{i-1}^*, \Lambda^*, \Lambda_1^*, \dots, \Lambda_{i-1}^*) \end{pmatrix}, \quad (30.17)$$

where φ_i, ψ_i will be described below.

The regularity conditions in the generative problem imply that the regularity conditions are also fulfilled with a sufficiently small ε in the perturbative problem. Therefore, with these ε there exists the Kuhn-Tucker vector $\Lambda^*(\varepsilon)$ satisfying the equations:

$$\begin{gathered} \nabla f_0(x^*(\varepsilon)) + \varepsilon\nabla f_1(x^*(\varepsilon)) = \sum_{l\in w} \lambda_l^*(\varepsilon)(\nabla g_{l_0}(x^*(\varepsilon)) + \varepsilon\nabla g_{l_1}(x^*(\varepsilon)), \\ g_{l_0}(x^*(\varepsilon)) + \varepsilon g_{l_1}(x^*(\varepsilon)) = 0, \quad l \in w, \end{gathered} \quad (30.18)$$

where w is the set of constraint indices fulfilled on x^* as strict equalities. This system can be regarded as an implicit specification of the function

$x^*(\varepsilon)$, $\lambda_l^*(\varepsilon)$, $l \in w$ in the neighborhood of the point $(x^*, \lambda_l^*,\ l \in w,\ \varepsilon = 0)$ and, considering that $\det(\nabla\nabla\mathcal{L}(x^*, \Lambda^*)) \neq 0$, on the strength of the well-known theorem of the implicitly specified functions we come to expansions (30.16). And the relations of (30.17) as well as the explicit expression for φ_i, ψ_i can be obtained by substituting (30.16) in (30.18) and expanding appropriate functions in a series according to degrees ε. Thus, in the case when the set of constraint indices of the perturbative problem fulfilled on the solution (with a sufficiently small ε) as strict equalities is known, the procedure of searching expansion coefficients is described. But, in general, when analyzing the generative problem the set of the identified cannot be available. In this situation we substitute the formal expansion of (30.16) in the functional of the perturbative problem to obtain

$$f_0(x^*)+\varepsilon(\nabla f_0(x^*))'x_1 + \varepsilon^2\left(\frac{1}{2}x_1^T\nabla\nabla f_0(x^*)x_1+(\nabla f_1(x^*))^Tx_1+(\nabla f_0(x^*))'x_2\right)+\cdots . \tag{30.19}$$

Evidently, x_1^* must be the solution of the coefficient maximization problem with ε

$$\max\{(\nabla f_0(x^*))^Tx_1 | (\nabla g_{l_0}(x^*))'x_1 \le -g_{l_1}(x^*),\ l \in \Omega(x^*)\}\,. \tag{30.20}$$

At the same time, the solution of (30.20) is constituted by the vector x_1 satisfying the relations

$$\begin{aligned} &(\nabla g_{l_0}(x^*))'x_1 = -g_{l_1}(x^*), \quad l \in w\,,\\ &(\nabla g_{l_0}(x^*))'x_1 \le -g_{l_1}(x^*), \quad l \in \Omega(x^*) \setminus w\,, \end{aligned} \tag{30.21}$$

where w is the set of the positive component indices of the Kuhn-Tucker vector, $\Omega(x^*)$ is the index set of the constraints fulfilled on x^* as strict equalities. Denote by X_1^* the vector set satisfying (30.21) and $\Omega_1(x_1)$ the set of constraint indices in (30.21) fulfilled on $x_1 \in X_1^*$ as strict equalities. Then x_1^*, x_2^* represent the solution of the problem

$$\begin{aligned} \max\Bigg\{&\frac{1}{2}x_1^T\nabla\nabla f_0(x^*)x_1+(\nabla f_1(x^*))^Tx_1+(\nabla f_0(x^*))^Tx_2 | x_1 \in X_1^*\,,\\ &\nabla g_{l_0}(x^*)x_2 \le -\frac{1}{2}x_1^T\nabla\nabla g_{l_0}(x^*)x_1 - (\nabla g_{l_1}(x^*))^Tx_1,\ l \in \Omega_1(x_1)\Bigg\}\,, \end{aligned} \tag{30.22}$$

providing the coefficient maximum with ε^2 in the perturbative problem functional writing. It is evident that

$$\max\Bigg\{(\nabla f_0(x^*))' x_2 | \nabla g_{l_0}(x^*) x_2$$

$$\leq -\frac{1}{2} x_1^T \nabla\nabla g_{l_0}(x^*) x_1 - (\nabla g_{l_1}(x^*))^T x_1,\ l \in \Omega_1(x_1)\Bigg\}$$

$$= (\Lambda^*)^T \Bigg[-\frac{1}{2} x_1^T \nabla\nabla g_{l_0}(x^*) x_1 - (\nabla g_{l_1}(x^*))^T x_1 \Bigg] . \qquad (30.23)$$

And, therefore, problem (30.22) can be written as

$$\max\Bigg\{\frac{1}{2} x_1^T \nabla_x \nabla_x \mathcal{L}(x^*, \Lambda^*) x_1$$

$$+ \Bigg(\nabla f_1(x^*) - \sum_{l \in w} \lambda_l^* \nabla g_{l_1}(x^*)\Bigg)' x_1 | x_1 \in X_1^* \Bigg\} . \qquad (30.24)$$

If the problem solution is unique, then it coincides with x_1^*. Furthermore, the Kuhn-Tucker vector Λ_1^* of problem (30.24) coincides with the coefficient with ε in (30.16). To obtain x_2^*, Λ_2^*, we need, by utilizing transformation similarly to (30.23), to reduce the coefficient maximization problem with x_2 to the problem of maximizing the functional which depends only on x_2^*.

§**30.4.** Consider application of the perturbations method to LP problems.

The nonuniqueness of the generative problem solution introduces changes into the procedure of constructing approximate solutions, since the problem arises with respect to isolating that of the generative solutions to which the perturbative tends as the parameter ε decreases. The nonuniqueness is particularly characteristic for linear problems.

Thus, we have the generative problem

$$\max c_0 x_0 ,$$
$$A_0 x = b_0, \quad x \geq 0 \qquad (30.25)$$

and the perturbative problem

$$\begin{gathered}\max(c_0 + \varepsilon c_1)x\,, \\ (A_0 + \varepsilon A_1)x = b_0 + \varepsilon b_1\,, \\ x \geq 0\,,\end{gathered} \tag{30.26}$$

where A_0, A_1 are the $(m \times n)$ matrices, c_0, c_1, x are the n vectors, b_0, b_1 are the m vectors. We assume that the rows of the matrix A_0 are linearly independent. Then the regularity condition for problem (30.25) is equivalent to the assumption that the basic components of each optimal solution of (30.25) are positive. In this case, the dual problem for (30.25) has a unique solution Λ^*. Consider the auxiliary problem

$$\begin{gathered}\max\{(c_1 x) - (\Lambda^*, A_1 x)\}\,, \\ x \in X^*\,,\end{gathered} \tag{30.27}$$

where X^* is the solution set of (30.25).

We will rearrange for our case theorem 30.6.

Theorem 30.11: Let the set $X^*(\varepsilon)$ of the perturbative problem solutions be constrained and the optimal solutions of the generative problem be nondegenerate. Then:

(1) the set X^{**} of limit points of the perturbative problem elements $X^*(\varepsilon)$ is contained in the solution set of the auxiliary problem, i.e.

$$X^{**} \subset \text{Arg max}\,\{c_1 x - (\Lambda^*, A_1 x) | x \in X^*\}\,;$$

(2) for the optimal value $\mathcal{F}^*(\varepsilon)$ of the perturbative problem objective functional the following expansion holds

$$\mathcal{F}^*(\varepsilon) = \mathcal{F}^* + \varepsilon(\mathcal{F}_1^* + \Lambda^{*^T} b_1) + O(\varepsilon^2)\,;$$

(3) if x^{**} is an arbitrary basic solution of the auxiliary problem and A_b is its associated basic submatrix of the matrix A, provided the columns corresponding to the nonbasic variables x^{**} are replaced by nonzero columns, then with all sufficiently small ε the basic program $\bar{x}(\varepsilon) = (\bar{x}_b(\varepsilon), 0)$, $\bar{x}_b(\varepsilon) = (A_{0b} + \varepsilon A_{1b})^{-1}\ (b_0 + \varepsilon b_1)$ is feasible for the

perturbative problem and the functional value on it differs from the optimal on $O(\varepsilon^2)$. Formulate the perturbations method algorithm for the LP problem.

Step 1: Define the solution set X^* of the generative problem (30.25).

Step 2: Obtain the solutions of the auxiliary problem (30.27) and fix the indices corresponding to basic variables.

Step 3: As the solution "suspicious" for optimality, in the originally (perturbative) problem (30.26), we take for the basic program the basic variable indices which coincide with those for the auxiliary problem.

Step 4: The solution is tested for admissibility to problem (30.26).

As it follows from theorem (30.11), if ε is sufficiently small, then it is feasible. Otherwise, the method of perturbations cannot be applied.

Step 5: Calculate characteristic differences and test the program obtained for optimality.

Step 6: If the program is not optimal, i.e. either the auxiliary problem solution is nonunique or insufficiently small, then the basic program obtained can be employed for further calculations that can be carried out, say, with the use of the simplex method.

§**30.5.** By using the algorithm described, we will provide several examples and demonstrate beneficial properties of separate problem types.

(A) Consider the following LP problem

$$\max\left\{\sum_{l=1}^{\mathcal{L}}\sum_{j\in\Omega_l} c_j x_j\right\},$$
$$(A+\varepsilon B)x = b, \quad x \geq 0, \tag{30.28}$$

where A is the matrix of a block diagonal structure and A_j $(j=\overline{1,\mathcal{L}})$ are its blocks, the vector b consists of m_j-dimensional vectors b_j, the vector $x=(x_1,\ldots,x_{\mathcal{L}})$. Note that the generative problem for (30.28) is of the immediately decomposable type (30.3).

The algorithm of the perturbations method for problem (30.28) is implemented as follows:

(1) Solve $\mathcal{L}$ subproblems of the type (30.4)

$$\max\left\{\sum_{j\in\Omega_l} c_j x_j | A_l x_l = b_l,\, x_l \geq 0\right\}, \quad l = \overline{1,\mathcal{L}}. \tag{30.29}$$

If the solution of these problems is unique, then we get the solution of (30.28) accurate to $O(\varepsilon)$. To refine the solution obtained, we fix the optimal basic sets of the indices σ_l ($l = \overline{1,\mathcal{L}}$) of each subproblem and carry out transition to point (3). If in some problem of (30.29) the solution is nonunique, which appears in even one of the nonbasic characteristic differences being equal to zero, then the optimal solution sets X_l^* of these problems are isolated and the vectors Λ_l, $l = \overline{1,\mathcal{L}}$ of the optimal dual variables, which are under the regularity assumption are unique and are fixed.

(2) Solve the auxiliary problem which falls into $\mathcal{L}$ independent subproblems

$$\min\left\{\sum_{s=1}^{M}(\Lambda^s, B_{sl}x_l)|x_l \in X_l^*\right\}, \quad l = \overline{1,\mathcal{L}}, \tag{30.30}$$

where B_{sl} is the submatrix B in which the rows correspond to the components b^s, $s = \overline{1,M}$ of the vector b, and the columns correspond to x_l. Moreover, Λ^s, $s = \overline{1,M}$ are here the components of the vector $\Lambda = (\Lambda_1, \dots, \Lambda_{\mathcal{L}})$ and $M = \sum_{j=1}^{\mathcal{L}} m_j$. Assuming that all the problems of form (30.30) have unique solutions and denoting them by x_l^*, we can take as the approximate (accurate to $O(\varepsilon)$) solution of (30.28) the vector

$$x^* = \{x_l^*, \quad l = \overline{1,\mathcal{L}}\}.$$

To refine the solution, the index set of the basic optimal bases σ_l is fixed in problems (30.30).

(3) We set in the system $(A + \varepsilon B)x = b$ all nonbasic variables being equal to zero, i.e.

$$x_k = 0, \quad k \notin \bigcup_{l=1}^{\mathcal{L}} \sigma_l \tag{30.31}$$

whereupon it takes the form

$$(\tilde{A} + \varepsilon \tilde{B})\tilde{x} = b\,, \tag{30.32}$$

where $\tilde{x}$ is the vector composed of variables with basic indices. If system (30.32) with the specified ε has the non-negative solution, then it, together with (30.31), forms a feasible solution of the original problem. To test the original problem solution for optimality, we solve the system of equations for the estimates

$$(\tilde{A} + \varepsilon \tilde{B})\Lambda = \tilde{c}\,,$$

where

$$\tilde{c} = \left\{ c_k, k \in \bigcup_{l=1}^{\mathcal{L}} \sigma_l \right\}.$$

Then we test fulfilling the inequalities $(A + \varepsilon B)\Lambda \geq c$, which enables us to see whether the fixed ε belongs to the segment indicated in the conditions of (30.11).

(B) Consider one more problem of the form

$$\begin{gathered} \max \sum_{l=1}^{\mathcal{L}} c_l x_l\,, \\ A_l x_l + \varepsilon B_l x = b_l + u_l, \quad \sum_{l=1}^{\mathcal{L}} u_l \leq \varepsilon b_0\,, \\ x_l \in X_l,\ u_l \geq 0,\ l = \overline{1, \mathcal{L}}\,, \end{gathered} \tag{30.33}$$

where $x = (x_1, \ldots, x_{\mathcal{L}})$, u_j, $j = \overline{1, \mathcal{L}}$ are the vectors of the unknowns A_l, B_l, $b_l\,(l = \overline{1, \mathcal{L}})$, b_0 are the matrices and vectors of the correlated dimensionalities, and X_l are polyhedrons. One cannot apply the algorithm of the perturbations method immediately to (30.33) since, due to the structure of the constraint first group, the generative (with $\varepsilon = 0$) problem obviously is irregular. By changing the variables $u_l = \varepsilon v_l$, we get the problem

$$\begin{gathered} \max \sum_{l=1}^{\mathcal{L}} c_l x_l\,, \\ A_l x_l + \varepsilon B_l x = b_l + \varepsilon v_l, \quad \sum_{l=1}^{\mathcal{L}} v_l \leq b_0\,, \\ x_l \in X_l, v_l \geq 0,\ l = \overline{1, \mathcal{L}}\,. \end{gathered} \tag{30.34}$$

The generative problem for (30.34) has the form

$$\max \sum_{l=1}^{\mathcal{L}} c_l x_l \,,$$
$$A_l x_l = b_l, \quad x_l \in X_l \,, \tag{30.35}$$
$$\sum_{l=1}^{\mathcal{L}} v_l \le b_0,\ v_l \ge 0,\ l = \overline{1, \mathcal{L}} \,.$$

Problem (30.35) has no unique solution. Indeed, even though all the problems

$$\max\{c_l x_l | A_l x_l = b_l, x_l \in X_l\} \tag{30.36}$$

have the unique solutions x_l^*, the variables v_l in (30.35) remain indeterminate and, to secure their approximate value v_l^*, we need to solve the following problem:

$$\max \sum_{l=1}^{\mathcal{L}} \Lambda_l v_l \,,$$
$$\sum_{l=1}^{\mathcal{L}} v_l \le b_0,\ v_l \ge 0 \,, \tag{30.37}$$

where Λ_l are the solutions of the problems dual with respect to (30.36). And problem (30.37) has the simple solution

$$v_l^{s*} = \begin{cases} b_0^s, & l = p_s, \\ 0, & l \ne p_s, \quad p_s = \arg\max\limits_l \lambda_l^s, \end{cases}$$

where λ_l^s are the components of Λ_j, b_0^s are the components of b_0. Thus, in the initial approximation, the resource S is transferred to the subsystem which employs it on the program x_l^* most efficiently. Next, following the method of perturbations, one solves the system

$$\tilde{A}_l \tilde{x}_l + \varepsilon \tilde{B}_l \tilde{x} = b_l + \varepsilon v_l^* \,, \tag{30.38}$$

where $\bar{x}_l$ is the vector involving the variables x_l with the index from the optimal basic set of (30.36), and $\tilde{A}_l$, $\tilde{B}_l$ are the appropriate submatrices of the matrices A_l, B_l. Evidently, as before, if system (30.38) admits a unique non-negative solution with the specified concrete ε, then it is also a feasible solution of the original problem.

(C) And, finally, we will consider the problem close to the directly aggregative (30.5):

$$\max \sum_{l=1}^{\mathcal{L}} \sum_{j \in \Omega_l} (c_l + \varepsilon \beta_l^j) x_j \,,$$
$$\sum_{l=1}^{\mathcal{L}} \sum_{j \in \Omega_l} (A_l + \varepsilon B_l^j) x_j \le b,\ x = (x_j) \ge 0\,, \tag{30.39}$$

where x_j, c_l, β_l^j are scalar vectors, A_l, B_l^j, b are m vectors. Introduce the notations $z_l = \sum_{j \in \Omega_l} x_j$ and rewrite problem (30.39) as

$$\max \left(\sum_{l=1}^{\mathcal{L}} c_l z_l + \varepsilon \sum_{l=1}^{\mathcal{L}} \sum_{j \in \Omega_l} \beta_l^j x_j \right),$$
$$\sum_{l=1}^{\mathcal{L}} A_l z_l + \varepsilon \sum_{l=1}^{\mathcal{L}} \sum_{j \in \Omega_l} \beta_l^j x_j \le b;\ x \ge 0;\ z \ge 0\,. \tag{30.40}$$

Let (z_l^*) be a unique solution of the aggregative problem

$$\max \sum_{l=1}^{\mathcal{L}} c_l z_l \,,$$
$$\sum_{l=1}^{\mathcal{L}} A_l z_l \le b, \quad z_l \ge 0\,. \tag{30.41}$$

Then the optimal solution set X^* of the generative problem is determined by the conditions

$$\sum_{j \in \Omega_l} x_j = z_l^*,\ x_j \ge 0,\ j \in \Omega_l,\ l \in \sigma\,,$$
$$x_j = 0, \quad j \in \Omega_l, \quad l \bar{\in} \sigma\,,$$

where σ is the variable index set entering into the optimal basis of problem (30.41). Let Λ be the solution of the problem dual with respect to (30.41). Based on theorem (30.11), we can state that the solution sought coincides, accurate to $O(\varepsilon^2)$ (by the functional), with the solution of the auxiliary problem

$$\max\left\{\sum_{l\in\sigma}\sum_{j\in\Omega_l}(\beta_l^j - \Lambda' B_l^j)x_j | x \in X^*\right\}, \tag{30.42}$$

which, due to separability, falls into $|\sigma|$ element problems ($l = \overline{1,\sigma}$):

$$\max\left\{\sum_{j\in\Omega_l}(\beta_l^j - \Lambda' B_l^j)x_j | x \in X^*\right\},$$

$$\sum_{j\in\Omega_l} x_j = z_l^*,\ x_j \geq 0\,.$$

Let us formulate the perturbations method algorithm as applied to problem (30.39)

<u>*Step 1*</u>: Obtain the solutions z^*, Λ of problem (30.41) and its dual problem.

<u>*Step 2*</u>: Let

$$j_l = \text{Arg}\max_j\,(\beta_l^j - \Lambda' B_l^j),\ l \in \sigma.$$

<u>*Step 3*</u>: If all j_l are uniquely determined, then the solution accurate to $O(\varepsilon)$ is equal

$$x_j = \begin{cases} z_l, & j = j_l, \quad l \in \sigma, \\ 0, & j \neq j_l, \quad l \in \sigma; \quad j \in \Omega_l, \quad l \in \sigma; \end{cases}$$

<u>*Step 4*</u>: Since the optimal basic set of (30.42) is formed from the indices j_l, $l \in \sigma$, then the nonzero components of the locally exact solution can be obtained from solving the system

$$\sum_{l\in\sigma}(\tilde{A}_l + \varepsilon\tilde{B}_l^{js})x_{js} = \tilde{b}\,, \tag{30.43}$$

where $\tilde{A}_l$, $\tilde{B}_l$ are the submatrices of the corresponding matrices from which all the rows with the indices of inequalities in (30.41) not transforming into equalities on z^* are removed.

If j_l are determined ambiguously, then the difference of the functional from the optimal on $O(\varepsilon^2)$ can be ensured on the solution of (30.43).

§31. DECOMPOSITION METHOD BASED ON AGGREGATION OF VARIABLES FROM DIFFERENT BLOCKS

We will describe the method[40] applicable to a wide class of extremal problems, in which the variables entering into different blocks are aggregated, as contrasted, say, to the method described in §29 of the present chapter where macrovariables correspond to each block.

§31.1. Let us describe the sector planning model taken as an example to consider the aggregation method.

Let a sector (industry) be specified by the following combination of sets: $\bar{k} = [1 : K]$ is the set of the sector plant numbers, $\bar{M} = [1 : M]$ is the number of nomenclature of the items produced at the sector plants, k_j is the set of numbers of the equipment groups available at the plant j, $I_j \subset \bar{I}$ is the number set of the complete items produced at the plant j. Assume that the technological chain of producing a particular article at a particular plant is rigidly fixed. Then the production capacities of each plant within the model of employing the noninterchangeable groups of equipment are described by the following inequalities

$$\sum_{i \in I_j} T_j^{ik} x_j^i \leq \Phi_j^k, \quad k \in k_j, \quad x_j^i \geq 0, \quad i \in I_j\,, \tag{31.1}$$

where x_j^i is the yearly output of the articles i at the plant j, Φ_j^k is the yearly fund of operation time of the equipment group k at the plant j, T_j^{ik} denote the expenditures of operation time of the equipment group k at the plant j for producing one article i. From inequalities (31.1) we see that the values x_j^i are determined for each plant j only for $i \in I_j$. Furthermore it is convenient to define them for all $i \in \bar{I}$ setting x_j^i for $i \bar{\in} \bar{I} \setminus I_j$. The connection between the output of complete articles and the sector final output is described by the relations

$$\sum_{j=1}^{k} (x_j^i + w_j^i) = \sum_{m=1}^{M} l^{i_m} Y^m (1 + \delta^m)\,, \tag{31.2}$$

where Y^m is the yearly final output of the form m, w_j^i is the inventory of the articles i at the plant j at the beginning of the planned period, l^{i_m}

is the number of the form i entering into the output of the form m (the complating coefficient), δ^m is the standard of the output inventory of the form m shifting to the next planned period (in unit fractions).

Assume that the set of the final output numbers $\bar{M}$ is divided into two nonintersecting subsets:

$$\bar{M} = \bar{M}_1 \cup \bar{M}_2, \quad \bar{M} = [1 : M_1], \quad \bar{M}_2 = [M_1 + 1 : M] .$$

The products from the set $\bar{M}_1$ are the most important for the sector from the standpoint of the national economic demand, the output of these products is compulsory and their volumes are either fixed or specified by constraints from below

$$Y^m \geq \bar{Y}^m, \quad m \in [1 : M_1] . \tag{31.3}$$

Assume that there exists the vector $\{\bar{Y}^{M_1+1}, \ldots, \bar{Y}^M\}$ such that the vector $\{\bar{Y}^1, \ldots, \bar{Y}^{M_1}, \bar{Y}^{M_1+1}, \ldots, Y^M\}$ satisfies (31.1), (31.2) for some x^i_j. As far as the products from the set $\bar{M}_2$ are concerned the decision making with respect to their volumes is within the competence of the planning body of the sector (industry) itself. This part of the nomenclature is called free.

Let the planning body set the free nomenclature vector

$$\{\bar{Y}^{M_1+1}, \ldots, \bar{Y}^M\} . \tag{31.4}$$

Two situations are possible here.

(1) Vector (31.4) is unfeasible, i.e. there are no values such that conditions (31.1), (31.2) are fulfilled.

(2) Vector (31.4) is feasible. Here, generally speaking, an infinite number of techniques is available at the plants and a particular part of the sector capacities may not be involved.

In both cases, the planning body of the sector encounters the necessity to reconsider vector (31.4). If vector (31.4) is regarded as the notion of the wanted structure of the free nomenclature of the sector final output, then the final plan must be sought for as

$$Y^{M_1+1} = \Theta \bar{Y}^{M_1+1}, \ldots, Y^M = \Theta \bar{Y}^M . \tag{31.5}$$

The components of vector (31.4) are called the assortment relations, and the magnitude Θ is identified as the number of assortment of M sets. Thus, if vector (31.4) is assigned, then we need to solve the following optimization problem: Θ has to be maximized under conditions (31.1) – (31.2). We rewrite (31.2) as

$$\sum_{j=1}^{k}(x_j^i + w_j^i) = \sum_{m=1}^{M_1} l^{i_m} Y^m (1+\delta^m) \quad \sum_{m=M_1+1}^{M} l^{i_m} Y^m (1+\delta^m),\ i = \overline{1,I}$$

or, utilizing (31.5), we get

$$\sum_{j=1}^{k} x_j^i + W^i = a^i\Theta + H^i, \quad i = \overline{1,I}\,,$$

where $W^i = \sum_{j=1}^{k} w_j^i$ is the summary pool of the article i in the sector at the beginning of the planned period, $H^i = \sum_{m=1}^{M_1} l^{i_m}\bar{Y}^m(1+\delta^m)$ is the summary volume of the article i needed to produce the final product of the compulsory nomenclature, $a^i = \sum_{m=M_1+1}^{M} l^{i_m}\bar{Y}^m(1+\delta^m)$ is the ith component of the assortment relation of the complete article output. Denoting $V^i = W^i - H^i$, $i = \overline{1,I}$; $b_j^{i_k} = T_j^{i_k}|\Phi_j^*$, $j = \overline{1,k}$; $k \in k_j$; $i \in I_j$ we will finally write the optimization problem as

$$\max \Theta\,, \tag{31.6}$$

$$\sum_{i\in I_j} b_j^{i_k} x_j^i \le 1,\ k \in k_j\,, \tag{31.7}$$

$$x_j^i \ge 0,\ i \in I_j,\ x_j^i = 0,\ i \in \bar{I}\setminus I_j;\ j \in [1:k]\,, \tag{31.8}$$

$$a^j\Theta - \sum_{j=1}^{k} x_j^i = V^i,\ i \in [1,I]\,, \tag{31.9}$$

$$\Theta \ge 0\,. \tag{31.10}$$

Problem (31.6) – (31.10) is an LP problem the dimensions of which are enormous in the real problems of sectoral planning. At the same time, the problem has a block structure: constraints (31.7) – (31.8) refer to separate plants and are block constraints, constraints (31.9) are sectoral or connecting.

§31.2. Let us describe the concept of the decomposition method for problem (31.6) – (31.10).

We introduce new variables – complete outputs as per all plants:

$$X^i = \sum_{j=1}^{k} x_j^i, \quad i = \overline{1, I}. \tag{31.11}$$

The values X^i are identified as aggregative variables. We also introduce specific outputs of the article i at plant j

$$\alpha_j^i = x_j^i / X^i, \quad j \in [1, k], \quad i \in [1 : I]$$

and call the values α_j^i the aggregation weights. Evidently, the conditions

$$\sum_{j=1}^{k} \alpha_j^i = 1, \quad i \in [1 : I],$$
$$\alpha_j^i \geq 0, \quad i \in I_j, \quad \alpha_j^i = 0, \quad i \in \bar{I} \setminus I_j, \quad j \in [1 : k], \tag{31.12}$$

are fulfilled.

Assume that the aggregation weights satisfying (31.12) are specified. Then, substituting in (31.6) – (31.8) the variables $x_j^i - \alpha_j^i\ X^i$, $j \in [1 : k]$; $i \in [1 : I]$ and denoting $B_j^{i_k} = b_j^{i_k} \alpha_j^i$, $j \in [1 : k]$, $k \in k_j$, $i \in I_j$, we write problem (31.6) – (31.10) in the form

$$\max \Theta, \tag{31.13}$$

$$\sum_{j=1}^{k} B_j^{i_k} X^i \leq 1, \quad j \in [1 : k], \quad k \in k_j, \tag{31.14}$$

$$a^i \Theta - X^i = V^i, \tag{31.15}$$

$$X^i \geq 0, \quad i \in [1 : I], \tag{31.16}$$

$$\Theta \geq 0. \tag{31.17}$$

We call problem (31.13) – (31.17) the problem in aggregative variables or the macroproblem.

The dual problem for (31.6) – (31.10) has the form

$$\min \varphi = \sum_{j=1}^{k} \sum_{k \in k_j} \xi_j^k + \sum_{i=1}^{I} V^i \mu^i ,$$

$$\sum_{k \in k_j} b_j^{i_k} \xi_j^k - \mu^i \geq 0, \quad j \in [1:k], \quad i \in I_j ;$$

$$\sum_{i=1}^{I} a^i \mu^i \geq 1 , \tag{31.18}$$

$$\xi_j^k \geq 0, \quad j \in [1:k], \quad k \in k_j .$$

And the dual problem for (31.13) – (31.17) is as follows

$$\min \psi = \sum_{j=1}^{k} \sum_{k \in k_j} \eta_j^k + \sum_{i=1}^{I} V^i \lambda^i ,$$

$$\sum_{j=1}^{k} \sum_{k \in k_j} B_j^{i_k} \eta_j^k - \lambda^i \geq 0, \quad i \in [1:I] , \tag{31.19}$$

$$\sum_{i=1}^{I} a^i \lambda^i \geq 1, \quad \eta_j^k \geq 0, \quad j \in [1:k], \quad k \in k_j .$$

We denote the optimal solutions of the problems by the sign $*$ above the corresponding letters. Let $\overset{*}{\lambda}{}^i$, $i \in [1:I]$ be unique optimal solutions of problem (31.19). For each fixed index j we formulate the block (plant) problems

$$\max h_j = \sum_{i \in I_j} \overset{*}{\lambda}{}^i x_j^i ,$$

$$\sum_{i \in I_j} b_j^{i_k} x_j^i \leq 1, \quad k \in k_j , \tag{31.20}$$

$$x_j^i \geq 0, \quad i \in I_j, \quad x_j^i = 0, \quad i \in I \setminus I_j .$$

The dual problems for (31.20) with each $j \in [1:k]$ are written as

$$\min \beta_j = \sum_{k \in k_j} \xi_j^k ,$$

$$\sum_{k \in k_j} b_j^{i_k} \xi_j^k - \overset{*}{\lambda}{}^i \geq 0, \quad i \in I_j , \tag{31.21}$$

$$\xi_j^k \geq 0, \quad k \in k_j .$$

The iterative process is constructed in the following way. With some fixed weights of α_j^i satisfying (31.12), we solve the problem in aggregative variables (31.13) – (31.17). It will be shown below that the problem, under some assumptions, is solved analytically. Further, by using the dual estimates $\overset{*}{\lambda}{}^i$, $i \in [1:I]$, the functionals of block problems (31.20) are formed and these problems are solved. Let $\hat{x}_j^i$, $j \in [1:k]$, $i \in [1:I]$ be their optimal solutions. Introduce the variables

$$\overset{*}{x}{}_j^i = \alpha_j^i \overset{*}{X}{}^i, \quad j \in [1:k], \quad i \in [1:I],$$

where $\overset{*}{X}{}^i$, $i \in [1:I]$ are the optimal solutions of problem (31.13) – (31.17). The values $\overset{*}{x}{}_j^i$ are termed the disaggregative solutions. Now aggregation weights are defined as the function $\alpha_j^i(p_j)$ according to the relation

$$\alpha_j^i(p_j) = \begin{cases} \overset{*}{x}{}_j^i + p_j(\hat{x}_j^i - \overset{*}{x}{}_j^i) \left[\sum_{j=1}^{k} [\overset{*}{x}{}_j^i + p_j(\hat{x}_j^i - \overset{*}{x}{}_j^i)] \right]^{-1}, i \in I_j , \\ 0, \; i \in I \setminus I_j , \end{cases} \tag{31.22}$$

where $0 \leq p_j \leq 1$, $j \in [1:k]$. Evidently, conditions (31.12) for the weights $\alpha_j^i(p_j)$ are met. If we consider problem (31.13) – (31.17) with weights (31.22), then the optimal value of the functional Θ in it represents a function from the parameters $p_j - \overset{*}{\Theta}(p_j)$. The problem arises with respect to maximizing the function $\overset{*}{\Theta}(p_j)$ under conditions $0 \leq p_j \leq 1$, $j \in [1:k]$. Let the maximum be achieved with some $\overset{*}{p}_j$, $j \in [1:k]$. Then the weights for the next step of the iterative process are determined from formula (31.22) with $p_j = \overset{*}{p}_j$. Thus, the algorithm forms a sequence of disaggregative solutions $\overset{*}{x}{}^i_j$ which together with $\overset{*}{\Theta}$ are admissible to the original problem (31.6) – (31.10).

§**31.3.** Establish the optimality criterion of the solution $\{\overset{*}{x}{}^i_j, \overset{*}{\Theta}\}$ for problem (31.6) – (31.10).

Assume that with some weights of α^i_j satisfying (31.12) we have obtained the optimal solution $\{\overset{*}{X}{}^i, \overset{*}{\Theta}\}$ of the problem in aggregative variables (31.13) – (31.17), and $\overset{*}{x}{}^i_j$ is the appropriate disaggregative solution. Let $\overset{*}{\lambda}{}^j$ be a unique optimal solution of the dual problem (31.19). Denote

$$\hat{h}_j = \sum_{i \in I_j} \overset{*}{\lambda}{}^i \hat{x}^i_j \,, \tag{31.23}$$

$$\overset{*}{h}_j = \sum_{i \in I_j} \overset{*}{\lambda}{}^i \overset{*}{x}{}^i_j \,. \tag{31.24}$$

The values $\hat{h}_j$, $\overset{*}{h}_j$ constitute the functional values of block problems (31.20) on the optimal and the disaggregative solution, respectively.

Let $\hat{h} = \sum_{j=1}^{k} \hat{h}_j$, $\overset{*}{h} = \sum_{j=1}^{k} \overset{*}{h}_j$. The following theorem gives an optimality condition of the disaggregative solution.

Theorem 31.1[40]**:** Suppose that for the optimal solutions $\{\overset{*}{X}{}^i, \overset{*}{\Theta}\}$, λ_i, $i \in [1 : I]$ of direct and dual macroproblems and the optimal solutions $\hat{x}^i_j$, $j \in [1 : k]$, $i \in I_j$ of block problems the following equalities are fulfilled

$$\hat{h}_j = \overset{*}{h}_j, \quad j \in [1 : k] \,. \tag{31.25}$$

Then the disaggregative solution $\{\overset{*}{x}{}^i_j, \overset{*}{\Theta}\}$, where $\overset{*}{x}{}^i_j = \alpha^i_j \overset{*}{X}{}^i$; $j \in [1 : k]$, $i \in I_j$ is an optimal program of the original problem.

§**31.4.** Since the central moment of the method described is represented by repeated solution of the problem in aggregative variables (31.13) – (31.17), then it is essential to ascertain the properties of the problem solution. We express X^i in terms of Θ according to (31.15):

$$X^i = a^i\Theta - V^i, \quad i \in [1:I]. \tag{31.26}$$

By substituting (31.26) in (31.14), obtain

$$\Theta \sum_{i=1}^{I} B_j^{i_k} a_i - \sum_{i=1}^{I} B_j^{i_k} V^i \le 1, \quad j \in [1:k], \quad k \in k_j.$$

From these inequalities we derive the expression for an optimal value of the variable Θ:

$$\overset{*}{\Theta} = \min_{\substack{j\in[1:k]\\ k\in k_j}} \left(1 + \sum_{i=1}^{I} B_j^{i_k} V^i\right) \Big/ \left(\sum_{i=1}^{I} B_j^{i_k} a_i\right). \tag{31.27}$$

And the optimal values $\overset{*}{X}{}^i$, $i \in [1:I]$ are determined in terms of $\overset{*}{\Theta}$ from formula (31.26). Consider now the dual problem (31.19). Assume that the output of each article is strictly positive:

$$\overset{*}{X}{}^i > 0, \quad i \in [1:I]. \tag{31.28}$$

According to the second theorem of duality, if the adjoint problems are solvable, then in each pair of their dual conditions one condition is free and another is fixed. Hence and due to (31.28) the first group of relations of (31.19) is fulfilled on the optimal solution as equalities, i.e.

$$\overset{*}{\lambda}{}^i = \sum_{j=1}^{k} \sum_{k\in k_j} B_j^{i_k} \overset{*}{\eta}{}_j^k. \tag{31.29}$$

Substituting (31.29) in the functional and the second group of constraints of problem (31.19), we come to the problem

$$\min \psi = \sum_{j=1}^{k} \sum_{k\in k_j} \overset{*}{\eta}{}_j^k \left(1 + \sum_{i=1}^{I} B_j^{i_k} V^i\right), \tag{31.30}$$

$$\sum_{j=1}^{k} \sum_{k\in k_j} \overset{*}{\eta}{}_j^k \left(\sum_{i=1}^{I} B_j^{i_k} a^i\right) = 1. \tag{31.31}$$

Let the minimum in (31.27) be attained on a unique pair of indices (j^*, k^*), i.e.

$$\overset{*}{\Theta} = \left(1 + \sum_{i=1}^{I} B_{j^*}^{i_{k^*}} V^i\right) \Big/ \left(\sum_{i=1}^{I} B_{j^*}^{i_{k^*}} a^i\right). \tag{31.32}$$

In this case problem (31.30) – (31.31) has a unique solution.

$$\overset{*}{\eta} k_{j^*}^{*} = 1 \Big/ \left(\sum_{i=1}^{I} B_{j^*}^{i_{k^*}} a^i\right), \quad \overset{*}{\eta}{}_j^k = 0, \quad (j,k) \neq (j^*, k^*). \tag{31.33}$$

And the values $\overset{*}{\lambda}{}^i$, according to (31.29) and (31.33) are equal

$$\overset{*}{\lambda}{}^i = B_{j^*}^{i_{k^*}} \Big/ \left(\sum_{i=1}^{I} B_{j^*}^{i_{k^*}} a^i\right), \; i \in [1:I] \tag{31.34}$$

and determined unambiguously. This is a simple case. Let the minimum in (31.27) be attained on a particular set R of index pairs (j,k). Then (31.27) can be written as

$$\overset{*}{\Theta} = \left(1 + \sum_{i=1}^{I} B_j^{i_k} V^i\right) \Big/ \left(\sum_{i=1}^{I} B_j^{i_k} a^i\right), \quad (j,k) \in R.$$

For the values $\overset{*}{\eta}{}_j^k$ we have the following problem

$$\sum_{(j,k)\in R} \overset{*}{\eta}{}_j^k \left(1 + \sum_{i=1}^{I} B_j^{i_k} V^i\right) \to \min \tag{31.35}$$

$$\sum_{(j,k)\in R} \overset{*}{\eta}{}_j^k \left(\sum_{i=1}^{I} B_j^{i_k} a^i\right) = 1, \tag{31.36}$$

$$\overset{*}{\eta}{}_j^k = 0, \quad (j,k) \notin R.$$

The problem has a nonunique solution, and the question arises with respect to determining the values $\overset{*}{\lambda}_i$ which form the functionals of block problems.

This case is referred to as degenerate. Thus, consider problem (31.35) – (31.36) when degenerating. A set of its solutions belongs to a polyhedron with nodes

$$\overset{*}{\eta}{}^{k^0}_{j^0} = 1 \Bigg/ \left(\sum_{i=1}^{I} B^{i_{k^0}}_{j^0} a^i \right), \quad (j^0, k^0) \in R. \tag{31.37}$$

Enumerate successively all the elements of the set R and denote the resultant set of numbers by $\bar{s} = [1 : s]$. Substitute (31.37) in (31.39):

$$\overset{*}{\lambda}{}^i_s = B^{i_k}_j \Bigg/ \left(\sum_{i=1}^{I} B^{i_k}_j a^i \right), \quad (j, k) \in R, \quad s \in \bar{s}.$$

Thus, the solutions of the dual macroproblem can be presented as

$$\overset{*}{\lambda}{}^i = \sum_{s=1}^{s} \gamma_s \overset{*}{\lambda}{}^l_s, \tag{31.38}$$

where the values γ_s belong to the set

$$G = \left\{ \gamma = (\gamma_1, \ldots, \gamma_s), \sum_{s=1}^{s} \gamma_s = 1, \gamma_s \geq 0, s \in [1 : s] \right\}.$$

Denote by M^j with each $j \in [1 : k]$ the sets of feasible solutions of block problems (31.20).

Consider the following problem:

$$\max_{\substack{x^i_j \in M^j \\ j \in [1:k]}} \min_{\gamma \in G} (h - \overset{*}{h}) = \max_{\substack{x^i_j \in M^j \\ j \in [1:k]}} \min_{\gamma \in G} \sum_{j=1}^{k} \sum_{i \in I_j} \sum_{s=1}^{s} \gamma_s \overset{*}{\lambda}{}^i_s (x^i_j - \overset{*}{x}{}^i_j). \tag{31.39}$$

If M^j, $j \in [1 : k]$ are the bounded sets, then problem (31.39) has the saddle point $(\bar{x}^i_j, \bar{\gamma}_s)$. Here for each j the values $\bar{x}^i_j$, $i \in I_j$ are the optimal solutions of block problems (31.20) with the functionals in which

$$\overset{*}{\lambda}{}^i = \sum_{s=1}^{s} \bar{\gamma}_s \overset{*}{\lambda}{}^i_s.$$

Denote in terms of $\bar{h}$ the value h for the saddle point $\bar{x}_j^i$. Thus we come to the following optimality criterion for the disaggregative solution $\overset{*}{x}{}_j^i$ in the degeneracy case.

Theorem[40]: The sufficient condition for optimality of the disaggregative solution $\{\overset{*}{x}{}_j^i, \overset{*}{\Theta}\}$ for the original problem (31.6) – (31.10) in the degeneracy case constitutes the equality

$$\bar{h} = \overset{*}{h}\,. \tag{31.40}$$

If problem (31.6) – (31.10) has a solution, and the program $\{\overset{*}{x}{}_j^i, \overset{*}{\Theta}\}$ is not optimal for the problem, then $\bar{h} > \overset{*}{h}$. Note that the non-uniqueness of solutions of dual macroproblems (31.19) presents one of the difficulties encountered when implementing the iterative scheme described. Reference 40 emphasizes the fact that degeneration of problems (31.19) becomes substantial when approaching the optimal solution of the original problem, and in the numerical experiments conducted the optimal dual variables $\overset{*}{\lambda}{}^i$ were calculated proceeding from the fact that the minimum in (31.27) is attained on a unique pair of indices even when approaching the optimum of the original problem.

§**31.5.** Let us formulate the algorithm of the decomposition method described for solving problem (31.6) – (31.10).

<u>*Step 1*</u>: Introduce some fixed aggregation weights α_j^i satisfying the conditions

$$\sum_{j=1}^{k} \alpha_j^i = 1, \quad i \in [1:I]\,,$$

$$\alpha_j^i \geq 0, \quad i \in I_j, \quad \alpha_j^i = 0, \quad i \in \bar{I} \setminus I_j, \quad j \in [1:k]\,.$$

Let the iteration number $l = 1$.

<u>*Step 2*</u>: Solve the problem in aggregative variables (31.13) – (31.17). We know (§31.4) is the analytic solution of the problem

$$\overset{*}{\Theta} = \min_{\substack{j\in[1:k]\\ k\in k_j}} \left(1+\sum_{i=1}^{I} B_j^{i_k} V^i\right) \Big/ \left(\sum_{i=1}^{I} B_j^{i_k} a^i\right), \tag{31.41}$$

$$\overset{*}{X}{}^i = a^i \overset{*}{\Theta} - V^i, \quad i \in [1:I].$$

Step 3: Obtain the dual estimates of $\overset{*}{\lambda}{}^i$, $i \in [1:I]$ to form the functionals of block problems (31.20) from formula (31.34)

$$\overset{*}{\lambda}{}^i = B_{j^*}^{i_{k^*}} \Big/ \left(\sum_{i=1}^{I} B_{j^*}^{i_{k^*}} a^i\right), \; i \in [1:I],$$

where $(j^*,\ k^*)$ is the pair of indices on which the minimum in (31.41) is achieved, and in the degenerate duality case of the macroproblem, i.e. when $\overset{*}{\lambda}{}^i$ is nonunique, any one of the optimal solutions $\overset{*}{\lambda}{}^i$ is taken. Let $\hat{x}_j^i$, $j \in [1:k]$, $i \in [1:I]$ be the optimal solutions of problems (31.20).

Step 4: Obtain the disaggregative solutions

$$\overset{*}{x}{}_j^i = \alpha_j^i \overset{*}{X}{}^i, \quad j \in [1:k], \quad i \in [1:I].$$

Step 5: Test the optimality criterion of the solution $\{\overset{*}{x}{}_j^i, \overset{*}{\Theta}\}$ of the original problem. If

$$\hat{h}_j = \overset{*}{h}_j, \quad j \in [1:k],$$

where $\hat{h}_j$, $\overset{*}{h}_j$ are calculated from formulae (31.23), (31.24), then the solution is optimal. Go to step 8.

Step 6: By using formula (31.22), we form the conditions of problem (31.13) – (31.17), the functional of which is a function from the parameters p_j, $j \subset [1:k]$, and solve the problem with additional constraints $0 \le p_j \le 1$, $j \in [1:k]$. Let $\overset{*}{p}_j$, $j \in [1:k]$ be the optimal solution.

Step 7: From formula (31.22) we define new aggregation weights $\alpha_j^i(p_j^*)$. Set $l = l+1$ and return to step 2.

Step 8: Terminate.

In conclusion it should be noted that the above decomposition method based on aggregating variables from different blocks can be applied to a wide class of extremal problems, e.g. problems of linear and square programming with a block diagonal structure of part of constraints, the optimal control problems with mixed constraints, etc.

Chapter 8

APPLICATION OF SOLUTION TECHNIQUES FOR LARGE DIMENSION PROBLEMS TO GRAIN FARMING OPTIMIZATION

§32. GRAIN CHAIN

§32.1. The Russian government is facing, say, the problem of developing the long-range plan of an appropriate complex by structural products including the national economic indicators prescribed from above and satisfying the national economic needs with final products of this diversified complex with minimal expenses.

It is impossible to construct a model embracing all the problems of agriculture and food processing industry with all the existing relations with other national economic industries. We can only develop a system of models, wherein each model represents the mathematical formulation of one or several closely interrelated problems abstracting from all other factors which actually affect the problem.

The system of models is constructed so that, proceeding from the prediction and general economic limitations on investments and labor force, the optimal distribution of basic funds and the labor force according to individual structural products and the limitations on imports are determined and the national economic demand for these products is established. Parallel planning is then carried out by all structural products: meat, milk, sugar and starch, grain, etc. The obtained results are balanced with each other and with other national economic industries, the efficiency of such a plan is determined and these new results enter again into the system as initial data. The entire process starts again. In other words, the process of planning is the iterative one.

One of the main problems is to carry out the process of planning structural products by mathematical methods, i.e. construction of the so-called chains. In this case the grain chain is an example for constructing other chains.

§**32.2.** The grain chain describes grain production and its subsequent consumption in agriculture itself and in industry before the first processing stage. Grain production and consumption may be represented schematically as follows: Here

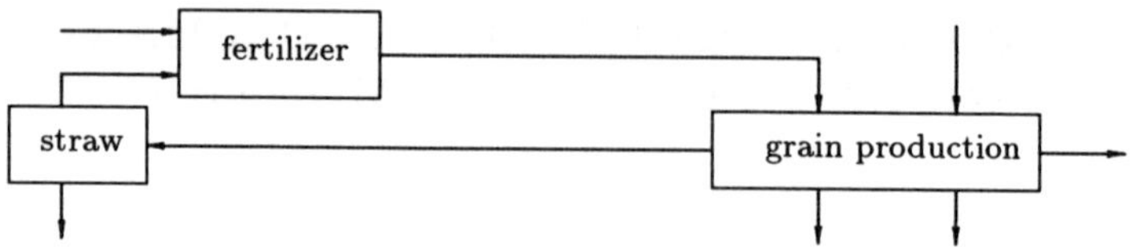

means the following: Not all the straw obtained from grain production as a by-product is needed for litter and industry. The remaining part is left cut into pieces in the field. After ploughing it turns into fertilizer.

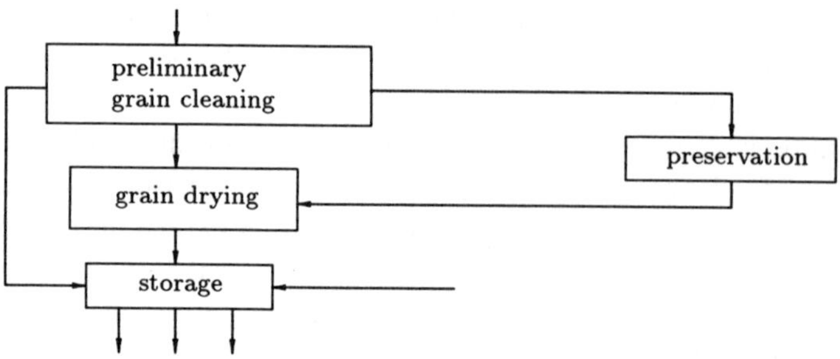

These processing stages are determined by specific agricultural conditions. Upon storage the natural properties of grain are preserved by reducing vital activity of grain. Humidity and relatively high temperatures are the essential factors affecting the grain's vital activity. If the water content in grain is more than 14.5 percent, its vital activity increases markedly and its quality simultaneously deteriorates. The cost of drying units is rather high, and they are used during harvesting. Because of this, from the viewpoint of the main fund utilization it may be beneficial to extend drying time by preserving grain. Three preservation techniques are considered here:

Grain Production and Consumption

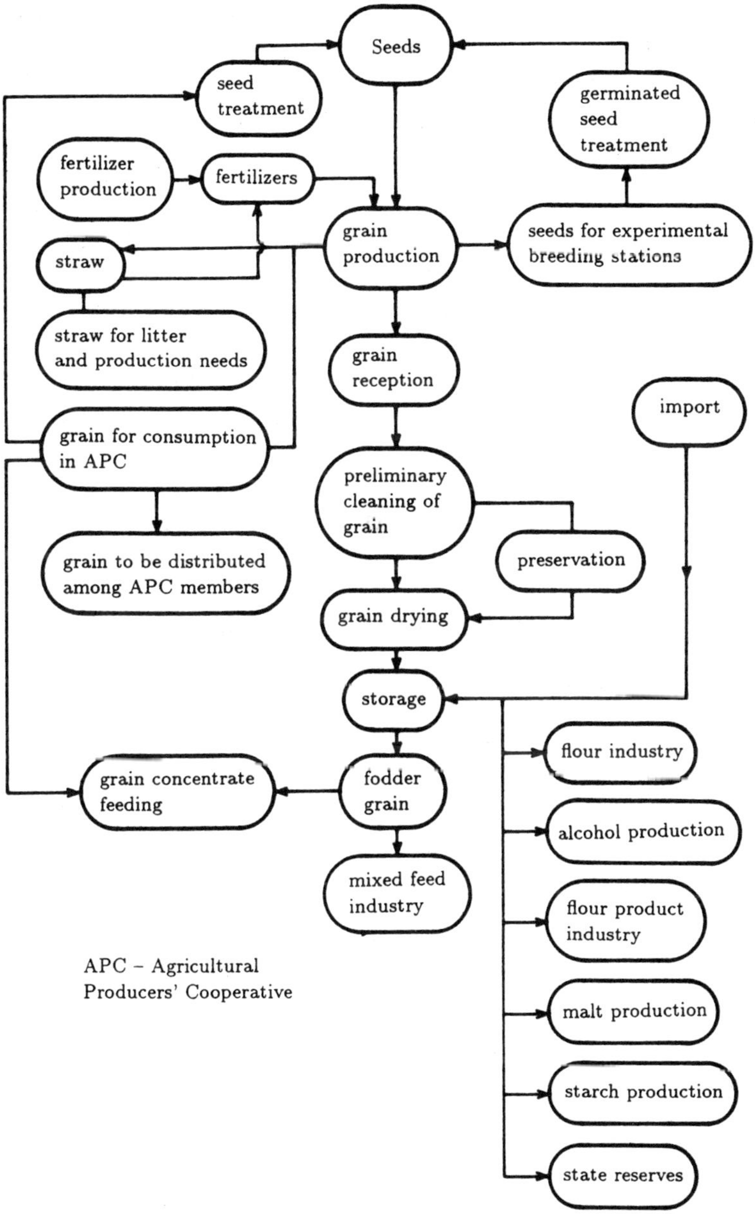

(1) Grain is aerated in suitable rooms by cold air at a temperature of $-1°$ to $+1°$. This is what is known as preservation by air supercooling (maximal storage life is 30 days).
(2) Grain is aerated by external air. This is what is known as simple preservation (maximal storage life is 14 days).
(3) Grain is in the open air. It is not treated. (Storage life is 7 days).

§**32.3.** The national economic demand for grain will grow progressively at the cost of increase in the livestock output. The need of population for cereal food products remains approximately the same. Russia has to import large quantities of grain from abroad, since its own grain production cannot satisfy this need. Thus, the problem arises with respect to growing high-yield crops, increase of fertility at the expense of land improvement, increase in drying and preservation capacities, increase in storage volume, enlargement and improvement of the fleet of combines, tractors and lorries, etc. This aims at providing Russia with its own grain.

We can construct the following grain chain:

— Proceeding from forecast, we determine the need for fodder. For the last year of the long-range plan we establish a fodder balance which results in obtaining the need for grain fodder.
— Based on the forecast, we determine the needs of the population for food products for the last year of the long-range plan. Proceeding from the knowledge of such needs, we have to construct the models optimizing satisfaction of needs by different criteria since a large number of food products are interchangeable. Starch may be taken as an example here.

Though starch is produced from potato or grain, in a sense women are indifferent to the kind of starch used in the kitchen, irrespective of whether it is produced from potato or grain. Grain is the weakest point in the Russian national economy. We would like to determine whether a particular portion of starch produced from grain can be made from potato and what expenses are associated with the problem, since equipment will require readjustment.

Implementation of such models will result in the knowledge of demand for cereal food products.

— The forecast touches upon requirements for growing new grades of grain. Here we deal with the requirement for yield capacity, quality of the grain itself, immunity against diseases and the like.
— Field capacity depends on the type and quality of fertilizer introduced per 1 hectare and on the extent to which land is improved. It is also determined by the application of pesticides to protect plants and by many other factors, which cannot be estimated qualitatively.

§**32.4.** The greater the yield capacity, the greater is the income obtained from 1 hectare. It can be presented by a concave function according to these indicators. Moreover, the amount of fertilizers, pesticides for plant protection and the sum of investments for land improvements are constrained by grain farming. Hence it is wise to construct production functions for each grade determining the dependence of yield capacity upon the three factors, i.e. construct

$\varphi_i(x_i, y_i, z_i)$ – the ith grade yield capacity, where

x_i – the amount of fertilizers introduced per 1 ha,
y_i – the sum of investments per 1 ha employed for land improvement,
z_i – the amount of toxic chemicals utilized,

and we have to determine with their use the optimal yield capacity under given conditions.

Let p_i be a procurement price of 1 centner of the ith grade grain,

q – the price of 1 centner of fertilizers,
v – the price of 1 centner of pesticides,
b_i – the average quantity of fertilizers introduced per 1 ha of the ith grade grain,
A_i – the area sown with the ith grade grain,
A – the area sown with grain,
c_i – the amount of investments for land improvement, which accounts, on the average, for 1 ha of the area sown with the ith grade grain,
α_i – the quantity of mineral fertilizers accounting, on the average, for 1 ha of the area sown with the ith grade grain.

The yield capacity optimization problem then takes the form:

$$\sum_{i=1}^{n} (p_i\varphi_i(x_i, y_i, z_i) - qx_i - y_i - vz_i) \to \max$$

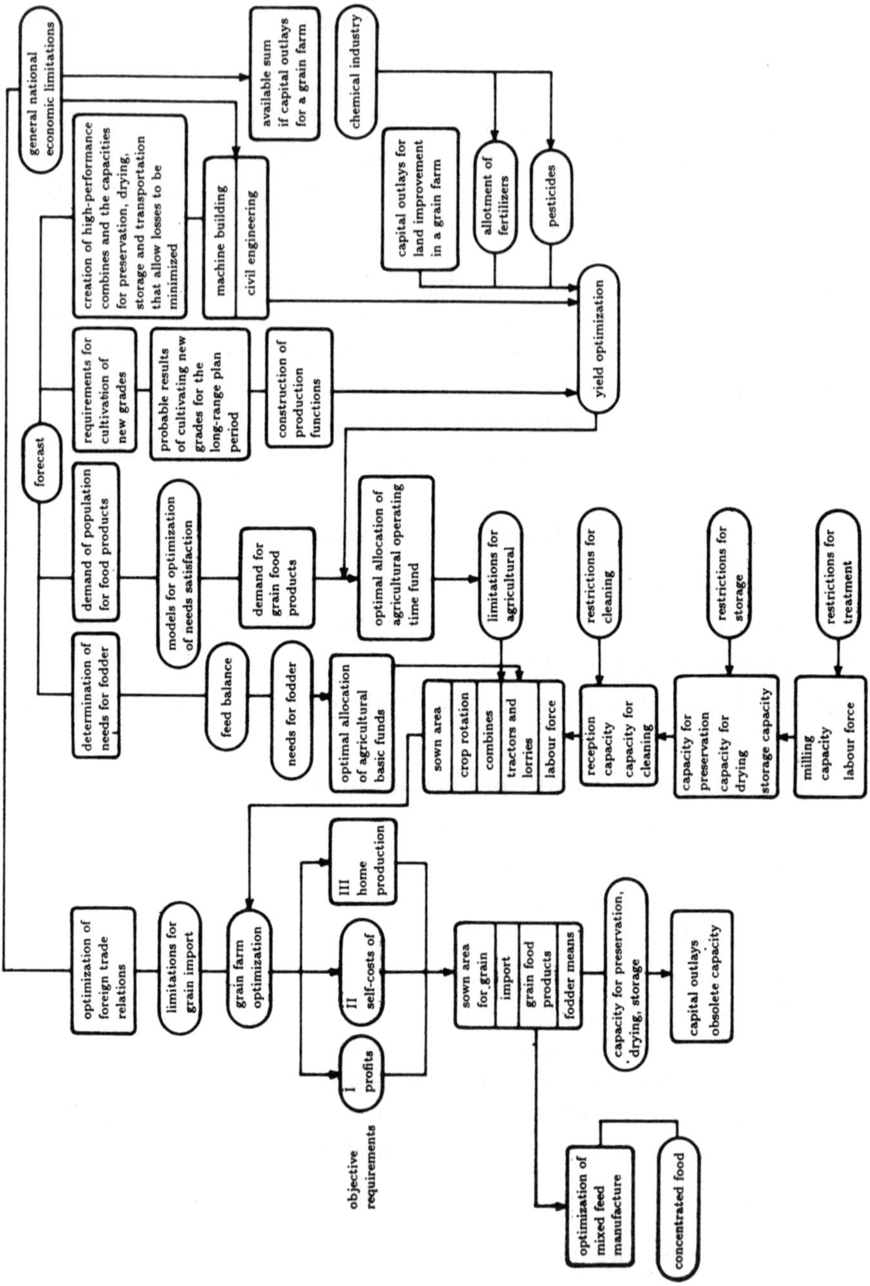
general national economic limitations
available sum if capital outlays for a grain farm
chemical industry
creation of high-performance combines and the capacities for preservation, drying, storage and transportation that allow losses to be minimized
machine building
civil engineering
capital outlays for land improvement in a grain farm
allotment of fertilizers
pesticides
yield optimization
forecast
requirements for cultivation of new grades
probable results of cultivating new grades for the long-range plan period
construction of production functions
demand of population for food products
models for optimization of needs satisfaction
demand for grain food products
optimal allocation of agricultural operating time fund
limitations for agricultural
restrictions for cleaning
restrictions for storage
restrictions for treatment
determination of needs for fodder
feed balance
needs for fodder
optimal allocation of agricultural basic funds
sown area
crop rotation
combines
tractors and lorries
labour force
reception capacity
capacity for cleaning
capacity for preservation
capacity for drying
storage capacity
milling capacity
labour force
optimization of foreign trade relations
limitations for grain import
grain farm optimization
III home production
II self-costs of
I profits
objective requirements
sown area for grain
import
grain food products
fodder means
capacity for preservation, drying, storage
capital outlays
obsolete capacity
optimization of mixed feed manufacture
concentrated food

thus

$$\sum_{i=1}^{n} x_i = X\,, \qquad \sum_{i=1}^{n} y_i = Y\,,$$

$$\sum_{i=1}^{n} z_i = Z\,, \qquad X = \frac{\sum_{i=1}^{n} b_i A_i}{A} n\,,$$

$$Y = \frac{\sum_{i=1}^{n} c_i A_i}{A} n\,,$$

where

$$Z = \frac{\sum_{i=1}^{n} \alpha_i A_i}{A} n\,.$$

We need data on the dependence of yield capacity of different grades on the amount of fertilizers and investments to carry out land improvement. Based on these data, we have to construct production functions and calculate optimal yielding capacity which enters into the grain farm optimization problem as coefficients. The above problem for the two-dimensional case can be solved by the method of dynamic programming.

By employing the models for optimal allocation of major funds and for that of labor (this includes training), we determine major funds and the operating time fund for the grain farming.

From the optimization model for foreign trade relations we can derive constraints on grain import.

An important constituent of the grain chain is the problem of mixed feed manufacture optimization. The present text does not discuss the problem in detail.

§33. GRAIN FARMING OPTIMIZATION PROBLEM

§33.1. The problem of grain farming optimization is a dominant link of the grain chain. The solution of the problem gives the possibility of obtaining information on the relation of home grain production and grain import, the optimal satisfaction of needs, the cost account of preservation, drying and storage of grain, the investments needed to achieve a particular level of production and the efficiency with which they are utilized. The problem of grain farming optimization represents the deterministic problem of nonlinear programming with 84 variables and 74 constraints. It must be solved by 3 objective functions for the long-range plan period. Its solution can be utilized in developing the final plan for grain farming.

The model was constructed under the following assumptions:

(1) The amount of grain in the state reserve does not change. Each year the state reserve is replenished with new grain, therefore the problem does not involve the variable which denotes the grain intended for the state reserve.

(2) The self-cost for 1 ha of the area sown with grain includes the expenses for transportation of grain to a reception center (5 km).

Transportation network optimization is the problem of local planning bodies.

(3) Harvesting continues for 20 days. Grain arrives uniformly at a reception center.

(4) The model is calculated for the long-range plan period. New construction and capacities obtained due to employment of investments immediately come into service.

(5) 10 percent of the entire grain harvest do not need drying. Water content in grain does not exceed 14 percent.

§33.2. Objective Functions.

We take the following as objective functions:

(1) maximum profits;
(2) minimum self-cost;
(3) maximum domestic production.

Maximum profits:

At present, one of the national economic development criteria is assumed to be the national income increment, and to isolate the part of the planned increment of the national income which accounts for grain and its products is extremely difficult. Profit is the economic category close to that of the national income increment. It is this category which has been taken to estimate grain farming and in the objective function we estimate by positive coefficients complex: food products, import grain, and fodder grain. The profit obtained at the earlier processing stages is proportionally included in these coefficients.

Minimum self-cost:

In this objective function, all variables are estimated. Of special interest here is the cost development of the complex intended for preservation, drying and storage of grain.

Maximum domestic production:

The main problem of the national economy consists in providing the country with an adequate quantity of raw material. Domestic production of grain is insufficient to satisfy the needs. Of vital importance is the knowledge of the quantity of grain the farming industry can produce, dry and store, and available investments (though it may be useless in the aspect of maximum profits and minimum self-cost).

§33.3. Variables.

The problem employs 84 variables, out of which 82 variables have economic meaning and 2 variables are auxiliary, assuming the value to be either 0 or 1.

All variables are subdivided into several groups. The first group determines grain production by individual cultures on lands of different quality, since the same culture yields differently on different lands and requires, accordingly, different expenditures of operating and machine time, etc.

By convention we consider 3 kinds of land:

— good land
— medium land
— bad land.

The entire land is estimated by a hundred-mark grading system. The best land obtains 100 marks. Each plot of land is graded by the following data:

— soil composition
— geological origin
— soil condition (e.g. slow changes in soil such as dewatering, decrease of lime content in soil composition)
— local climate
— characteristics of the place
— local hydrological conditions

and it is graded from 1 to 100 marks. Good land is assumed to be that with 60 – 100 marks, medium with 30 – 60 marks and bad land is credited with 1 – 30 marks. It should be noted here that agricultural treatment is worthwhile on the plots of land estimated with more than 16 marks.

For each culture it is advisable to till only two kinds of land: either good and medium or medium and bad. It is unjustified to cultivate, say, wheat on bad land, where its yield is poorer than that of rye, and, besides, it will require more expenditure. A unit of measurement for these variables is a hectare. We then get information on how many hectares of land that should be sown with a specific culture.

A full list of variables is given below.

x_{19} is the straw obtained from grain production;

x_{20} is the portion of straw left in the field. By employing this technology (leaving a portion of straw in the field), we reduce the expenditure of working and machine time per 1 ha of the plot sown with grain.

Variables x_{20}–x_{28} denote the tons of different grades of grain imported from developing and developed countries.

The following group of variables determines the capacities of preservation, drying and storage employed out of those available at the beginning of the last year of the long-range plan period. Apart from preservation in the open air and drying, the preservation capacity features only technical devices. The area needed for it, enters into the storage capacity which is vacant at that moment. The units of measure for the variables denoting preservation and storage are exact and those for the variables which denote drying in the units of different types are exact for a 20-hour day.

The drying capacity available at sugar mills are not designed for grain drying. It is held in reserve when, due to poor weather conditions, we have to dry during harvesting an extremely large amount of grain with high water content.

The group of variables determines the amount of individual food products made in the last year of the long-range plan period. Flour manufacturing involves two technologies of processing grain: in 12-stone mills (a routine practice) and in 6-stone mills. In the latter case the expenditure of working and machine time are taken into account. This is identified as the rapid milling procedure, the introduction of which requires large investments. The problem also contains such variables as the rye flour made by standard milling procedure and the rye flour made by rapid milling procedure.

The unit of measure for rye and wheat alcohol is a hectoliter, while for other food products – tons.

The following group of variables represents fodder grain, which is subdivided according to cultures, and by-products of the food industry also intended for fodder. The units of measure used are tons.

A large group of variables denotes the new capacities of preservation, drying, storage, reception, cleaning and machinery needed for a continuous process of grain reproduction and those available due to employment of investments. For every variable we take into account the existing three kinds of investments, namely, investments for an increase in the available capacity, i.e. for new construction works, investment for expanding the available capacity and those for reconstruction. At the cost of the first kind of investments we obtain the capacity for simple preservation, air supercooling preservation, drying in portable units and silage towers with the carrying capacity of 40 tons per hour, reception, cleaning, outdoor preservation, storage in granaries, in a complex of silage towers, in silage towers (80 thousand tons), in silage towers, (with the capacity of 38 thousand tons), designated by appropriate variables. At the cost of the second kind of investments we obtain the capacity for drying grain in dryers (32 t/h), for storing grain in silage towers (10–20 thousand tons) and their annexes.

The ton is taken as the unit of measure of the capacity for preservation and storage, the tons per day are defined as the units of measure of the capacity for reception, cleaning and drying, and the hours are the units of measure of the machine time fund for tractors, lorries and combines.

The last group of variables designates the obsolete capacity of flour milling by the standard procedure, the annual fund of machine time for obsolete tractors, lorries and combines, and the period of implementation of the long-range plan. In a sense these variables express obsolescence. By way of example, we consider the combines. Let the annual fund of machine time for the combines that have been in service since the beginning of the long-range plan period be, say 1000 hours. The planned number of the combines worn-out at the moment has been already subtracted. The time expenditure norm for old combines per 1 hectare is a. By employing investments, we obtain the auxiliary fund of machine time for new combines x_H which are, say twice as productive, i.e. their norm of time expenditure for 1 ha is $a/2$. Let the number of hectares be X, then the limitation on employment of the machine time fund for combines may be written as follows:

$$aX \leq 1000 + 2x_H \,,$$

or

$$aX - 2x_H \leq 1000 \,.$$

The available time fund (1000 hours) is not utilized in full, since from the standpoint of self-cost, profit and working time expenditures it is more useful to employ new combines, though they are associated with investment expenditures. The old combines are idle, they are obsolescent, but we have to pay for them. Because of this, we include the variable x_c denoting the machine time of the old combines which are not utilized due to obsolescence. The variable x_c enters into the objective function profit maximum with a negative coefficient of magnitude equal to the payment for funds calculated for 1 hour. This variable also enters into the objective function – self-cost minimum with the same magnitude, but positive sign. The limitation on employment of the machine time fund for combines then has to be written as

$$aX - 2x_H + x_c = 1000 \,.$$

List of Variables

x_1 – winter wheat on good land
x_2 – winter wheat on medium land
x_3 – spring wheat on good land
x_4 – spring wheat on medium land
x_5 – winter rye on medium land
x_6 – winter rye on poor land
x_7 – spring rye on medium land
x_8 – spring rye on poor land
x_9 – winter barley on good land
x_{10} – winter barley on medium land
x_{11} – spring barley on good land
x_{12} – spring barley on medium land
x_{13} – brewer's spring barley on good land
x_{14} – brewer's spring barley on medium land
x_{15} – oat on medium land
x_{16} – oat on poor land
x_{17} – mixture on medium land
x_{18} – mixture on poor land
x_{19} – (total amount of) straw
x_{20} – the straw left in the field

x_{21} – "durum wheat"
x_{22} – wheat
x_{23} – barley
x_{24} – maize
x_{25} – oat
} imports from developing countries

x_{26} – maize
x_{27} – millet
x_{28} – oat
} imports from developed countries

x_{29} – barn storage
x_{30} – granary storage
x_{31} – storage in a complex of silage towers
} available capacity

products of food industry:

x_{32} – oaten products
x_{33} – barley flour products
x_{34} – flour made from durum wheat
x_{35} – by standard milling procedure;
x_{36} – flour made from durum wheat by rapid milling procedure;
x_{37} – rye alcohol
x_{38} – wheat alcohol
x_{39} – wheat malt
x_{40} – brewer's barley malt
x_{41} – barley treated for brewing
x_{42} – maize starch
x_{43} – wheat starch
x_{44} – oat
x_{45} – brewer's barley
x_{46} – barley

grain and secondary products for feeding:

x_{47} – rye
x_{48} – wheat
x_{49} – millet
x_{50} – maize
x_{51} – wheat for feed meal
x_{52} – rye for feed meal
x_{53} – flour product manufacture waste

grain and secondary products for feeding:

x_{54} – rye bran
x_{55} – wheat bran
x_{56} – maize processing waste
x_{57} – reception capacity
x_{58} – cleaning capacity
x_{59} – granary storage

x_{60} – storage in a complex of silage towers
x_{61} – storage in silage towers, capacity 80 thousand tons

capacity obtained due to investment unitilization

x_{62} – storage in silage towers, capacity 38 thousand tons
x_{63} – storage in silage towers, capacity 10–20 thousand tons
x_{64} – storage in silage tower annexes
x_{65} – machine time fund for tractors and lorries
x_{66} – machine time fund for combines
x_{67} – milling capacity (standard milling procedure)

outdated capacity

x_{68} – machine time fund for outdated combines
x_{69} – machine time fund for tractors and lorries
x_{70} – drying in stationary driers

available capacity

x_{71} – drying in portable driers
x_{72} – drying in driers (32 ton/h)
x_{73} – drying in portable driers
x_{74} – drying in silage towers (40 ton/h)
x_{75} – preservation by air supercooling (new capacity)
x_{76} – simple preservation (available capacity)

new capacity

x_{77} – simple preservation (new capacity)
x_{78} – outdoor preservation (available capacity)
x_{79} – outdoor preservation (new capacity)

§33.4. Constraints and coefficients.

The problem has 74 constraints: agricultural, constraints for grain cleaning, preservation and treatment, general economic constraints and constraints, which characterize satisfaction of needs for food products and grain fodder.

The first 6 constraints bound above and below the sown area are subdivided into good, medium and poor land. Constant terms of these inequalities can be obtained from the models for optimal allocation of major funds. The lower bound for each kind of land is employed to compel farms to sow a particular area of each kind of land (which is vital to crop rotation) and thereby achieve a certain level of production. These constraints play an important part in solving the problems for maximum profit and minimum self-cost. Thus we have the following inequalities:

$$x_1 + x_3 + x_9 + x_{11} + x_{13} \leq b_1 ,$$

$$x_1 + x_3 + x_9 + x_{11} + x_{13} \geq b_2 ,$$

$$x_2 + x_4 + x_5 + x_7 + x_{10} + x_{12} + x_{14} + x_{15} + x_{17} \leq b_3 ,$$

$$x_2 + x_4 + x_5 + x_7 + x_{10} + x_{12} + x_{14} + x_{15} + x_{17} \geq b_4 ,$$

$$x_6 + x_8 + x_{16} + x_{18} \leq b_5 ,$$

$$x_6 + x_8 + x_{16} + x_{18} \geq b_6 .$$

Arable land in the country is limited, therefore we have to be careful not to allow separate plots to become lixiviated by the same crop. For this purpose, each farm needs to arrange for crop rotation, which is an advantageous succession in cultivating different crops on separate plots of the sown area over a particular number of years with the object of achieving a constant high yield while preserving fertility. Utilization, preservation and increase of soil fertility are determined by rotating crops on plots yearly. In this aspect it might be useful to cultivate in one year only one crop at each farm. But it is impossible from the economic point of view, i.e. from the standpoint of utilizing basic and current assets as well as the working time fund since the work to be done by all participants in production would be unequal. When constructing crop succession we need to comply with the following:

(1) Accommodation of a plant to itself and other plants. If the plants are accommodating, then we do not observe a drop in yield; if they are not accommodating, then there is a danger of pest accumulation.
(2) We should rotate crops which either improve or deteriorate soil structure.

(3) We should choose, for a plot, crops which secure an increase in the yield of the subsequent culture.

For individual grain crops, we have, say, the following data:

useful rotation:	winter wheat	– oat
	oat	– winter wheat
	winter wheat	– winter rye
	winter rye	– winter wheat
possible rotation:	winter wheat	– winter barley
	winter wheat	– spring barley
	oat	– winter barley
	winter rye	– winter wheat
impossible rotation:	spring barley	– winter barley
	winter barley	– spring barley
	spring barley	– winter wheat

Crop rotation cannot be confined to any one farm, it has to be carried out throughout the country. Therefore the problem involves crop rotation limitations on the amount of the cultivated wheat, barley, oat, spring grain as a whole and spring barley. Here we take into account the above accommodation of grain crops to themselves and other agricultural crops. For example, the number of hectares sown with wheat and barley must not exceed 65% of the entire area sown with grain. The exact percentage is obtained from the model of optimal allocation of the basic agricultural assets.

The constraints for imports are derived from the model of foreign relations optimization. On the one hand, the amount of grain imported from developing countries is bounded from above and, on the other, the same is with the sum of currency roubles for purchasing grain in developed countries. The coefficients with variables x_{26}, x_{27}, x_{28} are equal to the price of 1 ton of appropriate grain grade in currency roubles.

Constraints 13 and 14 are related to straw. Equation (13) evaluates the amount of straw to be produced in the planned year. We know the planned yield of each crop on different lands as well as the planned ratio, grain:

straw, which is, say

for winter wheat	1 : 1.2
for spring wheat	1 : 1.1
for winter rye	1 : 1.4, etc.

If, say, the planned yield of winter rye, which is cleared from the harvesting and drying losses, on medium land is 3 tons, then the planned yield of winter rye straw on medium land is 4.35 tons. Then we have equation (13):

$$\sum_{j=1}^{18} a_{13j}x_j - x_{19} = 0\,.$$

Inequality 14 secures satisfaction of the requirements in the planned year, in the straw intended for litter and industrial needs, i.e.

$$x_{19} - x_{20} \geq b_{14}\,.$$

b_{14} will become known to us from the fodder balance and the balance of light industry.

Equations (15)–(26) secure satisfaction of the needs for products of food industry, i.e. the needs for flour products made from oat and barley, for durum wheat flour (which exhibits especially high quality), flour from wheat and rye, alcohol from rye and wheat, malt from wheat and barley for brewing, brewer's treated barley, starch from maize and wheat. The needs of the population for grain food products remain nearly the same for subsequent years. The grain, from which these products were made, might have been more efficiently fed to livestock and used to satisfy the needs of the population. Therefore all constraints with the exception of those for flour manufacture are given say, as the equality:

$$x_{32} = b_{15}\,.$$

Now these x's have ceased to be variables. They are fixed to one value. We can eliminate them and thereby considerably diminish the problem's dimensions by evaluating through the use of the coefficients of material expenditure the requirements in separate grain grades for the food industry.

The amount of secondary products obtained from production and intended for fodder, are readily determined by coefficients. The problem retains these variables since it is the main link of the grain chain, and the grain chain may provide an example for constructing other chains. Subsequently, we need to diminish the dimensions of the problem, where the grain for the needs of the food industry will enter into appropriate equations as a constant term.

Accomplishment of constraint 27 secures satisfaction of the needs for fodder grain obtained from the fodder balance. It takes the following form

$$2.86x_7 + 2.17x_8 + 3.16x_{17} + 2.52x_{18} + 0.72x_{13} + 0.67x_{14} + x_{47} + x_{48} + x_{49} + x_{50} + x_{51} \geq b_{27}\,.$$

The needs for fodder maize and millet will be obtained from the fodder balance separately. We have therefore the inequalities

$$x_{52} \geq b_{28}\,,$$
$$x_{53} \geq b_{29}\,.$$

It may be said that one fifth of the brewer's spring barley cultivated, does not exhibit the properties needed for further processing. This portion of barley is used for feeding. The planned yield (cleared from losses) of this special spring barley on good land is 3.85 tons and on medium land, 3.6 tons. If we subtract from this yield the needs for seeds per 1 ha for the next year (which is on the average 0.2 t) and the needs for 35% of storage losses, then the "pure yield" is 3.6 tons on good land and 3.35 tons on medium land

$$\frac{3.6}{5} = 0.75\,, \qquad \frac{3.35}{5} = 0.67\,.$$

It is assumed that storage losses account for 1.5% of the stored grain. But during harvesting and in the process of drying, 1/12 of the annual output is consumed, i.e. only 11/12 of the entire grain output is stored for some time, and this accounts for 1.5% of storage losses. Thus it may be stated that storage losses of the entire output are

$$1.35\% \approx 1.5\% - \frac{1.5}{12}\%\,.$$

In equation (27), x_{49} denotes the amount of brewer's spring barley, which, apart from these 20%, is used for feeding. Spring rye is cultivated only when winter rye is destroyed by frost. It is completely used for feeding. The rye is incorporated in the problem, since the weather conditions for the planned year are not known, i.e. we do not know whether we have to cultivate this rye or not. Evidently, we need then to incorporate in the problem the additional constraints on the area sown with spring rye on medium and poor land, which should not exceed a particular percentage of the area sown with winter rye:

$$x_7 - a_1 x_5 \leq 0\,,$$

$$x_8 - a_1 x_6 \leq 0\,.$$

Mixture is completely used for feeding, too. Therefore, spring rye and mixture can be directly included in the fodder grain equation in the same manner as 20% of a brewer's spring barley.

Equations (32)–(37) are the balance equations for the secondary products of food industry. Thus we have:

$$0.1x_{34} + 0.1x_{35} + 0.1x_{36} + 0.1x_{37} - x_{54} = 0\,,$$

i.e. we obtain 0.1 t of feed-grade flour wheat from the manufacture of 1 t of flour from wheat. These constraints fall out of the problem in the above reduction of the problem dimension.

Constraints (38)–(45) are the balance equations for the manufacture and allocation of wheat, rye, durum wheat, barley, brewer's barley, oat, millet and maize; thus wheat provides an example of such equations:

$$\begin{aligned}4.05x_1 + 3.16x_2 + 3.95x_3 + 3.06x_4 + x_{22} - 1.33x_{36} - 1.33x_{37}\\ - 0.26x_{41} - 1.238x_{42} - 2x_{46} - x_{51} = 0\,.\end{aligned}$$

Here the coefficients with the variables designating grain production are equal to the planned yield cleared from the losses of yield and drying and 1.35% of storage losses, and the needs for seeds per 1 ha. The remaining coefficients are the coefficients of material expenditure for producing 1 ton of separate food products. Inasmuch as the needs for wheat alcohol (let it be 1000 hectoliters), wheat malt (say, 1000 tons) and wheat starch (let it be 2000 tons) are known, the constraint is transformed into

$$4.05x_1 + 3.16x_2 + 3.95x_3 + 3.06x_4 + x_{22} - 1.33x_{36} - 1.33x_{37} - x_{51} = 5506\,,$$

i.e. grain for the needs of the food industry with the exception of flour manufacture shows itself in the balance equation as a constant term. It is the same with the balance equations for rye, barley, brewer's barley, oat and maize (durum wheat is milled into flour, millet is used for feeding).

Equations (46)–(52) involve the upper restrictions of the capacity being available at the beginning of the last year of the long-range plan, for simple preservation, drying in stationary driers (tons per day), drying in portable driers (tons per day), storing in barns, granaries and in a complex of silage towers and for outdoor preservation. These equations are obtained from the capacity available at the beginning of the long-range plan minus physical wear of the capacity for the long-range plan period. Thus, for the problem we took the capacities available at the end of the preceeding period less the planned physical wear in the earlier years (here the capacity of driers has already been recalculated for 8% of dewatering).

Assume that the harvesting period is 20 days and let, say, 66.6% of the grain output be sold to the state. We need to clean the grain and dry the remaining 90%, i.e. about 60% of the entire yield. The remaining 33.4% are left and treated at farms. Their treatment is not centrally planned.

Assume that harvesting proceeds uniformly during these 20 days, then we have to process 1/20 of the entire yield each day. We denote the variables by

E	60% of the entire grain yield
b	the capacity for drying per day
V_1	the capacity for preservation by air supercooling
V_2	the capacity for simple preservation
V_3	the capacity for outdoor preservation.

Then for each day we must have

$$\frac{E}{20} - b = \frac{V_1}{20} + \frac{V_2}{20} + \frac{V_3}{20}\,.$$

As stated above, in the case of preservation by air supercooling the grain, storage time is 30 days, and when the time is up the grain dies. The grain thus preserved does not die during harvesting. Each day we observe an increase precisely by 1/20 of the entire capacity V_1 (It may be assumed that an increase in each kind of preservation is uniform, since, even though the cost structure changes each day, it remains the same as a whole). The second preservation technique surveys the grain storage time being 14 days before the grain dies. In this case, grain uniformly increases during the 14 days. After this period we have to dry the grain which has arrived for preservation on the first day. We release thus 1/14 of the capacity having been employed until now or 1/20 of the entire capacity V_2 which is replenished in its turn with the grain harvested on the 15th day. On this day a double amount of grain arrives for preservation, since the drying capacity has actually decreased by $V_2/20$. The same is with outdoor preservation, though in this case the grain storage time before drying is 7 days at most.

To illustrate, we take the following example. Let the harvesting period be 14 days ($T = 14$). Thus we have one preservation technique. In this case the maximal time of grain storage is 5 days ($c = 5$).

Let $E = 532$ tons

$b = 28$ tons per day

then $$\frac{E}{14} = 38 \text{ tons}$$

and $$\frac{E}{14} - b = 38 \text{ tons} - 28 \text{ tons} = 10 \text{ tons}$$

V should be 140 tons or 10 tons. On the sixth day 38 tons of the harvested grain arrive again. But only 18 tons out of these 38 tons can be dried, since on this day we have to dry those 10 tons which have arrived for preservation on the first day. Twenty tons of the incoming grain are preserved, 10 tons of the total amount ($= V/14$) occupy an empty place and another 10 tons ($= V/14$) take a new place. Thus, for the sixth day we also have the equality

$$\frac{E}{14} - b = \frac{V}{14}.$$

Furthermore, we need to secure the ratios provided that the grain stored by the outdoor preservation technique dries if only for 7 days

$$V_3 \leq 7b,$$

the grain stored by simple preservation dries for 14 days:

$$V_2 \leq 14b$$

and the grain generally stored by preservation dries for 30 days.

$$V_1 + V_2 + V_3 \leq 30b\,.$$

The minimal drying time is 20 days, and the maximal drying time is 50 days $(20 + 30)$. The real drying time obtained from the problem is equal to the ratio E/b, i.e. the overall drying capacity accounts for $(E/b) \cdot b$. The self-cost of drying 1 ton of grain in different driers is not the same and the self-cost of drying the overall amount of grain which constitutes thereby

$$\frac{\sum\limits_{j=1}^{18} d_j x_j}{\sum\limits_{j=73}^{77} x_j} \sum_{j=73}^{77} c_j x_j\,,$$

where $d_j = 0.9d'_j$ and d'_j is 66.6% of the planned amount (cleared from harvesting losses) of the demand for seeds per 1 hectare of the area.

We determine the self-cost of preservation of the entire amount of grain. The self-cost of preserving a ton of dry grain for 1 day by different techniques and in different devices is known, it accounts for c'_j $(j = 78, 79)$.

Hence (see the plot) the self-cost of preservation is equal to

$$10c'_0 + 20c'_0 + \cdots + 130c'_0 + 140c'_0 + 112c'_0 + 84c'_0 + 56c'_0 + 28c'_0 = \frac{19}{2} \cdot 140 \cdot c'_0$$

or, in a general form, it amounts to:

$$c'_0 \cdot V \cdot \frac{T + c}{2} = c_0 V(T + c)\,,$$

where $c_0 = c'_0/2$, though $T + c = E/b$, is the drying time.

The problem involves 3 preservation techniques, namely, for preservation by air supercooling $T + c = 50$, for simple preservation $T + c = 34$, for outdoor preservation $T + c = 27$, i.e. these capacities can be best utilized for 49, 33 and 26 days, respectively. The real time of their utilization depends on the magnitude

$$\frac{\sum\limits_{j=1}^{18} d_j x_j}{\sum\limits_{j=73}^{77} x_j} = \frac{E}{b}\,.$$

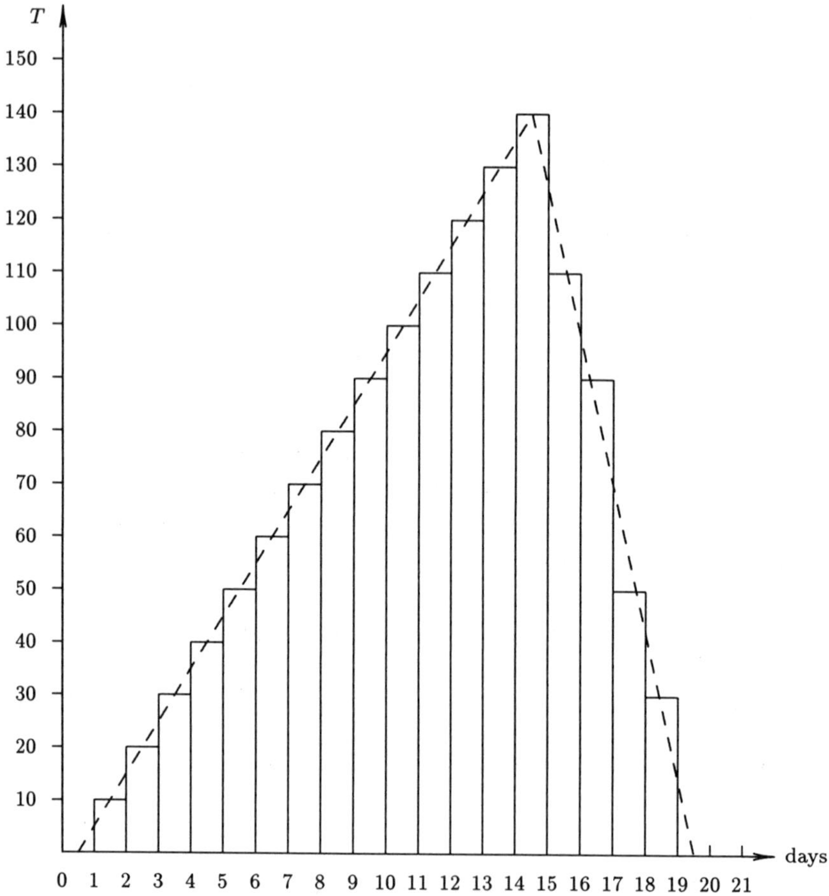

The first kind of preservation capacity is employed in any case for E/b – 1 day.

If $34 \leq E/b \leq 50$, then the second kind of preservation capacity is employed for 33 days and the third kind of preservation capacity for 26 days.

If $27 \leq E/b \leq 34$, the second kind of preservation capacity is employed for $E/b - 1$ day, and the third kind for 26 days.

If $20 \leq E/b \leq 27$, then the preservation capacity of the second and third kind is employed for $E/b - 1$ day.

Considering $c_j = c'_j/2$ $(j = 78, 79)$, we obtain that the self-cost of preservation accounts for

$$\frac{\sum_{j=1}^{18} d_j x_j}{\sum_{j=73}^{77} x_j} \cdot c_{78}x_{78} + \left[34s + (1-s)\frac{\sum_{j=1}^{18} d_j x_j}{\sum_{j=73}^{77} x_j}\right] c_{78}x_{78}$$

$$+ \left[27t + (1-t)\frac{\sum_{j=1}^{18} d_j x_j}{\sum_{j=73}^{77} x_j}\right] c_{79}x_{79}\,,$$

where

$$34s + (1-s)\frac{\sum_{j=1}^{18} d_j x_j}{\sum_{j=73}^{77} x_j} \leq \frac{34 + \dfrac{\sum_{j=1}^{18} d_j x_j}{\sum_{j=73}^{77} x_j}}{2}\,;$$

$$27t + (1-t)\frac{\sum_{j=1}^{18} d_j x_j}{\sum_{j=73}^{77} x_j} \leq \frac{27 + \dfrac{\sum_{j=1}^{18} d_j x_j}{\sum_{j=73}^{77} x_j}}{2}\,.$$

It is to be noted that preservation by air supercooling can proceed only in the silage towers with the capacity of 18 thousand tons and 38 thousand tons and simple preservation in granaries and in a complex of silage towers,

$$x_{78} \leq x_{64} + x_{65}\,,$$

$$x_{79} \leq x_{30} + x_{31} + x_{62} + x_{63}\,.$$

Furthermore, let it be planned that in that year if only 30% of the entire preservation capacity must be the capacity for preservation by air supercooling, then

$$V_1 \geq 0.3(V_1 + V_2 + V_3)\,.$$

The point is that if the weather conditions are poor throughout the harvesting period, then a large amount of humid grain has thus to be dried and preserved. Simple preservation then produces the desired effect, because the air by which grain is aerated exhibits a high percentage of humidity.

The silage towers with the capacity of 10–20 thousand tons and their annexes are added to the available complexes of silage towers. But this is not true for each place. Thus in some regions, where grain is little cultivated, a sufficient amount of storage capacities is available. The problem, therefore, includes the restriction that for the long-range plan period the silage towers (10–20 thousand tons) or their annexes can be added to not more than 30% of the available capacity for storing in complexes of silage towers. The number of these complexes of silage towers is

$$\frac{0.3x_{31}}{8300} \approx 0.00004x_{31} \qquad \text{(their capacity is 8300 tons)}.$$

The number of annexes to silage towers is

$$\frac{1}{11000}x_{67} \approx 0.0001x_{67}\,.$$

The number of silage towers (10–20 thousand tons):

$$\frac{1}{15000}x_{66} \approx 0.00007x_{66}\,.$$

Let us assume that, on average, 1.5 annexes or silage towers (10–20 thousand tons) can be added to every complex of silage towers. Constraint 60 thus takes the form:

$$0.00006x_{31} \geq 0.0001x_{67} + 0.00007x_{66}\,.$$

Constraint 61 is the balance equation for storage

$$\sum_{j=1}^{18} a_{61j}x_j + \frac{1}{6}\sum_{j=21}^{28} x_j - \sum_{j=29}^{31} x_j - \sum_{j=62}^{67} x_j = 0\,.$$

In this case we assume that during 2 months of grain harvesting and drying, 1/6 of the annual imports arrives for storage. The ratios a_{61j} constitute 11/12, i.e. 66.6% of the planned amount of grain (cleared from losses).

One twelfth of the entire yield is fed to livestock or processed by the food industry during harvesting.

Equation (62) restricts employment of the milling capacity (standard milling procedure) measured in machine-hours for the beginning of the last year of the long-range plan (the planned capacity for the beginning of the planned year minus physical wear). Let the data on this point be obtained from the forecast, thus we have

$$b_{62} = 0.667x_{34} + 0.667x_{36} + 0.667x_{38} + x_{70}\,.$$

Constraint 63 bounds from below the milling capacity (rapid milling procedure). The constraint allows a particular percentage of the milling capacity (standard milling procedure) to be reconstructed for the rapid milling procedure over the long-range plan period:

$$b_{63} \leq 0.3x_{35} + 0.3x_{37} + 0.3x_{39}\,.$$

The machine time fund b_{63} calculated for the standard milling procedure does not naturally enter into b_{62}.

The constraints on other capacities for making the remaining grain food products are not included in the problem. These constraints, much like the appropriate constraints on the machine time fund, enter into the models optimizing the satisfaction of consumers (the model optimizing the satisfaction of consumer's needs for flour is not constructed).

Equality (64) yields a restriction on the machine time fund for tractors and lorries

$$b_{64} = \sum_{j=1}^{18} a_{64j}x_j - a_{64,20}x_{20} - 1.5x_{68} + x_{72}\,,$$

where b_{64} is the machine time fund of tractors and lorries for a grain farm, which is available at the beginning of the first year of long-range plan minus physical wear before the last year of the long-range plan. The magnitude is obtained from the forecast: are the coefficients of machine time expenditures for old tractors and lorries (per 1 hectare) during a year. This includes all expenditures for preparing machines, for sowing, fertilizing and harvesting (apart from combines). $a_{64,20}$ is the amount of machine time saved by leaving 1 ton of straw in the field.

Equality (65) expresses a restriction on the machine time for combines

$$b_{65} = \sum_{j=1}^{18} a_{65j}x_j - 2x_{69} + x_{71} ,$$

where b_{65} is the machine time fund for old combines which had not been physically worn out before the last year of the long-range plan. The number of these combines is obtained from the forecast. We obtain the machine time fund by multiplying the number of combines by the harvesting time and the operating time of these combines.

The variables x_{68} and x_{69} in the objective function of the self-cost minimum are not estimated by positive coefficients. These expenditures enter into the planned self-cost of 1 hectare sown with a particular crop. Determination of these coefficients, particularly for different kinds of land, presents a serious problem. We should take into consideration the changes in technologies, the amount of mineral fertilizers applied, the changing expenditures of the direct and embodied labor, etc.

Each day 1/20 of the yield (equivalent to 66.6% of the overall output) arrives at the reception center. Restrictions for the reception capacity are given by inequality (66):

$$b_{66} \geq \frac{1}{20} \sum_{j=1}^{18} d'_j x_j - x_{60} .$$

(The reception capacity is the grain carrying capacity per day at the reception center. It is determined by the equipment of the center, namely, scales, conveyer systems, etc.) b_{66} is obtained from the forecast. It should be noted here that the present model does not consider special treatment of seeds. It takes into account only the place needed for seed storage.

The amount of grain arriving every day could be expressed also in terms of the daily output of combines:

$$\left[a \cdot \frac{b_{65}}{20} \cdot 0.5 + \frac{x_{69}}{20} \cdot 1.0 \cdot a - \frac{x_{71}}{20} \cdot 0.5 \cdot a \right] \cdot 0.666 ,$$

where 0.5 and 1.0 represent the number of hectares to be harvested for 1 hour by appropriate combines. But the present formula includes a the average planned yield which seems to differ from that obtained when solving the model. This may cause an error in the arrival of grain per day up to 150 000 tons.

In a sense, the outdoor preservation capacity may be thought of as a reception capacity, since the grain is actually "received" only when it is intended for drying, thus the equation includes the term.

The grain received is immediately cleaned, which means that the cleaning capacity must be, on one hand, larger than the reception capacity or equal to it and, on the other, 90% of the cleaning capacity must be larger or equal to the amount of grain which is dried or preserved each day by two procedures (to provide synchronism between receiving, cleaning, drying and preserving), i.e.

$$b_{67} \geq x_{60} - x_{61} ,$$

where $b_{67} = c_1 - b_{66}$, c_1 is the cleaning capacity available at the beginning of the last year of the long-range plan; and

$$0.9(c_1 + x_{61}) \geq \frac{V_1 + V_2}{20} + b ,$$

or

$$b_{68} = 0.9c_1 \geq \frac{V_1 + V_2}{20} + b - 0.9x_{61} .$$

The variables x_{60} and x_{61} in the objective function – self-cost minimum are not estimated. The self-cost of receiving and cleaning 1 ton of grain does not enter, as a constituent, into the production cost of 1 ton of grain.

We consider now a restriction on investments b_{69} as the available sum of investment for a grain farm. By subtracting from this sum the capital outlays for land improvement at the farm, it is obtained from general national economic restrictions for the given period of the long-range plan. As the coefficients with the variables for flour production by the rapid milling procedure we assumed the weighted sum of capital outlay for producing 1 ton of flour by the reconstructed capacity of three kinds of facilities with outputs of 31.5 tons/day, 63 tons/day, and 125 tons/day. Weights represent the planned percentage relation of their establishment. The coefficients with x_{60}, x_{61} and x_{75} – x_{77} denote the sum of capital outlay needed for construction of a new capacity for processing 1 ton of grain per day (i.e. receiving, cleaning, drying). The coefficients with the variables x_{62} – x_{67}

and x_{78} represent the sum of capital outlay needed for the construction of a new capacity for processing 1 ton of grain. The coefficients with x_{68} and x_{69} in the equation denote the sum of capital outlay needed for an increase in the annual machine time fund by 1 hour.

The latter three equations are the restrictions on the annual fund of operating time in the farming industry itself, intended for grain production, flour industry, preservation, drying and storage.

We first consider a constraint on the operating time fund for the farming industry:

$$b_{70} \geq \sum_{j=1}^{18} a_{70j}x_j - \bar{a}x_{20} - \bar{\bar{a}}x_{68} - \bar{\bar{\bar{a}}}x_{69}\,,$$

where a_{70j} are the summary expenditures of the operating time for 1 hectare sown with grain, considering that old technology and old machinery are employed.

When employing the technology, according to which straw is left in the field, we save $\bar{a}$ hours for 1 ton of the straw left in the field. When employing new machinery, we save $\bar{\bar{a}}$ hours for every hour of utilization of new tractors and lorries, and employment of combines saves $\bar{\bar{\bar{a}}}$ hours. Exact evaluation of those coefficients $(\bar{a}, \bar{\bar{a}}, \bar{\bar{\bar{a}}})$ is particularly difficult, since along with the employment of new facilities we also employ new technology, which cannot be included in the coefficients representing the average value for the entire farming industry. Evidently, the model here needs careful revision, particularly from the economic point of view.

Constraint 71:

$$b_{71} \geq \sum_{j=34}^{39} a_{71j}x_j\,,$$

where a_{71j} denotes the operating time expenditures for producing 1 ton of flour by different milling techniques.

Below is given the formula, where nonlinearity makes its appearance again just as in the objective function – self-cost minimum due to the variables for drying and preservation.

The coefficients a_{72j}, $j = 73, \ldots, 79$, are the expenditures of operating time for 0.5 ton of the grain being dried and preserved in different facilities. The remaining positive coefficients a_{72j} for $j = 29, 30, 31, 32, \ldots, 67$

represent the expenditures of operating time for storage of 1 ton of grain over 6 months, since utilization of grain for feeding and industrial needs is assumed to be uniform throughout the year. Therefore, on average, every ton of grain is not stored for more than 6 months. We have

$$b_{72} \geq \sum_{j=1}^{72} a_{72j}\, x_j + \frac{\sum_{j=1}^{18} d_j x_j}{\sum_{j=73}^{77} x_j} \sum_{j=73}^{78} a_{72j} x_j$$

$$+ \left[34s + (1-s) \frac{\sum_{j=1}^{18} d_j x_j}{\sum_{j=73}^{77} x_j} a_{72,78}\, x_{78} \right].$$

It should be noted that the present inequality involves double counting, since part of the operating time for drying and preservation enters into the operating time for storage, and to what extent it happens is not known for the present. To be more precise those expenditures are included in the two inequalities given below. But, on the other hand, definition of constant terms presents a serious problem.

The expenditures of operating time for receiving and cleaning enter into the coefficients of the operating time expenditures for producing 1 ton of grain.

b_{70}, b_{71}, and b_{72} are derived from the model of optimal distribution of the operating time fund of the farming industry.

§33.5. We consider the objective function coefficients. In the function – self-cost minimum, the coefficients are equal apart from the import of the obsolete capacity and the straw left in the field, the cost of processing, or the self-cost minus the expenditures for raw and grain materials (for import), are equal to the prevailing market price. The coefficient with x_{20} is equal to the decrease in the self-cost of producing 1 ton of grain from 1 ton of straw (due to the employment of a new technology). Fodders are not estimated here.

In the function-maximum of profit, the coefficients with x_{21} – x_{28} are equal to a difference between the selling price of imported grain inside the country and the market price. The coefficient with x_{20} is equal to c_{20}^2 with an inverse sign, though.

The coefficients in the objective function – home production maximum are equal to the net output, or the planned output which is cleared from the losses of harvesting, drying and storing as well as from the needs for seeds per 1 hectare of the sown area. The variables x_{19} – x_{73} are not estimated in the objective function.

§33.6. General formulation of the problem

$$\sum_{j=1}^{79} a_{ij}\, x_j \lesseqgtr b_i\,, \qquad i = 1, \ldots, 71\,,$$

$$\sum_{j=1}^{72} a_{72j}\, x_j + \frac{\sum_{j=1}^{18} d_j x_j}{\sum_{j=73}^{77}} \sum_{j=73}^{78} a_{72j}\, x_j + \left(34s + (1-s)\frac{\sum_{j=1}^{18} d_j x_j}{\sum_{j=73}^{77}} \right) a_{72,79}\, x_{79}\,,$$

$$34s + (1-s)\frac{\sum_{j=1}^{18} d_j x_j}{\sum_{j=73}^{77} x_j} \le \frac{34 + \frac{\sum_{j=1}^{18} d_j x_j}{\sum_{j=73}^{77} x_j}}{2}\,;$$

$$27t + (1-t)\frac{\sum_{j=1}^{18} d_j x_j}{\sum_{j=73}^{77} x_j} \le \frac{27 + \frac{\sum_{j=1}^{18} d_j x_j}{\sum_{j=73}^{77} x_j}}{2}\,;$$

$$s = 0,1;\ t = 0,1;\ x_j \ge 0,\ j = 1, \ldots, 79.$$

$$\text{I:} \quad \max \sum_{j=1}^{79} c_j^1 x_j\,;$$

$$\text{II:}\quad \min\left\{\sum_{j=1}^{72} c_j^2 x_j + \frac{\sum\limits_{j=1}^{18} d_j x_j}{\sum\limits_{j=73}^{77} x_j}\sum_{j=73}^{78} c_j^2 x_j + \left(34s + (1-s)\frac{\sum\limits_{j=1}^{18} d_j x_j}{\sum\limits_{j=73}^{77} x_j}\right) a_{72,79} x_{79}\right\},$$

$$\text{III:}\quad \max \sum_{j=1}^{18} c_j^3 x_j\,.$$

§34. SOLUTION TECHNIQUE

§34.1. We consider the problem

$$\min\left\{\sum_{j=1}^{n_1} c_j^1 x_j + \frac{\sum\limits_{j=1}^{n_1} d_j x_j}{\sum\limits_{j=1}^{n_2} y_j}\sum_{j=1}^{n_2} c_j^2 y_j\right\},$$

$$A(X,Y) \geq b\,,$$

$$x_j \geq 0\,; \qquad j = \overline{1,n_1}\,;$$

$$y_j \geq 0\,; \qquad j = \overline{1,n_2}\,, \tag{34.1}$$

where the matrix A is the matrix of dimension $m \times (n_1 + n_2)$ and b is the vector $(b_1, b_2, \ldots, b_m)$. The constraints on Y are such that $\sum\limits_{j=1}^{n_2} y_j > 0$ in any feasible program.

We set forth the technique of solving the problem by parametric programming.

A Parametric Form of the Problem

Problem (34.1) is equivalent to a parametric problem, where a parameter enters into the objective function and the coefficient matrix.

Definition 34.1: Extremal problem $(A, \mathcal{F})$ and $(A', \mathcal{F}')$ (both for a minimum), where A and A' are the sets of feasible programs and $\mathcal{F}$ and $\mathcal{F}'$ are objective functions, are identified as equivalent, if there are the mappings

$$\varphi : A \to A'$$

and

$$\psi : A' \to A\,,$$

satisfying the following properties:

1. $\mathcal{F}(a) \geq \mathcal{F}'[\varphi(a)]\,, \forall\, a \in A;$
2. $\mathcal{F}'(a') \geq \mathcal{F}[\psi(a')]\,, \forall\, a' \in A'.$

Theorem 34.1: Suppose we have the problems

$$\min \left\{ \sum_{j=1}^{n_1} c_j^1 x_j + \frac{\sum_{j=1}^{n_1} d_j x_j}{\sum_{j=1}^{n_2} y_j} \sum_{j=1}^{n_2} c_j^2 y_j \right\}, \tag{34.1}$$

$$A(X,Y) \geq b\,,$$
$$X \geq 0\,, \qquad Y \geq 0$$

and

$$\min_{\lambda} \min \left\{ \sum_{j=1}^{n_1} c_j^1 x_j + \lambda \sum_{j=1}^{n_2} c_j^2 y_j \right\},$$

$$A(X,Y) \geq b\,, \tag{34.2}$$
$$dX - \lambda_e Y = 0\,,$$
$$X \geq 0\,, \qquad Y \geq 0\,,$$

where

$$d = (d_1, d_2, \ldots, d_{n_1})\,,$$

$$l = (1, 1, \ldots, 1)\,.$$

Problems (34.1) and (34.2) are equivalent.

Proof:

I. Problem (34.2) can be regarded as

$$\min \left\{ \sum_{j=1}^{n_1} c_j^1 x_j + \lambda \sum_{j=1}^{n_2} c_j^2 y_j \right\},$$

$$A(X,Y) \geq b,$$
$$dX - \lambda_e Y = 0,$$
$$X \geq 0, \qquad Y \geq 0,$$
$$-\infty < \lambda < +\infty$$

taking λ here as a variable below the internal minimum we have

$$\min_{u,v} g(u,v) = \min_{u} \min_{v} g(u,v) = \min_{v} \min_{u} g(u,v).$$

II. Suppose we have two problems (Ω, f) and (Ω', f'), where

$$\Omega = \left\{ \begin{array}{c} (X,Y) : A(X,Y) \geq b \\ X \geq 0 \\ Y \geq 0 \end{array} \right\}$$

$$f(X,Y) = \sum_{j=1}^{n_1} c_j^1 x_j + \frac{\sum_{j=1}^{n_1} d_j x_j}{\sum_{j=1}^{n_2} y_j} \sum_{j=1}^{n_2} c_j^2 y_j,$$

$$\Omega' = \left\{ \begin{array}{c} (X',Y',\lambda) : A(X',Y') \geq b \\ dX' - \lambda_e Y' = 0 \\ X' \geq 0 \\ Y' \geq 0 \\ -\infty < \lambda < +\infty \end{array} \right\}$$

$$f'(X',Y',\lambda) = \sum_{j=1}^{n_1} c_j^1 x_j' + \lambda \sum_{j=1}^{n_2} c_j^2 y_j'.$$

We need to demonstrate that there exist such mappings

$$\varphi : \Omega \to \Omega' \quad \text{and} \quad \psi : \Omega' \to \Omega$$

exhibiting the properties

$$f(X,Y) \geq f'(\varphi(X,Y)),$$
$$f'(X',Y',\lambda) \geq f(\psi(X',Y',\lambda)).$$

Let $(X,Y) \in \Omega$ and $(X',Y',\lambda) \in \Omega'$. As φ, we take the mapping

$$\varphi(x_1,\ldots,x_{n_1};y_1,\ldots,y_{n_2}) = (x_1,\ldots,x_{n_1};y_1,\ldots,y_{n_2},\lambda),$$

where λ is calculated from the formula

$$\lambda = \frac{\sum_{j=1}^{n_1} d_j x_j}{\sum_{j=1}^{n_2} y_j} = \frac{dX}{eY},$$

and, as ψ, we take the mapping

$$\psi(x'_1,\ldots,x'_{n_1};y'_1,\ldots,y'_{n_2},\lambda) = (x'_1,\ldots,x'_{n_2};y'_1,\ldots,y'_{n_2}).$$

(1) We show that

$$\varphi : \Omega \to \Omega', \qquad \varphi(\Omega) \subseteq \Omega'.$$

Let

$$\forall (X,Y) \in \Omega,$$

then

$$A(X,Y) \geq b,$$
$$X \geq 0, \qquad Y \geq 0,$$
$$\varphi(x_1,\ldots,x_{n_1};y_1,\ldots,y_{n_2}) = (x_1,\ldots,x_{n_1};y_1,\ldots,y_{n_2},\lambda),$$

where

$$\lambda = \frac{dX}{eY},$$

i.e.,

$$dX - \lambda_e Y = 0$$

and

$$\lambda \in (-\infty, +\infty) .$$

Therefore, $\varphi(X,Y) \in \Omega'$.

(2) We demonstrate that

$$\psi : \Omega' \to \Omega \quad \text{i.e.} \quad \psi(\Omega') \subseteq \Omega .$$

Let

$$\forall (X', Y', \lambda) \in \Omega' ,$$

then

$$\begin{gathered} A(X', Y') \geq b , \\ dX' - \lambda_e Y' = 0 , \\ X' \geq 0 , \quad Y' \geq 0 , \\ -\infty < \lambda < +\infty , \\ \psi(x'_1, \ldots, x'_{n_1}; y'_1, \ldots, y'_{n_2}, \lambda) = (x'_1, \ldots, x'_{n_1}; y'_1, \ldots, y'_{n_2}) . \end{gathered}$$

Therefore, $\psi(X', Y', \lambda) \in \Omega$.

(3) We demonstrate that

$$f(X,Y) \geq f'\big(\varphi(X,Y)\big) .$$

Let

$$\begin{gathered} \forall (X,Y) \in \Omega , \\ f(X,Y) = c^1 X + \frac{dX}{eY} c^2 Y , \\ \varphi(X,Y) = (X,Y,\lambda) ; \quad f'\big(\varphi(X,Y)\big) = c^1 X + \lambda c^2 Y , \end{gathered}$$

but

$$\lambda = \frac{dX}{eY} .$$

Therefore

$$f'(\varphi(X,Y)) = c^1X + \lambda c^2Y = c^1X + \frac{dX}{eY}c^2Y = f(X,Y).$$

(4) We demonstrate that

$$f'(X',Y',\lambda) \geq f(\psi(X',Y',\lambda)).$$

Let

$$\forall (X',Y',\lambda) \in \Omega',$$
$$f'(X',Y',\lambda) = c^1X' + \lambda c^2Y',$$
$$\psi(X',Y',\lambda) = (X',Y'),$$
$$f(\psi(X',Y',\lambda)) = c^1X' + \frac{dX'}{eY'}c^2Y',$$

but since

$$(X',Y',\lambda) \in \Omega',$$
$$dX' - \lambda_e Y' = 0,$$

i.e.

$$\lambda = \frac{dX'}{eY'}.$$

Therefore,

$$f(\psi(X',Y',\lambda)) = c^1X' + \frac{dX'}{eY'}c^2Y' = c^1X' + \lambda c^2Y' = f'(X',Y',\lambda).$$

It is thereby demonstrated that problems (34.1) and (34.2) are equivalent.

Considering that the problems are equivalent, both of them simultaneously have or do not have feasible programs and the image of an optimal program would be the optimal program in the equivalent problem. As indicated in the demonstration, the problems

$$\min\left\{\sum_{j=1}^{n_1} c_j^1 x_j + \lambda \sum_{j=1}^{n_2} c_j^2 y_j\right\},$$

$$A(X,Y) \geq b,$$
$$dX - \lambda_e Y = 0,$$
$$X \geq 0, \qquad Y \geq 0,$$
$$-\infty < \lambda < +\infty$$

and

$$\min_{\lambda} \min_{\substack{A(X,Y)\geq b \\ dX-\lambda_e Y=0 \\ X\geq 0, \quad Y\geq 0}} \left\{ \sum_{j=1}^{n_1} c_j^1 x_j + \lambda \sum_{j=1}^{n_2} c_j^2 y_j \right\},$$

are equivalent.

By solving the internal parametric problem, we obtain the function $\varphi(\lambda)$, where each λ corresponds to the point

$$\big(x_1(\lambda), x_2(\lambda), \ldots, x_{n_1}(\lambda); y_1(\lambda), \ldots, y_{n_2}(\lambda)\big),$$

where the functions $x_j(\lambda)$ and $y_j(\lambda)$ are linear fractional, in which the minimum of the internal problem is achieved. Let $\varphi(\lambda^*) = \min \varphi(\lambda)$. Then the point $\big(x_1(\lambda^*), x_2(\lambda^*), \ldots, x_{n_1}(\lambda^*); y_1(\lambda^*), \ldots, y_{n_2}(\lambda^*)\big)$ constitutes the optimal program of problem (34.1).

Solve the nondegenerate parametric problem

$$\min \left\{ \sum_{j=1}^{n_1} c_j^1 x_j + \lambda \sum_{j=1}^{n_2} c_j^2 y_j \right\},$$

$$A(X,Y) = b,$$

$$dX - \lambda_e Y = 0,$$

$$X \geq 0, \qquad Y \geq 0.$$

§34.2. Definition of the Solution Stability Domain

We denote by A_1 the part of matrix A corresponding to X variables and by A_2 the part of matrix A, corresponding to Y variables

$$A_1 = \begin{pmatrix} a_{11}^1 & a_{12}^1 & \cdots & a_{1n_1}^1 \\ \hdashline a_{m1}^1 & a_{m2}^1 & \cdots & a_{mn_1}^1 \end{pmatrix},$$

$$A_2 = \begin{pmatrix} a_{11}^2 & a_{12}^2 & \cdots & a_{1n_2}^2 \\ \hdashline a_{m1}^2 & a_{m2}^2 & \cdots & a_{mn_2}^2 \end{pmatrix}.$$

In these designations the problem then takes the form

$$\begin{aligned}&\min(c^1X+\lambda c^2Y)\\&A_1X+A_2Y=b\,,\\&dX-\lambda_eY=0\,,\\&X\geq 0,\quad Y\geq 0\,.\end{aligned}\tag{34.3}$$

The dual problem to the above

$$\begin{aligned}&\max b\pi\\&\pi A_1+d\omega\leq c^1\,,\\&\pi A_2-\lambda\omega\leq c^2\,,\end{aligned}$$

where the dual estimate vector is represented as

$$(\pi,\omega)=(\pi_1,\pi_2,\dots,\pi_m,\omega)\,.$$

We solve problem (34.3) with $\lambda=\lambda_0$. When obtaining the optimal program $\bar{x}$, the following vectors enter into the basis $a^{1j_1},\dots,a^{1j_\alpha},a^{2j_1},\dots,a^{2j_\beta}$, where $\alpha+\beta=m+1$. According to the duality theory, an optimal program of the dual problem is derived from the system

$$\begin{aligned}&a^1_{1j_1}\pi_1+a^1_{2j_1}\pi_2+\dots+a^1_{mj_1}\pi_m+d_1\omega=c^1_{j_1}\,,\\&\text{-------------------------------------}\\&a^1_{1j_\alpha}\pi_1+a^1_{2j_\alpha}\pi_2+\dots+a^1_{mj_\alpha}\pi_m+d_{j_\alpha}\omega=c^1_{j_\alpha}\,,\\&a^2_{1j_1}\pi_1+a^2_{2j_1}\pi_2+\dots+a^2_{mj_1}\pi_m-\lambda_0\omega=\lambda_0c^2_{j_1}\,,\\&\text{-------------------------------------}\\&a^2_{1j_\beta}\pi_1+a^2_{2j_\beta}\pi_2+\dots+a^2_{mj_\beta}\pi_m-\lambda_0\omega=\lambda_0c^2_{j_\beta}\end{aligned}$$

by Kramer's rule.

Let us calculate the determinants:

$$\left\|\begin{matrix}a^1_{1j_1}&a^1_{2j_1}&\cdots&a^1_{mj_1}&d_{j_1}\\ \text{-----}&\text{-----}&\text{-----}&\text{-----}&\text{-----}\\a^1_{1j_\alpha}&a^1_{2j_\alpha}&\cdots&a^1_{mj_\alpha}&d_{j_\alpha}\\a^2_{1j_1}&a^2_{2j_1}&\cdots&a^2_{mj_1}&-\lambda_0\\ \text{-----}&\text{-----}&\text{-----}&\text{-----}&\text{-----}\\a^2_{1j_\beta}&a^2_{2j_\beta}&\cdots&a^2_{mj_\beta}&-\lambda_0\end{matrix}\right\|=\Delta=g(\lambda_0)=k\lambda_0+l\,.$$

Considering that the problem is nondegenerate, the determinant is nonzero.

For $1 \le i \le m$

$$\left\| \begin{array}{cccccc} a^1_{1j_1} & a^1_{2j_1} & \dots & c^1_{j_1} & \dots & d_{j_1} \\ \hdashline a^1_{1j_\alpha} & a^1_{2j_\alpha} & \dots & c^1_{j_\alpha} & \dots & d_{j_\alpha} \\ a^2_{1j_1} & a^2_{2j_1} & \dots & \lambda_0 c^2_{j_1} & \dots & -\lambda_0 \\ \hdashline a^2_{1j_\beta} & a^2_{2j_\beta} & \dots & \lambda_0 c^2_{j_\beta} & \dots & -\lambda_0 \end{array} \right\| \begin{array}{l} = \Delta_i = h_i(\lambda_0) \\ = p_i\lambda_0^2 + q_i\lambda_0 + r_i \end{array}$$

and for $i = m + 1$

$$\left\| \begin{array}{ccccc} a^1_{1j_1} & a^1_{2j_1} & \dots & a^1_{mj_1} & c^1_{j_1} \\ \hdashline a^1_{1j_\alpha} & a^1_{2j_\alpha} & \dots & a^1_{mj_\alpha} & c^1_{j_\alpha} \\ a^2_{1j_1} & a^2_{2j_1} & \dots & a^2_{mj_1} & \lambda_0 c^2_{j_1} \\ \hdashline a^2_{1j_\beta} & a^2_{2j_\beta} & \dots & a^2_{mj_\beta} & \lambda_0 c^2_{j_\beta} \end{array} \right\| \begin{array}{l} = \Delta_{m+1} = h_{m+1}(\lambda_0) \\ = q_{m+1}\lambda_0 + r_{m+1} \end{array}$$

and the dual estimates are equal

$$\pi_i = \frac{p_i\lambda_0^2 + q_i\lambda_0 + r_i}{k\lambda_0 + l}, \qquad i = 1, \dots, m,$$

$$\omega = \frac{q_{m+1}\lambda_0 + r_{m+1}}{k\lambda_0 + l}.$$

The vector $\bar{x}$ also depends on λ_0

$$\bar{x} = A_b^{-1} b$$

Calculate A_b^{-1} to find out in what way $\bar{x}$ depends on λ_0.

$$A \text{ basic} = \left(\begin{array}{cccccc} a^1_{1j_1} & \dots & a^1_{1j_\alpha} & a^2_{1j_1} & \dots & a^2_{1j_\beta} \\ a^1_{2j_1} & \dots & a^1_{2j_\alpha} & a^2_{2j_1} & \dots & a^2_{2j_\beta} \\ \hdashline a^1_{mj_1} & \dots & a^1_{mj_\alpha} & a^2_{mj_1} & \dots & a^2_{mj_\beta} \\ d_{j_1} & \dots & d_{j_\alpha} & -\lambda_0 & \dots & -\lambda_0 \end{array} \right)$$

$$\det A_b = \Delta = k\lambda_0 + l\,.$$

The inverse matrix elements are calculated from the formula

$$\alpha_{ij} = \frac{A_{ji}}{\det A_b} = \frac{g_{ji}(\lambda_0)}{k\lambda_0 + l} = \frac{k_{ji}\lambda_0 + l_{ji}}{k\lambda_0 + l}\,,$$

where A_{ji} is a cofactor to an a_{ji} element of the initial matrix. Thus A_b^{-1}:

$$\begin{pmatrix} \frac{k_{11}\lambda_0 + l_{11}}{k\lambda_0 + l} & \frac{k_{21}\lambda_0 + l_{21}}{k\lambda_0 + l} & \cdots & \frac{k_{m1}\lambda_0 + l_{m1}}{k\lambda_0 + l} & \frac{l_{m+1\,1}}{k\lambda_0 + l} \\ \frac{k_{12}\lambda_0 + l_{12}}{k\lambda_0 + l} & \frac{k_{22}\lambda_0 + l_{22}}{k\lambda_0 + l} & \cdots & \frac{k_{m2}\lambda_0 + l_{m2}}{k\lambda_0 + l} & \frac{l_{m+1\,2}}{k\lambda_0 + l} \\ \hdashline \frac{k_{1m+1}\lambda_0 + l_{1m+1}}{k\lambda_0 + l} & \frac{k_{2m+1}\lambda_0 + l_{2m+1}}{k\lambda_0 + l} & \cdots & \frac{k_{mm+1}\lambda_0 + l_{mm+1}}{k\lambda_0 + l} & \frac{l_{m+1\,m+1}}{k\lambda_0 + l} \end{pmatrix}$$

and

$$\bar{x}_J = \sum_{i=1}^{m} b_i \frac{k_{iJ}\lambda_0 + l_{iJ}}{k\lambda_0 + l} \qquad J \in \text{basis}\,,$$

i.e. $\bar{x}_J$ are the linear fractional functions from λ_0.

From the optimality criterion it follows that the vector $\bar{x}$ remains optimal for all λ, with which the following is accomplished:

$$a_{11}^1 \frac{p_1\lambda^2 + q_1\lambda + r_1}{k\lambda + l} + a_{21}^1 \frac{p_2\lambda^2 + q_2\lambda + r_2}{k\lambda + l} + \cdots$$
$$+ a_{m1}^1 \frac{p_m\lambda^2 + q_m\lambda + r_m}{k\lambda + l} + d_1 \frac{q_{m+1}\lambda + r_{m+1}}{k\lambda + l} \le c_1^1\,,$$

$$a_{1n_1}^1 \frac{p_1\lambda^2 + q_1\lambda + r_1}{k\lambda + l} + a_{2n_1}^1 \frac{p_2\lambda^2 + q_2\lambda + r_2}{k\lambda + l} + \cdots$$
$$+ a_{mn_1}^1 \frac{p_m\lambda^2 + q_m\lambda + r_m}{k\lambda + l} + d_{n_1} \frac{q_{m+1}\lambda + r_{m+1}}{k\lambda + l} \le c_{n_1}^1\,,$$

$$a_{11}^2 \frac{p_1\lambda^2 + q_1\lambda + r_1}{k\lambda + l} + a_{21}^2 \frac{p_2\lambda^2 + q_2\lambda + r_2}{k\lambda + l} + \cdots$$
$$+ a_{m1}^2 \frac{p_m\lambda^2 + q_m\lambda + r_m}{k\lambda + l} - \lambda \frac{q_{m+1}\lambda + r_{m+1}}{k\lambda + l} \le \lambda c_1^2\,,$$

$$a_{1n_2}^2 \frac{p_1\lambda^2 + q_1\lambda + r_1}{k\lambda + l} + a_{2n_2}^2 \frac{p_2\lambda^2 + q_2\lambda + r_2}{k\lambda + l} + \cdots$$

$$+ a^2_{mn_2} \frac{p_m\lambda^2 + q_m\lambda + r_m}{k\lambda + l} - \lambda \frac{q_{m+1}\lambda + r_{m+1}}{k\lambda + l} \le \lambda c^2_{n_2} ,$$

$$\sum_{i=1}^{m} b_i \frac{k_{iJ}\lambda + l_{iJ}}{k\lambda + l} \ge 0 \qquad J \in \text{basis}$$

Definition of the $\bar{x}$ solution stability domain:

(1) Let $k\lambda + l > 0$ and $k > 0$

The domain of stability is then determined by the inequalities

$$(k\lambda + l)\Delta_1 = \lambda^2 \sum_{i=1}^{m} a^1_{i1} p_i + \lambda \left[\sum_{i=1}^{m} a^1_{i1} q_i + d_1 q_{m+1} - c^1_1 k \right] + \sum_{i=1}^{m} a^1_{i1} r_i + d_1 r_{m+1} - c^1_1 l \le 0 ,$$

$$(k\lambda + l)\Delta_{n_1} = \lambda^2 \sum_{i=1}^{m} a^1_{in_1} p_i + \lambda \left[\sum_{i=1}^{m} a^1_{in_1} q_i + d_{n_1} q_{m+1} - c^1_{n_1} k \right] + \sum_{i=1}^{m} a^1_{in_1} r_i + d_{n_1} r_{m+1} - c^1_{n_1} l \le 0 ,$$

$$(k\lambda + l)\Delta_{n_1+1} = \lambda^2 \left[\sum_{i=1}^{m} a^2_{i1} p_i - q_{m+1} - c^2_1 k \right] + \lambda \left[\sum_{i=1}^{m} a^2_{i1} q_i - r_{m+1} - c^2_1 l \right] + \sum_{i=1}^{m} a^2_{i1} r_i \le 0 ,$$

$$(k\lambda + l)\Delta_{n_1+n_2} = \lambda^2 \left[\sum_{i=1}^{m} a_{in_2} p_i - q_{m+1} - c^2_{n_2} k \right] + \lambda \left[\sum_{i=1}^{m} a^2_{in_2} q_i - r_{m+1} - c^2_{n_2} l \right] + \sum_{i=1}^{m} a_{in_2} r_i \le 0 ,$$

$$\lambda \sum_{i=1}^{m} b_i k_{iJ} + \sum_{i=1}^{m} b_i l_{iJ} \ge 0 , \qquad \lambda > -\frac{l}{k}$$

or

$$\mathcal{L}_j \lambda^2 + M_j \lambda + N_j \le 0 \qquad j = 1, \dots, n$$

$$\lambda \sum_{i=1}^{m} b_i k_{iJ} + \sum_{i=1}^{m} b_i l_{iJ} \geq 0\,, \qquad \lambda > -\frac{l}{k}\,.$$

The system is consistent, since it is satisfied with $\lambda = \lambda_0$. Define

$$\lambda_1^j = \frac{-M_j - \sqrt{M_j^2 - 4\mathcal{L}_j N_j}}{2\mathcal{L}_j}\,,$$

$$\lambda_2^j = \frac{-M_j + \sqrt{M_j^2 - 4\mathcal{L}_j N_j}}{2\mathcal{L}_j}\,.$$

For those j, for which $\mathcal{L}_j > 0$, we have

$$\lambda_1 = \max_{\mathcal{L}_j > 0} \lambda_1^j\,,$$

$$\lambda_2 = \min_{\mathcal{L}_j > 0} \lambda_2^j\,.$$

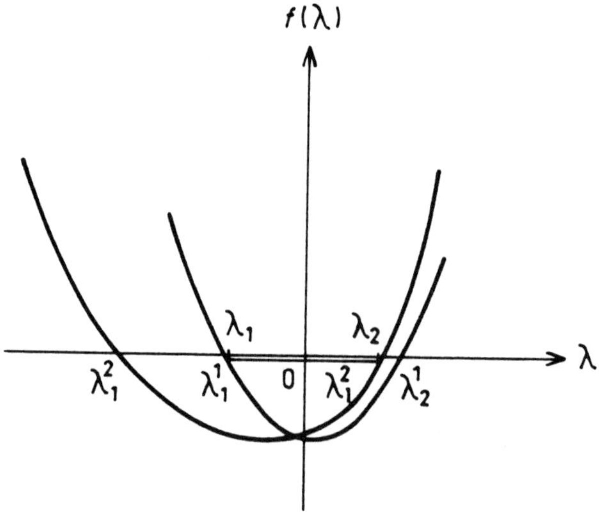

We have obtained the range of changes for $\lambda \in [\lambda_1, \lambda_2]$. If among the inequalities $\mathcal{L}_j\lambda^2 + M_j\lambda + N_j \leq 0$ there is no such value for which $\mathcal{L}_j > 0$, then we assume

$$\lambda_1 = -\infty\,,$$

$$\lambda_2 = +\infty \, .$$

For those j, for which $\mathcal{L}_j < 0$ we obtain:

$$\lambda_3 = \max_{\mathcal{L}_j<0} \lambda_1^j \, ,$$

$$\lambda_4 = \min_{\mathcal{L}_j<0} \lambda_2^j \, .$$

Here it is possible that for a particular j there are no real roots. The entire curve then lies under the axis λ without intersecting it. In this case we assume that $\lambda_1^j = -\infty$; $\lambda_2^j = +\infty$.

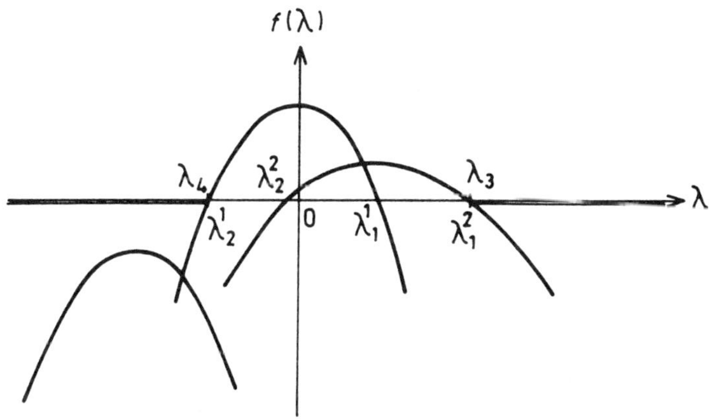

If for a particular j_0 $\lambda_1^{j_0} = \lambda_2^{j_0}$, then the curve also lies completely under the abscissa. In this case, as before, we assume that

$$\lambda_1^{j_0} = -\infty \, ,$$

$$\lambda_2^{j_0} = +\infty \, .$$

We have obtained the range of changes for λ $(-\infty, \lambda_4]$ and $[\lambda_3, +\infty)$. For those j, for which $\mathcal{L}_j = 0$, we calculate

$$\lambda^j = -\frac{N_j}{M_j} \, ,$$

define

$$\lambda_7 = \max_{\substack{\mathcal{L}_j=0 \\ M_j<0}} \lambda^j \, ,$$

$$\lambda_8 = \min_{\substack{\mathcal{L}_j=0 \\ M_j>0}} \lambda^j ,$$

and obtain the range of changes for λ $[\lambda_7, \lambda_8]$.

Let us consider the second group of inequalities

$$\lambda \sum_{i=1}^{m} b_i k_{iJ} + \sum_{i=1}^{m} b_i l_{iJ} \geq 0 , \qquad J \in \text{basis} .$$

Let for $J = 1, \ldots, \varepsilon$

$$\sum_{i=1}^{m} b_i k_{iJ} > 0$$

and for $J = \varepsilon + 1, \ldots, m+1$

$$\sum_{i=1}^{m} b_i k_{iJ} < 0 .$$

(The case $\sum_{i=1}^{m} b_i k_{iJ} = 0$ is not considered.) Calculate

$$\lambda^J = -\frac{\sum_{i=1}^{m} b_i l_{iJ}}{\sum_{i=1}^{m} b_i k_{iJ}}$$

Then there should be $\lambda \geq \lambda^J \quad J = 1, \ldots, \varepsilon$ and $\lambda \leq \lambda^J \quad J = \varepsilon+1, \ldots, m+1$. Define

$$\lambda_5 = \max_{\sum_{i=1}^{m} b_i k_{iJ} > 0} \lambda^J ,$$

$$\lambda_6 = \min_{\sum_{i=1}^{m} b_i k_{iJ} < 0} \lambda^J .$$

We have obtained the range of changes for λ $[\lambda_5, \lambda_6]$. The intersection of all the ranges obtained gives the domain of solution stability.

Define

$$\underline{\lambda} = \max\left\{\lambda_1; \lambda_5; \lambda_7; -\frac{l}{k}\right\},$$

$$\overline{\lambda} = \min\{\lambda_2; \lambda_4; \lambda_6; \lambda_8\},$$

$$\underline{\underline{\lambda}} = \max\left\{\lambda_1; \lambda_3; \lambda_5; \lambda_7; -\frac{l}{k}\right\},$$

$$\overline{\overline{\lambda}} = \min\{\lambda_2; \lambda_6; \lambda_8\}.$$

Then the $\bar{x}$ solution stability domain

$$[\underline{\lambda}, \overline{\lambda}] \cup \left[\underline{\underline{\lambda}}, \overline{\overline{\lambda}}\right]$$

(If $\underline{\lambda}$ or $\underline{\underline{\lambda}}$ is achieved on $-l/k$, then the bound is not included in the domain, e.g. $(\underline{\lambda}, \overline{\lambda}]$. If, say, $\overline{\lambda} < \underline{\lambda}$, then $[\underline{\lambda}, \overline{\lambda}] = \Lambda$). Therefore the domain of stability of the solution $\bar{x}$ is convex when either all $\mathcal{L}_j > 0$ or $\lambda_3 = -\infty$ and $\lambda_4 = +\infty$.

(2) Let $k\lambda + l < 0$ and $k > 0$. Then the domain of stability is determined by the inequalities

$$\left.\begin{aligned} &\mathcal{L}_j\lambda^2 + M_j\lambda + N_j \geq 0 && j \notin \text{basis} \\ &\lambda\sum_{i=1}^{m} b_i k_{iJ} + \sum_{i=1}^{m} b_i l_{ij} \leq 0 && J \in \text{basis} \end{aligned}\right\}$$

We determine from similar reasonings

$$\lambda_1' = \min_{\mathcal{L}_j > 0} \lambda_1^j,$$

$$\lambda_2' = \max_{\mathcal{L}_j > 0} \lambda_2^j,$$

$$\lambda_3' = \min_{\mathcal{L}_j < 0} \lambda_1^j,$$

$$\lambda_4' = \max_{\mathcal{L}_j < 0} \lambda_2^j,$$

$$\lambda_5' = \min_{\sum_{i=1}^{m} b_i k_{iJ} > 0} \lambda^J,$$

$$\lambda'_6 = \max_{\sum_{i=1}^{m} b_i k_{iJ} < 0} \lambda^J ,$$

$$\lambda'_7 = \min_{\substack{\mathcal{L}_j = 0 \\ M_j < 0}} \lambda^j ,$$

$$\lambda'_8 = \max_{\substack{\mathcal{L}_j = 0 \\ M_j > 0}} \lambda^j .$$

To obtain the solution stability domain, intersect the obtained line segments.

Define

$$\underline{\lambda}' = \max\{\lambda'_4; \lambda'_6; \lambda'_8\} ,$$

$$\overline{\lambda}' = \min\left\{-\frac{l}{k}; \lambda'_1; \lambda'_3; \lambda'_5; \lambda'_7\right\} ,$$

$$\underline{\underline{\lambda}}' = \max\{\lambda'_2; \lambda'_4; \lambda'_6; \lambda'_8\} ,$$

$$\overline{\overline{\lambda}}' = \min\left\{-\frac{l}{k}; \lambda'_3; \lambda'_5; \lambda'_7\right\} .$$

Then the domain of stability of the solution $\bar{x}$

$$[\,\underline{\lambda}', \overline{\lambda}'\,] \cup [\,\underline{\underline{\lambda}}', \overline{\overline{\lambda}}'\,]$$

(If $\overline{\lambda}'$ or $\overline{\overline{\lambda}}'$ is achieved on $-l/k$ then the bound here is not included in the domain, e.g. $[\underline{\lambda}', \overline{\lambda}')$. If, say $\overline{\lambda}' < \underline{\lambda}'$, then $[\underline{\lambda}', \overline{\lambda}'] = \Lambda$).

§34.3. Solution Algorithm

<u>*Step 1*:</u> Suppose we have the problem

$$\min \left\{ \sum_{j=1}^{n_1} c_j^1 x_j + \frac{\sum_{j=1}^{n_1} d_j x_j}{\sum_{j=1}^{n_2} y_j} \sum_{j=1}^{n_2} c_j^2 y_j \right\} ,$$

$$A(X, Y) = b ,$$

$$X \geq 0 , \qquad Y \geq 0 .$$

Replace it by the problem

$$\min_{\lambda} \min_{\substack{A_1X+A_2Y=b \\ dX-\lambda_eY=0}} \left(c^1X + \lambda c^2Y\right)$$

$$X \geq 0\,, \qquad Y \geq 0\,.$$

Step 2: Solve the parametric problem

$$\min\ \left(c^1X + \lambda c^2Y\right)\,,$$
$$A_1X + A_2Y = b\,,$$
$$dX - \lambda_eY = 0\,,$$
$$X \geq 0\,, \qquad Y \geq 0\,.$$

Step 2.1: Prescribe $\lambda = \lambda_0$.
Suppose for $\lambda = \lambda_0$ there is the basic initial program

$$\left(x_{J1}(\lambda_0), x_{J2}(\lambda_0), \ldots, x_{J\alpha}(\lambda_0), y_{J1}(\lambda_0), \ldots, y_{J\beta}(\lambda_0)\right)\,.$$

Calculate A_b^{-1} for the program and fill in the simplex table with the changing λ

N	c_J	x_J	x^1	...	x^{n_1}	x^{n_1+1}	...	$x^{n_1+n_2}$
J_1	c^1_{J1}	$x_{J1}(\lambda)$	$x^1_{J_11}(\lambda)$	...	$x^1_{J_1n_1}(\lambda)$	$x^1_{J_1n_1+1}(\lambda)$	...	$x^1_{J_1n_1+n_2}(\lambda)$
J_α	$c^1_{J\alpha}$	$x_{J\alpha}(\lambda)$	$x^1_{J_\alpha1}(\lambda)$	...	$x^1_{J_\alpha n_1}(\lambda)$	$x^1_{J_\alpha n_1+1}(\lambda)$	...	$x^1_{J_\alpha n_1+n_2}(\lambda)$
J_1	c^2_{J1}	$y_{J1}(\lambda)$	$x^2_{J_11}(\lambda)$	...	$x^2_{J_1n_1}(\lambda)$	$x^2_{J_1n_1+1}(\lambda)$	...	$x^2_{J_1n_1+n_2}(\lambda)$
J_β	$c^2_{J\beta}$	$y_{J\beta}(\lambda)$	$x^2_{J\beta_1}(\lambda)$	...	$x^2_{J_\beta n_1}(\lambda)$	$x^2_{J_\beta n_1+1}(\lambda)$	...	$x^2_{J\beta n_1+n_2}(\lambda)$
		$\varphi(\lambda)$	$\Delta_1(\lambda)$	...	$\Delta_{n_1}(\lambda)$	$\Delta_{n_1+1}(\lambda)$	...	$\Delta_{n_1+n_2}(\lambda)$

where

$$\varphi(\lambda) = \sum_{k=1}^{\alpha} c^1_{Jk}x_{Jk} + \sum_{k=1}^{\beta} c^2_{Jk}y_{Jk}\,,$$

$$x^j(\lambda) = A_b^{-1}(\lambda)A_1^j\,, \qquad j = 1, \ldots, n_1\,,$$
$$x^{n_1+j}(\lambda) = A_b^{-1}(\lambda)A_2^j\,, \qquad j = 1, \ldots, n_2\,,$$
$$\Delta_j(\lambda) = \sum_{k=1}^{\alpha} c_{Jk}^1 x_{J_k\,j}^1(\lambda) + \sum_{k=1}^{\beta} c_{Jk}^2 x_{J_k\,j}^2(\lambda)\,.$$

Solve the problem for $\lambda = \lambda_0$ by a direct simplex method. If the problem for $\lambda = \lambda_0$ is unsolvable, i.e. the linear form is not bounded, then we define the set of all λ for which the problem cannot be solved from the relations

$$\left.\begin{array}{ll} x_{J_k\,j}^1(\lambda) \le 0\,, & k = 1, \ldots, \alpha \\ x_{J_k\,j}^2(\lambda) \le 0\,, & k = 1, \ldots, \beta \end{array}\right\}$$

(the vector j should enter into the basis).
If we obtain the optimal program, then from the table of the relations obtained we determine the domain of stability of this program for all λ, for which

$$\Delta_j(\lambda) = \frac{\mathcal{L}_j\lambda^2 + M_j\lambda + N_j}{k\lambda + l} \le 0 \qquad j \notin \text{basis}$$
$$x_J(\lambda) = \sum_{i=1}^{m} b_i \frac{k_{iJ}\lambda + l_{iJ}}{k\lambda + l} \ge 0 \qquad J \in \text{basis.}$$

Step 2.2: Suppose that λ crosses one bound of the domain of stability. Then either one of x_J becomes less than zero or one of Δ_j becomes larger then zero. If one of x_J becomes less than zero, then the problem is further solved by the dual simplex method in one table.

The vector J originates from the basis, the vector, on which $\max\limits_{x_{Jj}<0} \left\{-\frac{\Delta_j}{x_{Jj}}\right\}$ is achieved, enters into it.

We determine the domain of those λ for which the problem has no programs from the relations

$$x_{Jj} \ge 0\,.$$

- If one of the characteristic differences Δ_j becomes larger than zero, then the problem is further solved by the simplex method in one table. The vector j enters into the basis.

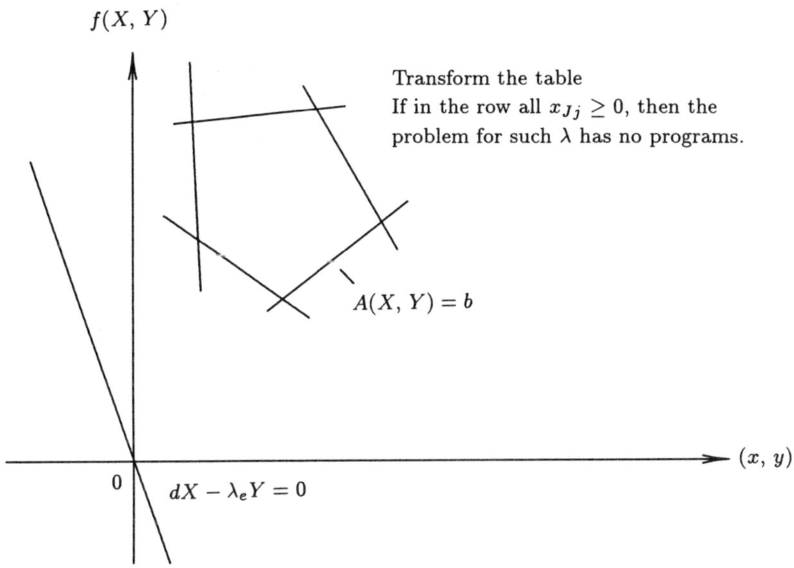

Fit $\theta = \min\limits_{x_{Jj}>0} \dfrac{x_J}{x_{Jj}} = \dfrac{x_k}{x_{kj}}$, the vector k leaves the basis. Transform the simplex table.

- If simultaneously one of x_J becomes less than zero and one of Δ_j becomes larger than zero, then the problem is further solved by the dual simplex method so that it would always be in a feasible region.

  Transform the table.

<u>*Step 2.3*</u>: The new problem obtained is tested for optimality.

- If the program is optimal, then we determine the stability of the domain once again.

  Go to 2.2.

- If the program is not optimal, the problem is further solved according to three possible cases till we get the optimal program. Determine its domain of stability.

  Go to step 2.2.

The problem solution is completed if for each λ from $(-\infty;+\infty)$ we obtain the optimal program or see that the problem is unsolvable.

Step 2.4: The solution of the parametric problem results in the function of the objective function values depending on λ, which looks as

$$\varphi(\lambda) = \begin{cases} \dfrac{a_1\lambda^2 + b_1\lambda + C_1}{d_1\lambda + l_1} & \lambda \in (-\infty, \beta_1] \\ \dfrac{a_2\lambda^2 + b_2\lambda + C_2}{d_2\lambda + l_2} & \lambda \in [\beta_1, \beta_2] \\ \text{-------------------------} & \\ \dfrac{a_t\lambda^2 + b_t\lambda + C_t}{d_t\lambda + l_t} & \lambda \in [\beta_{t-1}, +\infty) \end{cases}$$

or $\varphi_i(\lambda) = -\infty$, $\lambda \in [\beta_{i-1}, \beta_i]$, when the linear form is not bounded on the set of programs or for $\lambda \in [\beta_{i-1}, \beta_i]$ $\varphi_i(\lambda)$ does not exist, when for such λ the problem has no programs at all.

Step 3: We have $\min\limits_{\lambda} \varphi(\lambda)$ if among $\varphi_i(\lambda)$ there are no such that $\varphi_i(\lambda) = -\infty$.

(a) $d_i = 0$

- If $a_i/l_i > 0$, then the function has the minimum at the point $\lambda = -b_i/2a_i$. If

$$\lambda = -\frac{b_i}{2a_i} \notin [\beta_{i-1}, \beta_i],$$

then the function takes its minimal value at the ends of the segment line.

- If $a_i/l_i < 0$, then the function assumes its minimal value at the ends of the segment line.
- If $a_i = 0$, then it is equivalent to the case $b.1$.

(b) $d_i \neq 0$

The case, when the point

$$-\frac{l_i}{d_i} \in [\beta_{i-1}, \beta_i],$$

can not occur, since there would then be

$$\varphi_i\left(-\frac{l_i}{d_i}\right) = -\infty,$$

which means that the linear form is not bounded on the set of programs, and this shows itself even in the process of solving the problem.

Derive

$$\frac{a_i\lambda^2 + b_i\lambda + c_i}{d_i\lambda + l_i} = \gamma_i\lambda + \delta_i + \frac{\Omega_i}{d_i\lambda + l_i}.$$

(1) $\Omega_i = 0$ is a linear function, it assumes its minimal value at the extremities of the segment line.

(2) $\gamma_i = 0$ is a monotone function, it assumes its minimal value at the bound of the segment line.

(3) $\gamma_i \neq 0, \Omega_i \neq 0$.
- If $\gamma_i\Omega_i < 0$, then the function is monotonic, it assumes its minimal value at the bound of the segment line.
- If $\gamma_i\Omega_i > 0$, then the function has the minimum at the point

$$\lambda = \frac{-l_i\gamma_i + \sqrt{l_i^2\gamma_i^2 - \gamma_i(l_i^2\gamma_i - \Omega_i d_i)}}{\gamma_i d_i} = \frac{-l_i\gamma_i + \sqrt{\gamma_i\Omega_i d_i}}{\gamma_i d_i}.$$

If the derived point $\lambda \notin [\beta_{i-1}, \beta_i]$, then the function achieves its minimal value at the bound of the segment line.

Step 3.1: Calculate for each i the minimal value of the function $\varphi_i(\lambda)$ and its corresponding λ_i^*

$$\frac{a_i\lambda_i^{*2} + b_i\lambda_i^* + c_i}{d_i\lambda_i^* + l_i}$$

Step 3.2: Determine

$$\min_i \left\{ \frac{a_i\lambda_i^{*2} + b_i\lambda_i^* + c_i}{d_i\lambda_i^* + l_i} \right\} = \frac{a_{i_0}\lambda_{i_0}^{*\,2} + b_{i_0}\lambda_{i_0}^* + c_{i_0}}{d_{i_0}\lambda_{i_0}^* + l_{i_0}}.$$

Step 3.3: The obtained $\lambda_{i_0}^*$ corresponds to the optimal program $\bar{x}(\lambda_{i_0}^*)$.

Terminate. The program $\bar{x}(\lambda_{i_0}^*)$ is optimal for the original problem.

Remark: The more complex problems, where the parameter λ enters into more than one constraint, are not solved by the present algorithm, since we would have to solve then the inequality of higher degrees.

§34.4.

Block Scheme

Suppose we have the problem

$$\min \left\{ \sum_{j=1}^{n_1} c_j^1 x_j + \frac{\sum_{j=1}^{n_1} d_j x_j}{\sum_{j=1}^{n_2} y_j} \sum_{j=1}^{n_2} c_j^2 y_j \right\}$$

$$A(X, Y) = b$$

$$X \geq 0\,, \qquad Y \geq 0$$

↓

The above is replaced by the problem

$$\min_{\lambda} \min \left(c^1 X + \lambda c^2 Y \right)$$

$$A_1 X + A_2 Y = b$$

$$dX - \lambda e Y = 0$$

$$X \geq 0\,, \qquad Y \geq 0$$

↓

Suppose we have the original basic program of the problem

$$\min \left(c^1 X + \lambda c^2 Y \right)$$

$$A_1 X + A_2 Y = b$$

$$dX - \lambda e Y = 0$$

$$X \geq 0\,, \qquad Y \geq 0$$

$$\lambda = \lambda_0$$

↓

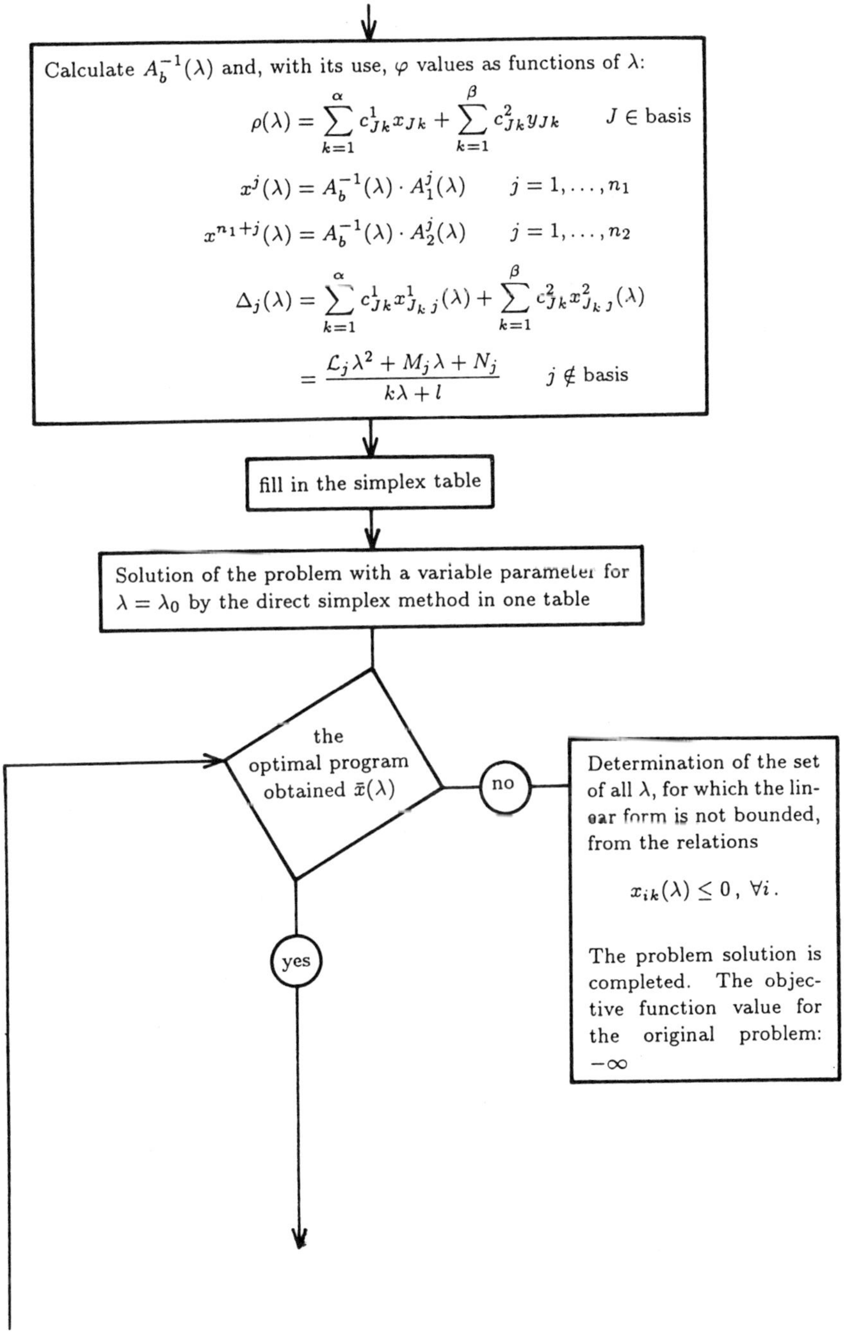
Calculate $A_b^{-1}(\lambda)$ and, with its use, φ values as functions of λ:
$\rho(\lambda) = \sum_{k=1}^{\alpha} c^1_{Jk} x_{Jk} + \sum_{k=1}^{\beta} c^2_{Jk} y_{Jk} \qquad J \in \text{basis}$
$x^j(\lambda) = A_b^{-1}(\lambda) \cdot A_1^j(\lambda) \qquad j = 1, \ldots, n_1$
$x^{n_1+j}(\lambda) = A_b^{-1}(\lambda) \cdot A_2^j(\lambda) \qquad j = 1, \ldots, n_2$
$\Delta_j(\lambda) = \sum_{k=1}^{\alpha} c^1_{Jk} x^1_{J_k j}(\lambda) + \sum_{k=1}^{\beta} c^2_{Jk} x^2_{J_k j}(\lambda) = \frac{\mathcal{L}_j \lambda^2 + M_j \lambda + N_j}{k\lambda + l} \qquad j \notin \text{basis}$
fill in the simplex table
Solution of the problem with a variable parameter for $\lambda = \lambda_0$ by the direct simplex method in one table
the optimal program obtained $\bar{x}(\lambda)$
no
Determination of the set of all λ, for which the linear form is not bounded, from the relations
$x_{ik}(\lambda) \leq 0 \,, \ \forall i \,.$
The problem solution is completed. The objective function value for the original problem: $-\infty$
yes

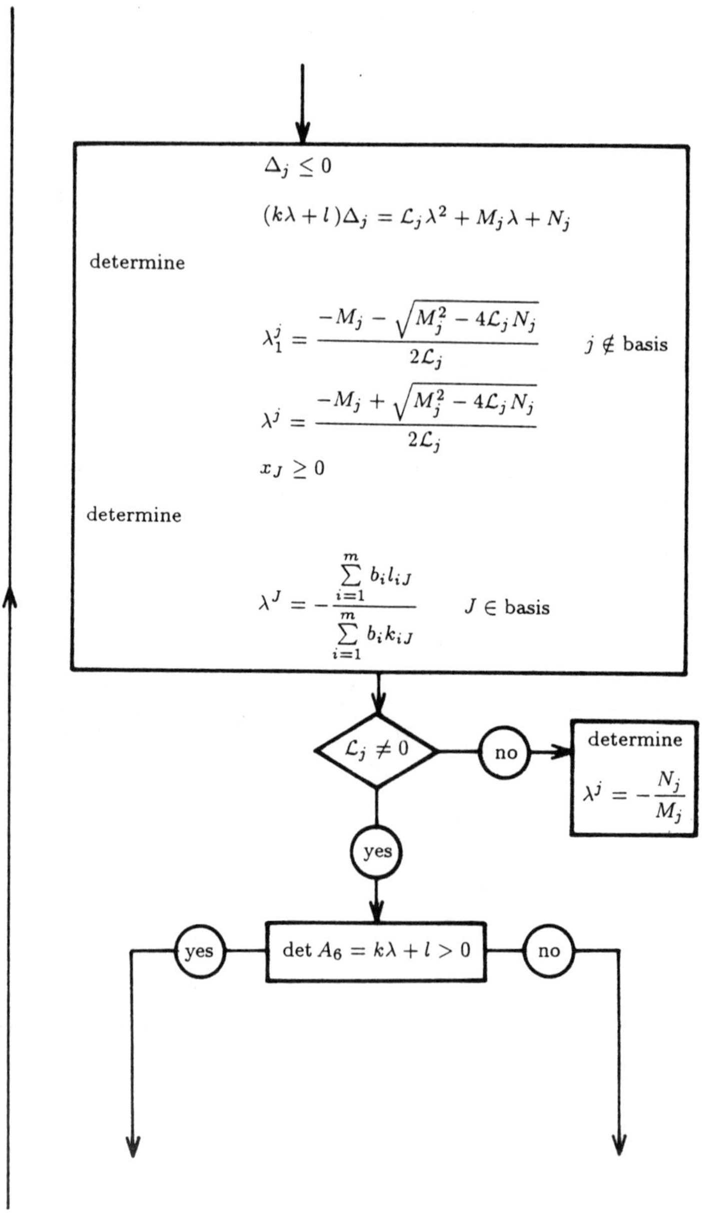
$\Delta_j \le 0$
$(k\lambda + l)\Delta_j = \mathcal{L}_j\lambda^2 + M_j\lambda + N_j$
determine
$\lambda_1^j = \dfrac{-M_j - \sqrt{M_j^2 - 4\mathcal{L}_j N_j}}{2\mathcal{L}_j}$
$j \notin$ basis
$\lambda^j = \dfrac{-M_j + \sqrt{M_j^2 - 4\mathcal{L}_j N_j}}{2\mathcal{L}_j}$
$x_J \ge 0$
determine
$\lambda^J = -\dfrac{\sum_{i=1}^m b_i l_{iJ}}{\sum_{i=1}^m b_i k_{iJ}}$
$J \in$ basis
$\mathcal{L}_j \neq 0$
no
determine
$\lambda^j = -\dfrac{N_j}{M_j}$
yes
yes
det $A_6 = k\lambda + l > 0$
no

$$\lambda_1 = \max_{\mathcal{L}_j > 0} \lambda_1^j$$

$$\lambda_2 = \min_{\mathcal{L}_j > 0} \lambda_2^j$$

If among the inequalities

$$\mathcal{L}_j \lambda^2 + M_j \lambda + N_j \leq 0$$

there are no inequalities for which $\mathcal{L}_j > 0$, then we assume $\lambda_1 = -\infty$, $\lambda_2 = +\infty$

$$\lambda_3 = \max \lambda_1^j$$

$$\lambda_4 = \min_{\mathcal{L}_j < 0} \lambda_2^j$$

If for a particular j, there are no real roots, then we assume $\mathcal{L}_j < 0$,

$$\lambda_1^j = -\infty\,,$$

$$\lambda_2^j = +\infty\,.$$

If for a particular j_0, for which $\mathcal{L}_{j_0} < 0$

$$\lambda_1^{j_0} = \lambda_2^{j_0}$$

then we assume

$$\lambda_1^{j_0} = -\infty\,,$$

$$\lambda_2^{j_0} = +\infty\,.$$

$$\lambda_5 = \max_{\sum_{i=1}^{m} b_i k_{iJ} > 0} \lambda^J$$

$$\lambda_6 = \min_{\sum_{i=1}^{m} b_i k_{iJ} < 0} \lambda^J$$

$$\lambda_7 = \max_{\substack{\mathcal{L}_j = 0 \\ M_j < 0}} \lambda^j$$

$$\lambda_8 = \min_{\substack{\mathcal{L}_j = 0 \\ M_j > 0}} \lambda^j$$

$$\lambda_1' = \min_{\mathcal{L}_j > 0} \lambda_1^j$$

$$\lambda_2' = \max_{\mathcal{L}_j > 0} \lambda_2^j$$

If for a particular j, for which $\mathcal{L}_j > 0$, there are no real roots, then we assume $\lambda_1^j = +\infty$, $\lambda_2^j = -\infty$

$$\lambda_3' = \min \lambda_1^j$$

$$\lambda_4' = \max_{\mathcal{L}_j < 0} \lambda_2^j$$

If among the inequalities

$$\mathcal{L}_j \lambda^2 + M_j \lambda + N_j \geq 0$$

there are no such λ, for which $\mathcal{L}_j < 0$, then we assume

$$\lambda_3' = +\infty\,,$$

$$\lambda_4' = -\infty\,.$$

If for a particular j_0, for which $\mathcal{L}_{j_0} > 0$

$$\lambda_1^{j_0} = \lambda_2^{j_0}$$

then we assume

$$\lambda_1^{j_0} = +\infty\,,$$

$$\lambda_2^{j_0} = -\infty\,.$$

$$\lambda_5' = \min_{\sum_{i=1}^{m} b_i k_{iJ} > 0} \lambda^J$$

$$\lambda_6' = \max_{\sum_{i=1}^{m} b_i k_{iJ} < 0} \lambda^J$$

$$\lambda_7' = \min_{\substack{\mathcal{L}_j = 0 \\ M_j < 0}} \lambda^j$$

$$\lambda_8' = \max_{\substack{\mathcal{L}_j = 0 \\ M_j > 0}} \lambda^j$$

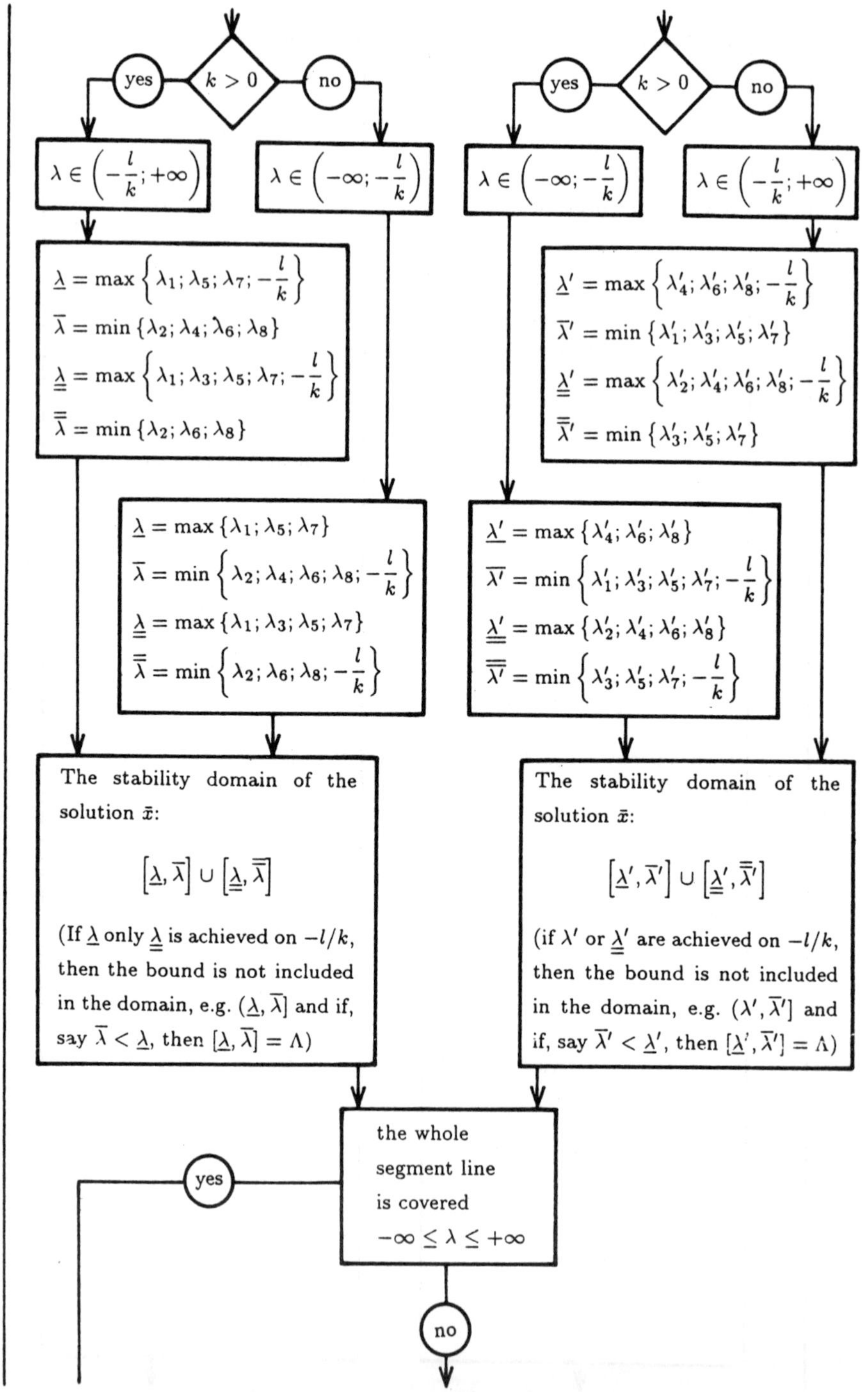

k > 0
yes
no
λ ∈ (−l/k; +∞)
λ ∈ (−∞; −l/k)
λ = max {λ1; λ5; λ7; −l/k}
λ̄ = min {λ2; λ4; λ6; λ8}
λ = max {λ1; λ3; λ5; λ7; −l/k}
λ̿ = min {λ2; λ6; λ8}
λ = max {λ1; λ5; λ7}
λ̄ = min {λ2; λ4; λ6; λ8; −l/k}
λ = max {λ1; λ3; λ5; λ7}
λ̿ = min {λ2; λ6; λ8; −l/k}
λ′ = max {λ′4; λ′6; λ′8; −l/k}
λ̄′ = min {λ′1; λ′3; λ′5; λ′7}
λ′ = max {λ′2; λ′4; λ′6; λ′8; −l/k}
λ̿′ = min {λ′3; λ′5; λ′7}
λ′ = max {λ′4; λ′6; λ′8}
λ̄′ = min {λ′1; λ′3; λ′5; λ′7; −l/k}
λ′ = max {λ′2; λ′4; λ′6; λ′8}
λ̿′ = min {λ′3; λ′5; λ′7; −l/k}
The stability domain of the solution x̄:
[λ, λ̄] ∪ [λ, λ̿]
(If λ only λ is achieved on −l/k, then the bound is not included in the domain, e.g. (λ, λ̄] and if, say λ̄ < λ, then [λ, λ̄] = Λ)
The stability domain of the solution x̄:
[λ′, λ̄′] ∪ [λ′, λ̿′]
(if λ′ or λ′ are achieved on −l/k, then the bound is not included in the domain, e.g. (λ′, λ̄′] and if, say λ̄′ < λ′, then [λ′, λ̄′] = Λ)
the whole segment line is covered −∞ ≤ λ ≤ +∞

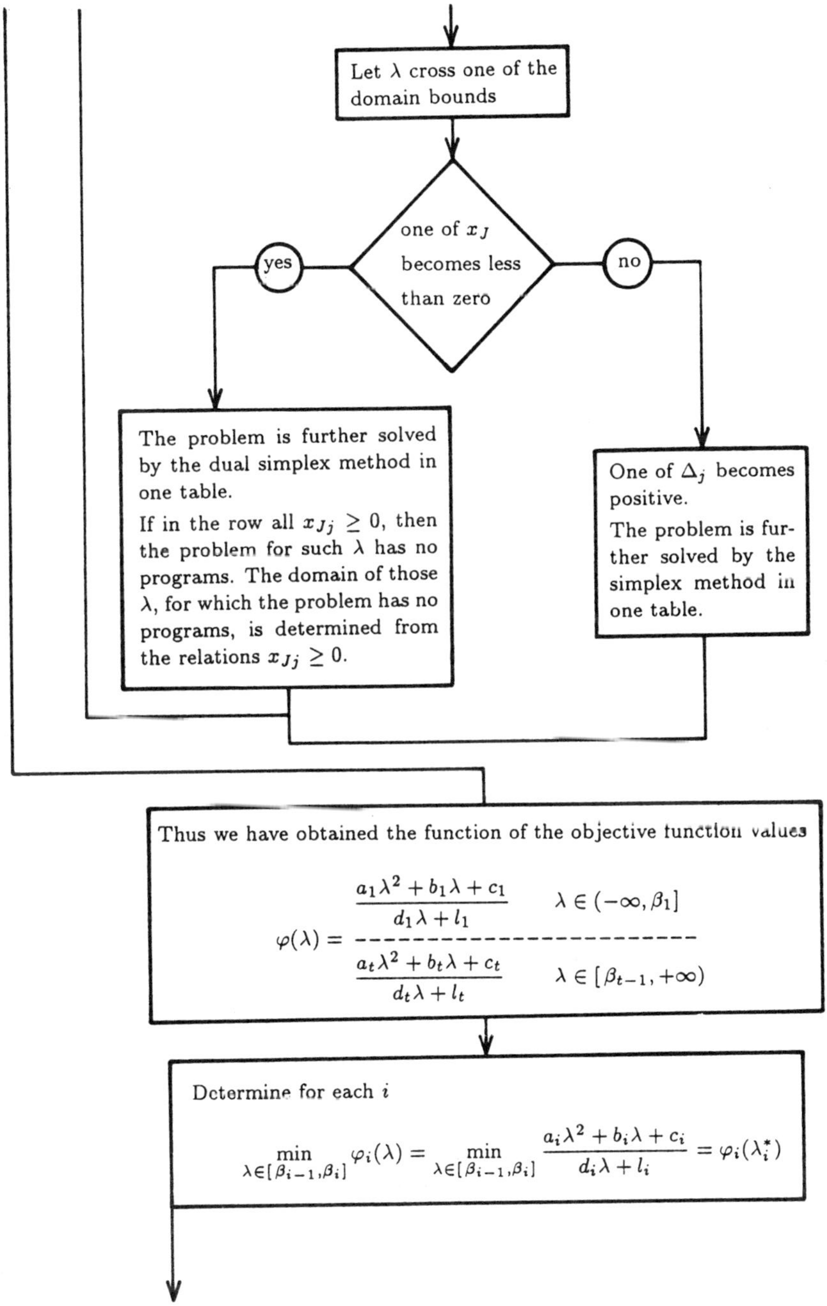
Let λ cross one of the domain bounds
one of x_J becomes less than zero
yes
no
The problem is further solved by the dual simplex method in one table.
If in the row all $x_{Jj} \geq 0$, then the problem for such λ has no programs. The domain of those λ, for which the problem has no programs, is determined from the relations $x_{Jj} \geq 0$.
One of Δ_j becomes positive.
The problem is further solved by the simplex method in one table.
Thus we have obtained the function of the objective function values
$\varphi(\lambda) = \frac{a_1\lambda^2 + b_1\lambda + c_1}{d_1\lambda + l_1} \quad \lambda \in (-\infty, \beta_1]$
$\frac{a_t\lambda^2 + b_t\lambda + c_t}{d_t\lambda + l_t} \quad \lambda \in [\beta_{t-1}, +\infty)$
Determine for each i
$\min_{\lambda\in[\beta_{i-1},\beta_i]} \varphi_i(\lambda) = \min_{\lambda\in[\beta_{i-1},\beta_i]} \frac{a_i\lambda^2 + b_i\lambda + c_i}{d_i\lambda + l_i} = \varphi_i(\lambda_i^*)$

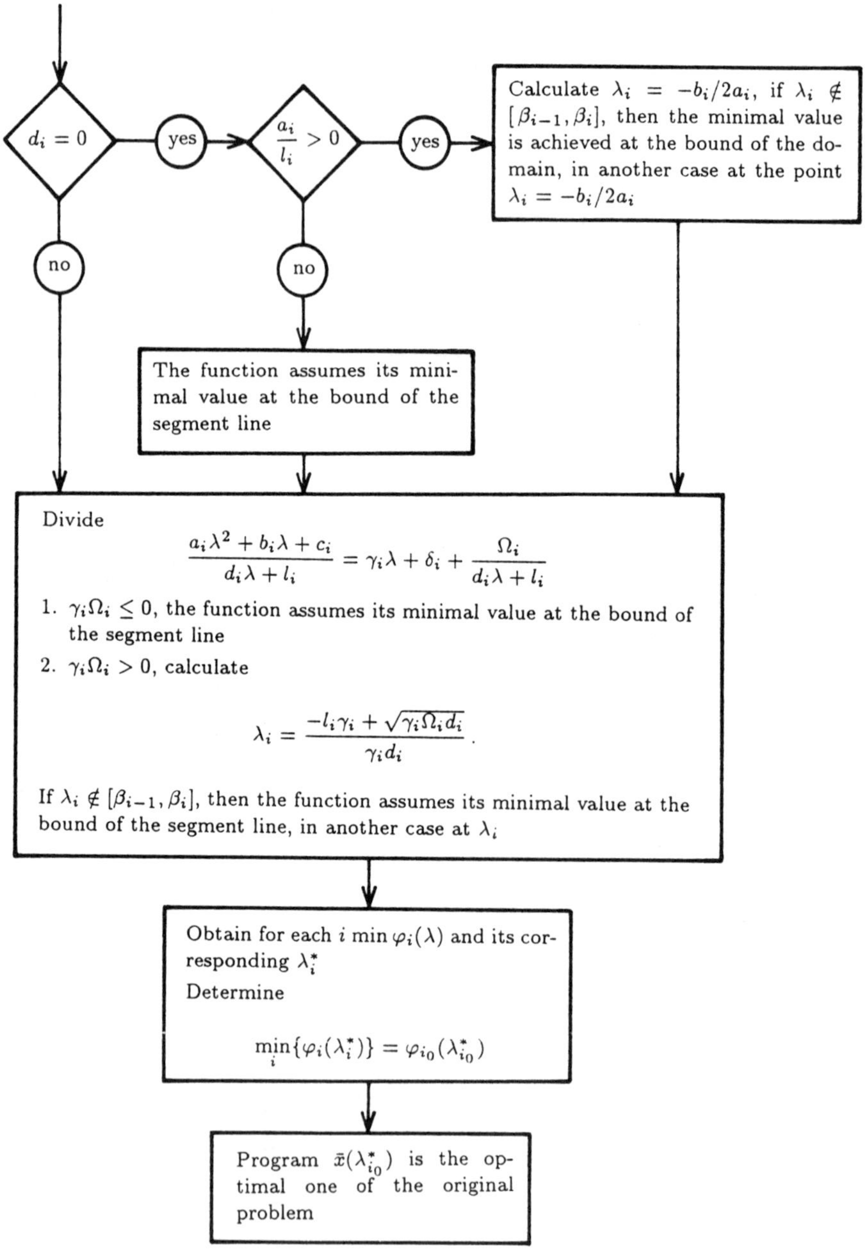

$d_i = 0$
yes
$\frac{a_i}{l_i} > 0$
yes
Calculate $\lambda_i = -b_i/2a_i$, if $\lambda_i \notin [\beta_{i-1}, \beta_i]$, then the minimal value is achieved at the bound of the domain, in another case at the point $\lambda_i = -b_i/2a_i$
no
no
The function assumes its minimal value at the bound of the segment line
Divide
$$\frac{a_i\lambda^2 + b_i\lambda + c_i}{d_i\lambda + l_i} = \gamma_i\lambda + \delta_i + \frac{\Omega_i}{d_i\lambda + l_i}$$
1. $\gamma_i\Omega_i \leq 0$, the function assumes its minimal value at the bound of the segment line
2. $\gamma_i\Omega_i > 0$, calculate
$$\lambda_i = \frac{-l_i\gamma_i + \sqrt{\gamma_i\Omega_i d_i}}{\gamma_i d_i}.$$
If $\lambda_i \notin [\beta_{i-1}, \beta_i]$, then the function assumes its minimal value at the bound of the segment line, in another case at λ_i
Obtain for each i $\min \varphi_i(\lambda)$ and its corresponding λ_i^*
Determine
$$\min_i\{\varphi_i(\lambda_i^*)\} = \varphi_{i_0}(\lambda_{i_0}^*)$$
Program $\bar{x}(\lambda_{i_0}^*)$ is the optimal one of the original problem

§35. APPLICATION OF THE ALGORITHM TO THE GRAIN FARMING OPTIMIZATION PROBLEM

§35.1. Immediate application of the above parametric programming method to the problem of grain farming optimization is possible due to the nonlinear character of constraint 72. But the inequality *can* be replaced by two inequalities (which is more exact), where one of them stands for employment of the annual fund of the operating time for storage, i.e.

$$b_{72} \geq \sum_{j=29}^{31} a_{72\,j} x_j + \sum_{j=62}^{67} a_{72\,j} x_j \,.$$

In the present equation, $a_{72\,j}$ denotes the operating time expenditures for storing 1 ton of grain for 6 months, and b_{72} the annual operating time fund for grain storage.

The second equation takes the form:

$$b_{73} \geq \sum_{j=73}^{79} a_{73\,j} x_j \,,$$

where $a_{73\,j}$ are the operating time expenditures for drying 1 ton of grain or for preserving 1 ton of grain for 1 day, and b_{73} is the daily operating time fund for drying and preservation. Evidently, it will be used at most in the last day of harvesting when the preservation capacity and the drying capacity are fully employed. The difficulty is encountered only in determining b_{72} and b_{73}, since these magnitudes separately have not so far entered into statistics.

The feasible region now is the convex polyhedron determined by 73 linear inequalities. With the objective function – maximum profit and the objective function – home production maximum, the problem was turned into a linear programming problem which can be solved by the simplex method. With the objective function – self-cost minimum, the problem was solved by the parametric programming method.

§35.2. We consider a problem in the self-cost minimum.

Denote by

$$\lambda = \frac{\sum_{j=1}^{18} d_j x_j}{\sum_{j=73}^{77} x_j},$$

$$r(\lambda) = 34s + (1-s)\frac{\sum_{j=1}^{18} d_j x_j}{\sum_{j=73}^{77} x_j} = 34s + (1-s)\lambda,$$

where $s = 0.1$.

$$\mu(\lambda) = 27t + (1-t)\frac{\sum_{j=1}^{18} d_j x_j}{\sum_{j=73}^{77} x_j} = 27t + (1-t)\lambda,$$

where $t = 0.1$. Then constraints 74 and 75 taken together with restrictions on s and t change to

$$r(\lambda) = \min\{34, \lambda\},$$
$$\mu(\lambda) = \min\{27, \lambda\}.$$

Let us formulate the problem in terms of new designations

$$\min\left\{\sum_{j=1}^{72} c_j^2 x_j + \lambda \sum_{j=73}^{78} c_j^2 x_j + r(\lambda) c_{79}^2 x_{79}\right\},$$

$$\sum_{j=1}^{79} a_{ij} x_j \geq b_i, \qquad i = 1, \dots, 73$$

(the inequalities with the sign $\leq$ are multiplied by -1)

$$\lambda = \frac{\sum_{j=1}^{18} d_j x_j}{\sum_{j=73}^{77} x_j}, \qquad r(\lambda) = \min\{34, \lambda\},$$
$$\mu(\lambda) = \min\{27, \lambda\},$$
$$x_j \geq 0, \qquad j = 1, \dots, 79.$$

The relation

$$\frac{\sum_{j=1}^{18} d_j x_j}{\sum_{j=73}^{77} x_j} = \lambda$$

can be varied according to its economic meaning only in the interval.[20,50]

§35.3. We substitute the problem with an equivalent parametric programming problem

$$\min_{20\le\lambda\le 50} \min \left\{ \sum_{j=1}^{72} c_j^2 x_j + \lambda \sum_{j=73}^{78} c_j^2 x_j + r(\lambda) c_{79}^2 x_{79} \right\},$$

$$\sum_{j=1}^{79} a_{ij} x_j \ge b_i\,, \qquad i = 1, \ldots, 73\,,$$

$$\sum_{j=1}^{18} d_j x_j - \lambda \sum_{j=73}^{77} x_j = 0\,,$$

$$r(\lambda) = \min\{34, \lambda\}\,,$$

$$\mu(\lambda) = \min\{27, \lambda\}\,,$$

$$x_j \ge 0\,, \qquad j = 1, \ldots, 79.$$

Considering that the parameters r and μ depend on λ, the process of solving the internal parametric problem has to be subdivided into three parts. We solve first the problem

$$\min \left\{ \sum_{j=1}^{72} c_j^2 x_j + \lambda \sum_{j=73}^{79} c_j^2 x_j \right\}$$

$$\sum_{j=1}^{79} a_{ij} x_j \ge b_i\,, \qquad i = 1, \ldots, 73\,, \tag{35.1}$$

$$\sum_{j=1}^{18} d_j x_j - \lambda \sum_{j=73}^{77} x_j = 0$$

$$x_j \ge 0\,, \qquad j = 1, \ldots, 79\,;$$

where $\lambda \in [20, 27]$.

As soon as λ crosses the upper bound and $\lambda \leq 34$ is satisfied we change the objective function coefficients, i.e. turn to solving the problem

$$\min\left\{\sum_{j=1}^{72} c_j^2 x_j + \lambda \sum_{j=73}^{79} c_j^2 x_j\right\}$$

$$\begin{aligned} &\sum_{j=1}^{79} a_{ij}x_j \geq b_i\,, \qquad i = 1, \ldots, 73\,, \\ &\sum_{j=1}^{18} d_j x_j - \lambda \sum_{j=73}^{77} x_j = 0 \\ &x_j \geq 0\,, \qquad j = 1, \ldots, 79. \end{aligned} \tag{35.2}$$

For this purpose, we calculate new characteristic differences of the optimal program $\bar{x}(27)$ of problem (35.1) from the formula

$$\Delta_j(\lambda) = \sum_{k=1}^{\alpha} c_{Jk}^1 x_{Jk_j}(\lambda) + \sum_{k=1}^{\beta} c_{Jk}^2 x_{Jk_j}^2(\lambda)\,,$$

where $c_{Jk}(\lambda)$ are the objective function coefficients of problem (35.2), and solve problem (35.2) until λ crosses the upper bound of the interval.[27,34] As soon as we have $\lambda > 34$, we similarly proceed to solving the parametric problem for $\lambda \in [34, 50]$

$$\min\left\{\sum_{j=1}^{72} c_j^2 x_j + \lambda \sum_{j=73}^{78} c_j^2 x_j + 34 c_j^2 x_j (j > 79)\right\},$$

$$\begin{aligned} &\sum_{j=1}^{79} a_{ij}x_j \geq b_i\,, \qquad i = 1, \ldots, 73\,, \\ &\sum_{j=1}^{18} d_j x_j - \lambda \sum_{j=73}^{77} x_j = 0 \\ &x_j \geq 0\,, \qquad j = 1, \ldots, 79. \end{aligned} \tag{35.3}$$

The regular solution of problems (35.1), (35.2), (35.3) results in the function of the objective function values $\varphi(\lambda)$. By the algorithm we define $\min_{\lambda \in [20,50]} \varphi(\lambda) = \varphi(\lambda^*)$ and thereby obtain the optimal program $x^*(\lambda^*)$ of the original problem.

It should be noted here that when solving the above problems we do not obtain as Δ_j the expressions involving λ with a degree larger than the second, though the parameter enters into the objective function with a larger number of variables than in the constraint case.

The above problem of grain farm optimization was solved experimentally for

$$\lambda = \frac{\sum\limits_{j=1}^{18} d_j x_j}{\sum\limits_{j=73}^{77} x_j} = 50$$

by the three objective functions. The constraints proved to be consistent. The obtained solutions showed that after careful treatment of matrix coefficients and constant terms the problem could give results well suited for the process of planning. As noted above, the coefficients $\bar{a}$, $\bar{\bar{a}}$, $\bar{\bar{\bar{a}}}$ and the operating time expenditures for preserving and drying grain particularly require a more exact definition. The problem has the shortcomings of a more general character. Thus, all the coefficients represent average values over the whole country. The present problem is a static model for the period of one long-range plan. The capacities obtained due to capital outlays, immediately go into operation. This is the starting point for the next study which aims at constructing a dynamic model of grain farming optimization wherein the investment freezing and changes in the coefficients of the matrix A and in the right-hand sides for the long-range plan period are included.

Chapter 9

MAJOR PROBLEMS OF MULTIOBJECTIVE OPTIMIZATION

§36. FORMULATION OF THE MULTIOBJECTIVE OPTIMIZATION PROBLEM

The multiobjective optimization problem is formulated as follows: Define an optimal choice of the element $x^0 \in X$ by a multiobjective indicator $f(x) = \{f(x)|y\}$, $y \in Y$, where X, Y are the sets of admissible elements $x \in \tilde{X}$, $y \in \tilde{Y}$ and $f \in F$ is an objective functional.

The multiobjective optimization problem amounts to the extremal problem for a sealar indicator, if the set Y consists of a unique element y^0 and an optimal choice of the element $x^0 \in X$, is derived from the condition for finding the extremum of $f(x) = f(x|y^0)$ by $x \in X$.

We call the multiobjective problem a vector optimization problem, if a set $Y = I$ consists of indices $1, 2, \ldots, m$ $(m < +\infty)$. For other cases with specified set Y, the multiobjective optimization problem will be regarded conceptually.

From the mathematical standpoint the above formulation of the multiobjective optimization problem is incorrect. Since attainment of an optimum of the multiobjective indicator $f(x)$ with $x \in X$ is impossible in that if the element $x^0 \in X$ is selected from the optimum condition of any of the components $f(x|y^0)$, for $y^0 \in Y$ of the multiobjective indicator $f(x) = \{f(x|y), y \in Y\}$, then no possibility is left for optimization (with the chosen $x^0 \in X$) of the remaining components. Therefore, formulation of the multiobjective indicator optimization problem has to be improved.

It is necessary to formulate the rule defining the concept of an optimum by a multiobjective indicator. It is customary to call a rule of that kind the principle choice of a solution to the multiobjective optimization problem.

Investigation of multiobjective optimization problems involves developing methods for specifying and incorporating determinative elements such as normalization, convolution, priority, etc., formulation of the axioms and properties satisfied by the principles of choice, consideration of concrete principles of choice and analysis of their properties, e.g. improvability, characterization, and stability. And, finally, it involves statement and development of methods for tackling multiobjective optimization problems.

Let us briefly define the main characteristics in use, proceeding from the assumption that components of the multiobjective indicator f are preset in a positive ingredient (when selecting the elements $x \in X$ their increase is desirable). Simultaneously, suppose that there exist and are finite the utilized operations of finding min and max both immediately by $x \in X$, $y \in Y$ and in all the others employed hereinafter.

Normalization will be taken as a unique representation of F in F which transforms an objective functional $f \in F$ into another element of a space F. Among the major normalization procedures, we shall indicate the following:

– Natural normalization $\mu_f(x|y)$ $\quad \dfrac{f(x|y) - \min\limits_{x \in X} f(x|y)}{\max\limits_{x \in X} f(x|y) - \min\limits_{x \in X} f(x|y)}$;

– Reduction to one dimension $\nu[\psi]$ $\quad \nu[f(x|y)] = \nu[p(y)] \times \nu[\psi]$;

– Transformation to measureless magnitudes $\quad f(x|y)/\nu[f(x|y)]$;

– Substitution of an ingredient $\quad -f(x|y); 1/f(x|y)$;

– Normalization of comparison $\quad f(x|y)/\max\limits_{x \in X} f(x|y)$;

– Savage normalization $\quad \max\limits_{x \in X} f(x|y) - f(x|y)$;

– Averaging normalization $\quad f(x|y) \Big/ \sum\limits_{y \in Y} f(x|y)$.

Convolution of components of the multiobjective indicator $f \in F$ will be termed the representation $G \in \{F \to R^1\}$ which transforms a totality of components of the multiobjective indicator f corresponding to objective terms $y \in Y$ into an objective scalar indicator $G(f(x)) = G[\{f(x|y)\}, y \in Y] \in R^1$. When tackling practical problems of multiobjective optimization we employ linear convolutions as well as convolutions of maximization, minimization, product and various Cobb–Douglas functions:

$$G(f(x)) = \sum_{y \in Y} P(y) f(x|y) ,$$

$$G(f(x)) = \min_{y \in Y} [P(y) f(x|y) + q(y)] ,$$

$$G(f(x)) = \max_{y \in Y} [P(y) f(x|y) + q(y)] ,$$

$$G(f(x)) = \prod_{y \in Y} P(y) f(x|y) ,$$

$$G(f(x)) = \prod_{y \in Y} [P(y) f(x|y)] q(y) .$$

The problems of obtaining and substantiating a choice of convolutions are studied within the theory of utility.

The study of the concept of priority in multiobjective optimization is based on comparing objective terms $y \in Y$, components $f(\cdot \,|\, y)$ of the multiobjective indicator $f \in F$ and values of $f(x) \in F$ of the multiobjective indicator $f \in F$ on elements $x \in X$. In this case it should be borne in mind that so far there has been a lack of precise definitions of priority corresponding to awareness of the essence of the multiobjective approach in the optimization theory based on a clear-cut differentiation of comparability by importance of objective terms, preferableness (significancy) of components $f(\cdot \,|\, y)$ and efficiency of values of $f(x) \in F$ with $x \in X$.

Comparison of elements z_1, z_2 of a set (space) z is regarded as fulfillment of some binary relation of order N in the sense that there exists one of the following conditions: z_1 is better than z_2 by $N(z_1 N z_2)$; z_2 is better than z_1 by $N(z_2 N z_1)$; z_1 is equivalent to z_2 by $N(z_1 \backsim z_2 = z_1 N z_2 \wedge z_2 N z_1)$; z_1 is not better than z_2 by $N(z_1 \bar{N} z_2)$; z_2 is not better than z_1 by $N(z_2 \bar{N} z_1)$; z_1 is not worse than z_2 by $N(z_1 N z_2 \vee z_1 \backsim z_2)$; z_2 is not worse than z_1 by $N(z_2 N z_1 \vee z_2 \backsim z_1)$; z_1 and z_2 are incomparable by $N(\overline{z_1 N z_2}) \wedge$

$(\overline{z_2 N z_1}) \wedge (\overline{z_1 \backsim z_2})$. The binary relation of order N is here determined by the following set of axioms:

(1) $zNz \quad \forall z \in Z$ (reflexivity);

(2) $(z_1 N z_2) \vee (z_2 N z_1) \quad \forall z_1, z_2 \in Z$ (asymmetry);

(3) $(z_1 N z_2) \wedge (z_2 N z_3) \Rightarrow z_1 N z_3$ (transitivity);

(4) $z_1 \backsim z_2 \iff (z_1 N z_2) \wedge (z_2 N z_1)$ (equivalence);

The binary relation N satisfying axioms 1, 2, 3; 2, 3 and 1, 2, 3, 4 is called, respectively, preference (a nonstrict order), regulation (a strict order) and order (a perfect order) over the elements of the set z. Hence, priority can be defined as follows:

The relation of priority over the set Y of objective terms is defined as a binary relation of order $N(Y)$ for Y and makes it possible to compare objective terms y_1, $y_2 \in Y$ by importance according to fulfillment of the condition $y_1 N(Y) y_2$, i.e. y_1 is more important than $y_2 \iff y_1 N(Y) y_2$, y_1 and y_2 are equally important $\iff y_1 \backsim y_2$ etc.

The priority relation on components $f(\cdot \,|\, y)$ of the multiobjective indicator with $y \in Y$ is defined as a binary relation of order N_f on the set $\{f(\cdot \,|\, y)\}$, $y \in Y$ and enables us to compare the components $f(\cdot \,|\, y^1)$ and $f(\cdot \,|\, y^2)$ with y_1, $y_2 \in Y$ by preference (significance) in compliance with fulfillment of the condition $f(\cdot \,|\, y_1) N_f f(\cdot \,|\, y_2)$, i.e. the y_1th component $f(\cdot \,|\, y_1)$ is preferable (more significant) as compared to y_2 – the component $f(\cdot \,|\, y_2)$ if and only if this condition is met, etc.

The priority relation on values of $f(x) \in F$ with $x \in X$ is defined as a binary relation of order $N(f)$ on the set $\mathcal{F} = \{f(x) : x \in X\}$ and enables us to compare values $f(x_1)$, $f(x_2) \in \mathcal{F}$ by efficiency (a degree of increase in f), i.e. the value $f(x_1)$ is more efficient than $f(x_2) \iff f(x_1) N(f) f(x_2)$, etc.

In practice, dealing with priority problems, only one of the possible versions of the priority relation is used due to the lack of clear-cut differentiation of importance, significance and efficiency when specifying multiobjective indicators. In so doing, it should be borne in mind that the priority relation can be defined not only as the binary relations of order $N(Y)$, N_f and $N(f)$ on Y, $\{f(\cdot \,|\, y)\}$, $y \in Y$ and $\mathcal{F}$, but as more complex relations of order both on these sets and on their direct product which, however, are induced by the initial binary orders $N(Y)$, N_f, $N(f)$, and while specifying them one should be careful because it is necessary to take account of the correlated statement of priority on the objective terms, components and values of the multiobjective indicator.

The principle of choice will be identified as the formation of the relation of order $N(f)$ on $\mathcal{F}$ with which normalization, convolutions and inclusion of the priority specification are chosen on the sets Y, $\{f(\cdot \mid y)\}$, $y \in Y$ and $\mathcal{F}$. Formulation of the principle of selecting $N(f)$ enables one to determine in what sense the solution of the problem for multiobjective optimization of the indicator $f(x)$ on the set X is understood and define a set of optimal elements. Notwithstanding that many principles of choice have been formulated, the problem of prescribing and taking account of priority on a set of objective terms and components of the multiobjective indicator, basically, has not yet been considered. Some principles of choice and their associated optimality conditions of $x^0 \in X$ are given below:

– Dominant result principle

$$\max_{y\in Y} f(x^0|y) \geq \max_{y\in Y} f(x|y) \quad \forall x \in X\,;$$

– Tradeoff principle

$$\exists p(y) \in P : \sum_{y\in Y} P(y) f(x^0|y) = \max_{x\in X} \sum_{y\in Y} P(y) f(x|y)\,;$$

– Nash principle

$$f^*(y) = 0, \quad \forall y \in Y, \quad \prod_{y\in Y} f(x^0|y) \geq \prod_{y\in Y} f(x|y), \quad \forall x \in X\,;$$

– Overall efficiency principle

$$\sum_{y\in Y} f(x^0|y) \geq \sum_{y\in Y} f(x|y)\,, \quad \forall x \in X\,;$$

– Equality principle

$$f(x^0|y') = f(x^0|y'')\,, \quad \forall y', y'' \in Y\,;$$

– Principle of Jeffry's efficiency inproper

$$\bar{\exists} x \in X : f(x|y) > f(x^0|y), \quad \forall y \in Y, \forall k > 0, \quad \exists y'' \in Y,$$

$$\exists x \in \{x \in X : f(x|y) > f(x^0|y'')\}$$

$$\frac{f(x|y'') - f(x^0|y'')}{f(x^0|y') - f(x|y')} < k, \quad \forall y' \in Y : f(x|y') < f(x^0|y');$$

– Principle of Jeffry's efficiency proper

$$\bar{\exists} x \in X : f(x|y) > f(x^0|y), \quad \forall y \in Y,$$

$$\exists k > 0 : \frac{f(x|y) - f(x^0|y)}{f(x^0|y') - f(x|y')} \leq k$$

$$\forall y \in Y, \quad \forall x \in \{x \in X : f(x|y) > f(x^0|y)\},$$

$$\exists y' \in \{y \in Y : f(x|y) < f(x^0|y)\};$$

– Slayter's principle

$$\bar{\exists} x \in X : f(x|y) > f(x^0|y), \quad \forall y \in Y;$$

– Pareto's principle N^{dom}

$$\bar{\exists} x \in X : \{f(x|y) \geq f(x^0|y) \ \ \forall y \in Y\} \wedge \{\exists y^0 \subset Y : f(x|y^0) > f(x^0|y^0)\},$$

– Partial dominance principle

$$\exists Y_1 \subseteq Y : f(x^0|y) \geq f(x|y) \quad \forall x \in X, \quad y \in Y_1;$$

– Sure result principle

$$\min_{y \in Y} f(x^0|y) \geq \min_{y \in Y} f(x|y) \quad \forall x \in X;$$

– Least bias principle

$$\|f(x^0) - f^*\| \leq \|f(x) - f^*\| \quad \forall x \in X$$

$$f^* = \{f^*(y)\}, y \in Y; \quad f^*(y) = \max_{x \in X} f(x|y) \quad \forall y \in Y;$$

– λ-criterion principle $N_\lambda(f)$

$$x^0 \in X_{\lambda^0} : \lambda^0 = \max_{0 \le \lambda \le 1} (\lambda | X_\lambda \ne \Lambda)$$

$$X_\lambda = \{x \in X : \lambda f(x|\lambda) \ge \lambda \quad \forall y \in Y\} ;$$

– Gurvitz α-criterion principle $N_\lambda^\alpha(f)$

$$x^0 \in X : \Big[\alpha \min_{y \in Y} \lambda f(x^0|y) + (1-\alpha) \max_{y \in Y} \lambda_f(x^0|y) \Big]$$
$$\ge \Big[\alpha \min_{y \in Y} \lambda f(x|y) + (1-\alpha) \min_{y \in Y} \lambda f(x|y) \Big] \quad \forall x \in X ;$$

– Uncertainty function maximum principle

$$x^0 \in X : H[f(x^0)] \ge H[f(x)] \quad \forall x \in X .$$

Define other concepts employed in solving multiobjective optimization problems.

The element $x^0 \in X$ is called Pareto-optimal (unimprovable) by the multiobjective indicator $f \in F$, if there is no $x \in X$ such that conditions $f(x|y) \ge f(x^0|y)$, $\forall y \in Y$ are met, if only for one $y' \in Y$ this inequality is strict. In such an event, $x^0 \in X$ is termed dominant (optimal according to the dominance principle) if inequalities $f(x^0|y) \ge f(x|y)$ $\forall x \in X$, $y \in Y$ are completed, i.e. the dominant element $x^0 \in X$ is Pareto-optimal.

Define a natural nonstrict order N^{dom} on $\mathcal{F}$ as an intersection $\bigcap_{y \in Y} N_Y$ of the binary relations of orders N_Y on $\{f(x|y) : x \in X\}$, $y \in Y$, i.e. $f(x_1)N^{\mathrm{dom}}f(x_2)$ with x_1, $x_2 \in X \iff f(x_1|y) \ge f(x_2|y)$, $\forall y \in Y$.

The natural nonstrict order $\hat{N}^{\mathrm{dom}}$ is called nonstrict order N^{dom}, such that $f(x_1)\hat{N}^{\mathrm{dom}}f(x_2)$, x_1, $x_2 \in X \iff f(x_1)N^{\mathrm{dom}}f(x_2)$ and there exists if only one $y' \in Y$ such that $f(x_1|y') > f(x_2|y')$. Since the relations of order are determined immediately on the values of the multiobjective indicator $f \in F$, they (N^{dom} and $\hat{N}^{\mathrm{dom}}$) are called natural. These orders are decisive in the principles of choice in the form of dominance N^{dom} and Pareto-optimality $(N^{\mathrm{dom}} \backslash \hat{N}^{\mathrm{dom}}) = N^{\mathrm{dom}} \wedge \hat{N}^{\mathrm{dom}}$.

The element $x^0 \in X$ is termed improvable by the multiobjective indicator $f \in F$ if there exists $x \in X$ such that inequalities $f(x|y) \ge f(x^0|y)$ $\forall y \in Y$ are completed, and if only for one $y' \in Y$ this inequality in strict, in other words, $x^0 \in X$ is improvable $\iff \exists x \in X : f(x)\hat{N}^{\mathrm{dom}}f(x^0)$.

Define the compromise region by $f \in F$ as follows

$$R(f) = \left\{ x^0 \in X : \sum_{y \in Y} p(y) f(x^0|y) = \max_{x \in X} \sum_{y \in Y} p(y) f(x|y) \quad \forall p(y) \in S \right\},$$

where S denotes the space of pseudoprobabilistic distributions $p = \{p(y)| y \in Y\}$ on Y such that $p(y) \geq 0$, $\forall y \in Y$, $\sum_{y \in Y} p(y) = 1$ is completed.

Definition of the tradeoff region $R(f)$ is conditioned by S. Karlin's results concerning establishment of relation between Pareto-optimality and linear convolution optimization in vector optimization problems with $Y = I = \{1, 2, \ldots, m\}$. Thus, if for some $x^0 \in X$, $p^0 = (p_1^0, \ldots, p_m^0) \in S$ the condition $\sum_{i=1}^{m} p_i^0 f_i(x_i^0) = \max_{x \in X} \sum_{i=1}^{m} p_i^0 f_i(x)$ is met, then x^0 is Pareto-optimal. If the set X is convex, the functions $f_i(x)$ are concave and $x^0 \in X$ is Pareto-optimal, then there exists $p^0 = (p_1^0, \ldots, p_m^0) \in S$, $p_i^0 > 0\,(i = \overline{1, m})$ for which the maximum condition holds for $x \in X$ of the linear form $\sum_{i=1}^{m} p_i^0 f_i(x)$.

Let us set forth the augmentation of Karlin's results obtained by V. V. Khomeniuk[7] for conditions of the conceptual formulation of the multiobjective optimization problem.

Theorem 36.1: For $x^0 \in X$ to be Pareto-optimal, it is necessary that for any $p(y) \in S$

$$\sum_{y \in Y} p(y) f(x^0|y) \geq \sum_{y \in Y} p(y) f(x|y) \quad \forall x \in X\,.$$

Theorem 36.2: For $x^0 \in X$ to be Pareto-optimal, it is necessary that there might exist $p^0(y) \in S$, for which

$$\sum_{y \in Y} p^0(y) f(x^0|y) = \max_{x \in X} \sum_{y \in Y} p^0(y) f(x|y)\,.$$

Theorem 36.3: For $x^0 \in X \backslash [R(f)]$ to be Pareto-optimal on a set $X \backslash [R(f)]$, where $[R(f)]$ is a closure of the set $R(f)$, it is sufficient that for some $p^0(y) \in S$

$$\max_{x \in X} \sum_{y \in Y} p^0(y) f(x|y) = \max_{p(y) \in S} \left[\max_{x \in X} \sum_{y \in Y} p(y) f(x|y) \right].$$

Corollary 36.1: The set of Pareto-optimal elements $x^0 \in X$ on X belongs to the tradeoff region $R(f)$.

Besides, theorem (36.3) formulates a sufficient condition for the existence of the Pareto-optimal element $x^0 \in X \backslash [R(f)]$ outside the closure of the trade-off region.

Theorem 36.4: For the set of Pareto-optimal elements $x^0 \in X$ on X to coincide with the trade-off region $R(f)$, it is necessary and sufficient that there would not exist $p^0(y) \in S$, for which the following condition is met

$$\max_{x \in X} \left[\sum_{y \in Y} p^0(y) f(x|y) \right] = \max_{p(y) \in S} \left[\max_{x \in X} \sum_{y \in Y} p(y) f(x|y) \right].$$

Consider axiomatic problems of a rational choice. As far as multiobjective optimization is concerned, notwithstanding numerous works addressing the problem, so far there has been a lack of clear and sufficiently rigorous studies on the development of the axiomatic approach. To this point there have been formulated principles for selecting the elements $f \in F$ optimal by the multiobjective indicator $x^0 \in X$, though it has been done, as a rule, without revealing the systems of axioms and properties that should be satisfied by these principles which are fulfilled. Within the multiobjective optimization theory there arise the problems remaining most commonly unsolved: what are the conditions (axioms) to be satisfied by the principle of choice as a relation of order $N(f)$? Do the principles of choice satisfying these axioms exist? What are the principles of choice to be taken as a guideline within the accepted system of axioms?

The basic systems of axioms have been stated in Arrow,[1,2] Arrow–Gurvitz,[3] V. V. Khomeniuk,[7] Milnor,[9] Nash,[10] as well as in Sen.[12,13] In this case it has been shown that these systems of axioms are satisfied accordingly by the following principles of choice in the form of specification of the relation of order $N(f)$: $N(f) = \Lambda$, $N(f) = \bigcap_{y \in Y} N_y$, $N(f) = \prod_{y \in Y} [f(\cdot \mid y) - f^*(y)]$, $N(f) = \min_{y \in Y} f(\cdot \mid y)$, $N(f) = N(\{\min_{y \in Y} f(\cdot \mid y), \max_{y \in Y} f(\cdot \mid y)\})$, where $f^*(y)$ is the status quo element of a set $\mathcal{F}^F = \{f(x) \; \forall x \in X : f \in F\}$, usually defined as $f^*(y) = \min_{x \in X} f(x|y) \; \forall y \in Y$ and $N(\{\min_{y \in Y} f(\cdot \mid y), \max_{y \in Y} f(\cdot \mid y)\})$ denotes the relation of order on the set R^2 of the pairs of real numbers (k, K)

satisfying the condition $(k_1 \geq k_2) \wedge (K_1 \geq K_2) \Rightarrow (k_1, K_2)N(\{\min_{y \in Y} f(\cdot \,|\, y), \max_{y \in Y} f(\cdot \,|\, y)\})(k_2, K_2)$.

V. V. Khomeniuk has formulated in his work[7] modification of the Arrow system of axioms satisfied by the principles of maximal efficiency by $N_\lambda(f)$ and those of parametric maximal efficiency $N_\lambda^p(f)$ with any $p \in [0, 1]$.

Arrow has formulated the following five natural axioms to be satisfied by order $N(f)$ of the principle of choice.

(1) Axiom of independence from positive linear transformation of the multiobjective indicator

$$N(f) \Longleftrightarrow N(\{[(m|y)f(\cdot \,|\, y) + n(y)]\}, y \in Y) \; \forall m(y) > 0, n(y) \in R^1 .$$

(2) Axiom of independence from the choice of elements $x \in X$, i.e. if for two multiobjective indicators

$$f_1(x) = \{f^1(x|y)\}, \quad y \in Y; \quad f_2(x) = \{f^2(x|y)\}, \quad y \in Y$$

elements x_1, $x_2 \in X$ are such that there are

$$f^1(x^1|y) \geq f^1(x^2|y), \quad f^2(x^1|y) \geq f^2(x^2|y) \; \forall y \in Y ,$$

and exist y_1, $y_2 \in Y$, for which $f^1(x^1|y^1) > f^1(x^2|y^1)$, $f^2(x^1|y^2) > f^2(x^2|y^2)$ then we get $x^1 N(f^1) x^2$, $x^1 N(f^2) x^2$.

(3) Axiom of universality of $N(f)$ by $f \in F$, i.e. the principle of choice of $N(f)$ must be defined for any multiobjective indicator $f \in F$.

(4) Axiom of the "dictator" absence: there is no $y^* \in Y$ such that

$$f(x^1|y^*) \geq f(x^2|y^*) \Rightarrow x^1 N(f) x^2 ,$$

x^1, $x^2 \in X$, i.e. there is no $y^* \in Y$ such that the condition $x^1 N(f) x^2$ with x^1, $x^2 \in X$ is determined by the y^*th component of the multiobjective indicator.

(5) Axiom of the Pareto-weakened optimality: the condition $N^{\text{dom}} = \bigcap_{y \in Y} N_y \subseteq N(f)$ is met, where N_y denotes the binary relation of order in R^1 established by the yth component of the multiobjective indicator f, i.e. if $x^0 \in X$ is optimal according to the dominance principle $f(x^0|y) \geq f(x|y) \; \forall x \in X$, $y \in Y$ then $x^0 \in X$ is optimal by $N(f) : x^0 N(f) x$, $\forall x \in X$.

The well-known Arrow paradox resides in the fact that there is no order $N(f)$ satisfying these five axioms. Naturally it produces the desire to reformulate those five axioms so that there might exist the relation of order $N(f)$ satisfying these new requirements.

Thus, $f^1(x)$, $f^2(x)$ are two multiobjective indicators of the element $x \in X$, whereas N_y^1, N_y^2 are the two relations of order in R^1 corresponding to the yth component $f^1(x)$, $f^2(x)$ with $y \in Y$, then the multiobjective indicator f^2 is termed the improvement of f^1, and f^1 the degradation of f^2, if $x^1 N_y^2 x^2$, $\forall y \in Y \Rightarrow x^1 N_y^1 x^2$, $\forall y \in Y$, i.e. if the condition $f^2(x^1|y) \geq f^2(x^2|y)$, $\forall y \in Y$ implies fulfillment of the inequalities $f^1(x^1|y) \geq f^1(x^2|y)$, $\forall y \in Y$.

The changes in axioms 1–5 suggested by Sen consist in strengthening 1, weakening 4 and abandoning 5 which becomes here a consequence of axioms 2, 3, 6 and 7, and axioms 6 and 7 are formulated as follows.

(6) Axiom of monotonocity (of positive correlation of the principle of choice): if, for elements x^1, $x^2 \in X$ the following conditions are fulfilled:
 (a) $x^1 N(f^1) x^2$ and f^2 constitute improvement of f^1, then $x^1 N(f^2) x^2$;
 (b) $x^1 \bar{N}(f^1) x^2$ and f^2 constitute degradation of f^1, then $x^1 \bar{N}(f^2) x^2$.
(7) Axiom of the free principle of choice: for any $x^1, x^2 \in X$ differing from one another there are the multiobjective indicators $f^1, f^2 \in F$ such that $x^1 N(f^1) x^2$, $x^1 \bar{N}(f^2) x^2$ exist.

In the above conditions it might be shown that axiom 5 is a consequence of axioms 2, 3, 6 and 7 and the system of Sen's axioms (axioms 2, 3, 6 and 7) holds for the principle of selecting $N(f)$ if and only if it is "trivial", i.e. if it is the dominance principle $N(f) = N^{\text{dom}} = \bigcap_{y \in Y} N_y$.

Thus, since the answer to the question about the presence of the principle of choice (with the exception of the dominance in the class of order relations was basically negative, attempts were undertaken to develop new systems of axioms and substantiate, for them, the presence of the principles of choice in the class of binary relations of order.

The system of axioms developed by Nash consists of axiom 1 (that of independence from positive linear transformation of f and f^*) and the following three axioms.

(8) Axiom of Pareto-optimality.
(9) Axiom of independence from nonexistent elements, i.e. if $X' \subseteq X$ $x^0 N(f) x$, $\forall x \in X$, $x^0 \in X'$ then $x^0 N(f) x$ $\forall x \in X'$.

(10) Axiom of symmetry $N(f) \backsim N(f_l)$, where f_l is obtained from rearrangement of components according to the operation of representation $l \in \{Y \to Y\}$.

If the element $f^*(y)$ is status quo transformed into $f_l^*(y)$ in compliance with the operation l, then this system of axioms is satisfied by the principle of choice as $N(f) = \prod_{y \in Y} [f(\cdot \,|\, y) - f^*(y)]$.

Milnor has formulated a system of the following three axioms.

(11) Axiom of incorporating the relation of order $N(f)$ into N^{dom}, i.e. $x^0 N^{\text{dom}} x^*$, x^0, $x^* \in X \Rightarrow x^0 N(f) x^*$.

(12) Axiom of independence of fulfilling the condition $x^1 N(f) x^2$ with x^1, $x^2 \in X$ from the expansion $X' \supseteq X$.

(13) Axiom of independence from the dominated elements: if for $x' \in X$ $\exists x^2 \in X : f(x^2|y) \geq f(x^1|y)\ \forall y \in Y$, then the dominated element cannot belong to the set $X^0 = \{x^0 \in X : x^0 N(f) x, \forall x \in X\}$ of the elements optimal according to the principle $N(f)$, i.e. the set X^0 does not depend on the removal of the dominated elements from X. The system of Milnor axioms is satisfied by the principle of selecting $N(f) = \min_{y \in Y} f(\cdot \,|\, y)$, axiom 8 being, basically, the requirement for the dominated elements not to belong to the set X^0 of the elements optimal according to the principle of selecting $N(f)$.
Arrow and Gurvitz advanced the system composed of axioms 10 (of symmetry) and 8 and, moreover, including the following two axioms.

(14) Axiom of independence from the removal of the recurring components $f \in F$, i.e. $N(f) \backsim N(f_M)$ where M is the compression of $Y \subseteq \tilde{Y}$ into $Y' \subseteq \tilde{Y}$ such that $Y' \subseteq Y$, $y^1 \neq y^2$, $\forall y^1$, $y^2 \in Y'$, $\forall y^1 \in Y$, $\exists y^2 \in Y' : y^1 = y^2$.

(15) Axiom of independence from the removal of the inessential and some essential elements, in other words, if $X_1 \subseteq X_2 \subseteq H = \tilde{X}$, $X_1 \cap X_2^0 \neq \Lambda$ then $X_1^0 = X_1 \cap X_2^0$, where

$$X_1^0 = \{x^0 \in X_1 : x^0 N(f) x \quad \forall x \in X_1\},$$
$$X_2^0 = \{x^0 \in X_2 : x^0 N(f) x \quad \forall x \in X_2\}.$$

The Arrow–Gurvitz system of axioms is satisfied by the principle of the guaranteed result, in which $N(f) = \min_{y \in Y} f(\cdot \,|\, y)$, i.e. $x^0 N(f) x$,

$\forall x \in X \iff \min_{y\in Y} f(x^0|y) = \max_{x\in X}\min_{y\in Y} f(x|y)$, a modified system of Arrow's axioms is composed of axioms 1–3 and the following axioms.

(16) Axiom of the "dictator" acceptability, i.e. there can exist $y^* \in Y$ such that fulfillment of the condition $x^1N(f)x^2$ with x^1, $x^2 \in X$ is determined by fulfillment of the inequality $f(x^1|y^*) \geq f(x^2|y^*)$ for the y^*th component of the multiobjective indicator f.

(17) Axiom of unimprovability of the elements optimal by $N(f)$ i.e. if for $x^* \in X$ the condition $x^*N(f)x\ \forall x \in X$ is fulfilled, then x^* is unimprovable (Pareto-optimal) on the set $X\backslash[X^0]$ where $[X^0]$ is a closure of the set $X^0 = \{x^0 \in X : x^0N(f)x\ \forall x \in X\}$.

The above system of axioms is satisfied by the maximal efficiency principle in the form of $N_\mu(f)$ and the parametric principle of maximal efficiency in the form of $N_\mu^p(f)$ with $p \in [0,1]$, and $N_\mu^p(f)$ satisfies the Arrow–Gurvitz system of axioms.

Although the principles of choice proceed from the employment of convolutions of the multiobjective indicator components, the formulated systems of axioms and the appropriate principles of choice do not touch upon the problems of definition, specification and incorporation of priority. Thus, in practical problems of multiobjective optimization one should be guided by the principle of maximal efficiency $N_\mu(f)$ and the parametrical principle of maximal efficiency $N_\mu^p(f)$.

Let us consider comparison of orders for various principles of choice.

The binary relation of order N on the elements $z \in Z$ of an arbitrary set will be referred to as extential if there exist elements z_1, $z_2 \in Z$, $z_1 \neq z_2$ such that z_1Nz_2. If the relation of order N is extential, then it will be denoted by N^{ext}. The relation of order N_1^{ext} is stronger than the relation of order N_2^{ext}, i.e. $N_2^{\text{ext}} \subseteq N_1^{\text{ext}}$ if the intersection of sets is nonempty

$$z_1 = \{z^0 \in Z : z^0N_1^{\text{ext}}z \quad \forall z \in Z\},$$
$$z_2 = \{z^0 \in Z : z^0N_2^{\text{ext}}z \quad \forall z \in Z\}$$

and if $z_1 \subseteq z_2$, i.e. there exists $N_1^{\text{ext}} \triangleright N_2^{\text{ext}} \iff N_2^{\text{ext}} \subseteq N_1^{\text{ext}} \iff \{(z_1 \cap z_2 \neq \Lambda) \wedge (z_1 \subseteq z_2)\}$.

It will be stated that the relation of order N^{ext} on z is extraextential if it is extential and the set $z^0 = \{z^0 \in Z : z^0N^{\text{ext}}z\ \forall z \in Z\} \neq \Lambda$ of the elements $z^0 \in Z$ being the best by N^{ext} is nonempty.

Since later we shall study only extraextential relations of orders the symbol ext for N^{ext} is omitted, hence $N^{\text{ext}} = N$.

Thus, the relation of order N is stronger than each N_m, $m \in R$ if $\{N \triangleright N_m\}$ and $\{z^0 \cap z_m\} \neq \Lambda$, i.e. $z^0 \subseteq z_m$ where we denote via z^0 and z_m the sets of the form

$$z_0 = \{z^0 \in Z : z^0 N z \quad \forall z \in Z\},$$

$$z_m = \{z^0 \in Z : z^0 N_m z \quad \forall z \in Z\}.$$

It will be stated that the relation of order N is stronger than a totality $\{N_m\}$, $m \in R$, i.e. $N \triangleright \{N_m\}$, $m \in R$, if $\{N \triangleright N_m\}$ $\forall m \in R$ and $z^0 \cap \{\bigcap_{m\in R} z_m\} \neq \Lambda$, $z^0 \subseteq \bigcap_{m\in R} z_m$. The totality of relations of orders $\{N_m\}$, $m \in R$ is stronger than N, i.e. $\{N_m\} \triangleright N$, $m \in R$ if $\{N_m \triangleright N\}$ $\forall m \in R$ and $\{\bigcap_{m\in R} z_m\} \cap z^0 \neq \Lambda$, i.e. if $\{\bigcap_{m\in R} z_m\} \subseteq z^0$.

The relation of order N^{dom} is extential for the prescribed X, Y, $f \in F$, i.e. there are dominant and dominated elements in X, but the problem of choice according to the multiobjective indicator arises due to the fact that the set $X^0 = \{x^0 \in X : x^0 N^{\text{dom}} x \ \forall x \bar{\in} X\}$ of the best by N^{dom} elements is generally empty. However, the relation of order $N^{\text{dom}} \backslash \hat{N}^{\text{dom}}$ close to N^{dom} is usually extraextential, since it is assumed that the multiobjective optimization problem is nondegenerate, in other words, the set of Pareto-optimal elements $x^0 \in X$ is nonempty, and the Pareto-optimal elements $x^0 \in X$ are exactly the best by $N^{\text{dom}} \backslash \hat{N}^{\text{dom}}$, because the problem of choosing the element optimal by the multiobjective indicator arises due to the fact that this set is sufficiently representative and, sometimes, coincides with X. One of the basic requirements for the principle of choice in the form of the relation of order $N(f)$ must reside in the fact that $N(f)$ would be extraextential. Moreover, it is natural to conjecture that $N(f)$ is stronger than $N^{\text{dom}} \backslash \hat{N}^{\text{dom}}$ in the sense of the given definition, specifically, $N_\mu(f) \triangleright (N^{\text{dom}} \backslash \hat{N}^{\text{dom}})$, $N_\mu^p(f) \triangleright (N^{\text{dom}} \backslash \hat{N}^{\text{dom}})$ $\forall p \in [0,1]$.

The above concept $\triangleright$ introduced on the relations of order N enables one to undertake comparison of the principles of choosing $N(f)$ as to whether there exist, for a specified principle $N(f)$, other principles of choice in which their corresponding relations of order are stronger than $N(f)$ or, conversely, $N(f)$ is stronger than they are. Furthermore, for the concrete specified pairs of the principles of choice (or sets) it might be pointed out which of them is stronger and it can be done for the purpose of finding the strongest relations of order $N(f)$ with the prescribed X, Y, f. According to the principle of choice as $N(f)$ we find for them the sets of the best elements which are assumed to be the solution of the multiobjective optimization problem.

The prescription of the principle of choice in the form of the extra extential relation of order $N(f)$ for particular X, Y, f is specified by the system of axioms (conditions) which should be satisfied by the principle and, thus, it determines the system of axioms which should be satisfied by the principle. Hence we have two problems: the primal, which is to determine for the specified relations of order $N_m(f)$ with $m \in R$, the system of axioms to be satisfied by $N(f) \triangleright \{N_m(f)\}$, $m \in R$, and how this system of axioms shall agree with the system of axioms for $\{N_m(f)\}$, $m \in R$, and the inverse, which is to determine for the given $\{N_m(f)\}$, $m \in R$, the system to be satisfied by $N(f) : \{N_m(f)\}_{m \in R} \triangleright N(f)$ and how this system shall agree with the systems of axioms for $\{N_m(f)\}$, $m \in R$.

So far the relation of order, $N(f)$, for different principles of choice has been determined by choosing a convolution $c_f(x) = \{c_f(x|y)\}$, $y \in Y$, with $x \in X$ from the space of mapping the set X into R^1, i.e. $c_f \in \{X \to R^1\}$, where $c_f(x)$ is a value of the functional c_f on the element $x \in X$. In this case, for the prescribed convolution, c_f, the relation of order $N_c(f) = N(c_f)$ is determined by a condition $x^1 N_c(f) x^2 | x^1, x^2 \in X \iff c_f(x^1) \geq c_f(x^2)$.

Hence prescribing a totality of the relations of order (principles of choice) $\{N_m(f)\}$, $m \in R$ is equivalent to prescribing a totality of functionals $\{c_f^m\}$, $m \in R$ and permits the definition of a linear shell $\mathcal{L}_0$ and a linear space $\mathcal{L}$ of the relations of order $N_c^p(f)$ with $p = \{p_m\}$, $m \in R$ as follows: $\mathcal{L}_0 = \{N_c^p(f)\}$, $p \in S(R)$, $\mathcal{L} = \{N_c^p(f)\}$, $p \in \tilde{S}(R)$ where $\tilde{S}(R)$ is a space of pseudoprobabilistic distributions on R

$$\tilde{S}(R) = \{p = \{p_m\}, m \in R : p_m \geq 0 \quad \forall m \in R\},$$

$$S(R) = \left\{ p \in \tilde{S}(R) : \sum_{m \in R} p_m = 1 \right\}.$$

The relation of order $N_c^p(f)$ with $p \in \tilde{S}(R)$ is derived by $c_f^p = \sum_{m \in R} p_m c_f^m$ in the form of $N_c^p(f) = \sum_{m \in R} p_m N_m(f)$, thus: $x^1 N_c^p(f) x^2 | x^1, x^2 \in X \iff c_f^p(x^1) \geq c_f^p(x^2)$.

Thus, by the systems of axioms A_m for the principles of choice $N_m(f)$ with $m \in R$, it is possible to define the systems of axioms A_c^p for the principles of choice in the form of $N_c^p(f)$ with $p \in \tilde{S}(R)$ as well as the more general systems $A(\mathcal{L}_0)$ and $A(\mathcal{L})$ as a totality of the systems of axioms $\{A_c^p\}$, $p \in S(R)$ and $\{A_c^p\}$, $p \in \tilde{S}(R)$ respectively. It is the choice $p \in S(R)$ or $p \in \tilde{S}(R) \backslash S(R)$ that forms concrete relations of order $N_c^p(f) \in \mathcal{L}_0$ and $N_c^p(f) \in \mathcal{L} \backslash \mathcal{L}_0$. Using the above definition of $\triangleright$ as the base, it permits revealing whether with a specified $p \in S(R)$ or $p \in \tilde{S}(R)$ the relation of order in the principle of choice $N_c^p(f)$ is stronger than $\{N_m(f)\}$, $m \in R$ and for which $p \in S(R)$ or $p \in \tilde{S}(R)$ the corresponding $N_c^p(f) \triangleright \{N_m(f)\}$, $m \in R$. Thus, following this way we can seek the solution of the stated primal and reverse problems. It is also possible to suggest other approaches to investigation of the above problems.

§37. THE STUDY OF THE MAXIMAL EFFICIENCY PRINCIPLE

Proceeding from the conceptual formulation of the multiobjective optimization problem, suppose that for any $y \in Y$ the values $f^0(y) = \min_{x \in X} f(x|y)$, $f^*(y) = \max_{x \in X} f(x|y)$ are finite. Employ natural normalization to eliminate different scales for measuring the components of the multiobjective indicator $f(x) = \{f(x|y)\}$, $y \in Y$, thus we have $\mu_f(x|y) = [f(x|y) - f^0(y)]/[f^*(y) - f^0(y)]$ where the inequalities $0 \leq \mu_f(x|y) \leq 1$, $\forall x \in X$, $y \in Y$ exist.

The value $\mu_f(x|y)$ with the fixed $y \in Y$ represents the degree of attaining an optimum, specifically, a maximum with positive ingredients of f at the point $x \in X$ by the yth component of the multiobjective indicator f. Let X_μ be a set of elements of the form $X_\mu = \{x : X : \mu_f(x|y) \geq \mu \forall y \in Y\}$, where μ is a prescribed value, $0 \leq \mu \leq 1$.

In this case, the choice of a particular level μ provides for the nonempty set X_μ the degree of attaining an optimum by any one of the multiobjective indicator components being not less than μ. It might be natural to define the maximal efficiency problem as the problem of obtaining a maximally possible level μ^0, i.e. we have to derive the solution of the μ-problem of the form $\mu^0 = \max_{0 \leq \mu \leq 1} (\mu | X_\mu \neq \Lambda)$.

Identify the element $x^0 \in X$ as optimal according to the maximal efficiency principle, if $x^0 \in X_{\mu^0}$, i.e. the conditions $\mu_f(x^0|y) \geq \mu^0$, $\forall y \in Y$ are fulfilled. The value of μ^0 is simultaneously the best possible approximation to an optimum by all components of the multiobjective indicator (the priority on components f is not specified).

In vector optimization, one of the well-known principles of choice is the principle of obtaining a guaranteed result or the maximin principle the essence of which, as far as the formulation closest to the maximal efficiency principle is concerned, resides in deriving the optimal element $x^0 \in X$ from the condition

$$\min_{y \in Y} \mu_f(x^0|y) = \max_{x \in X}\left[\min_{y \in Y} \mu_f(x|y)\right].$$

Thus, we have the following statements.

Theorem 37.1: The maximal efficiency principle coincides with the maximin principle, thus $\mu^0 = \max_{x \in X}\left[\min_{y \in Y} \mu_f(x|y)\right]$.

Lemma 37.1: We have the inequality

$$\mu^0 \geq \max_{x \in X}\left[\min_{y \in Y} \mu_f(x|y)\right].$$

Lemma 37.2: The inequality is fulfilled as

$$\max_{x \in X}\left[\min_{y \in Y} \mu_f(x|y)\right] \geq \mu^0.$$

The maximal efficiency principle may be considered, on one hand, as a particular interpretation of the guaranteed result principle (maximin), and moreover as an independent principle of choice, being more "handy" as compared to maximin, wherein the game problem of optimization is stated and solved, whereas the μ-problem is the extremal problem for optimization of deriving μ^0, $x^0 \in X$ from the condition

$$\mu^0 = \max_{\substack{0 \leq \mu \leq 1 \\ x \in X}} \{\mu | \mu_f(x|y) \geq \mu \quad \forall y \in Y\}.$$

Test fulfillment of the axioms of choice for the maximal efficiency principle. Thus, we have

Lemma 37.3: The degree of attaining an optimum $\mu_f(x|y)$ at the point $x \in X$ by the yth component of the multiobjective indicator is independent of the positive linear transformation for any $y \in Y$,

$$\mu_{[m(y)f+n(y)]}(x|y) = \mu_f(x|y) \ \forall m(y) > 0, \quad n(y) \in R^1.$$

Theorem 37.6: The maximal efficiency principle satisfies the axiom 1 of independence of the principle of choice from the positive linear transformation of the multiobjective indicator $f \in F$.

Lemma 37.4: The degree of success in attaining an optimum of $\mu_f(x|y)$ is independent of choosing the elements $x \in X$, i.e. if for two multiobjective indicators f^1, $f^2 \in F$ the elements x^1, $x^2 \in X$ are such that for any $y \in Y$ the inequalities $f^1(x^1|y) \geq f^1(x^2|y)$; $f^2(x^1|y) \geq f^2(x^2|y)$ are fulfilled and there exist y^1, $y^2 \in Y$, for which the inequalities are strict, then for all $y \in Y$

$$\mu_{f^1}(x^1|y) \geq \mu_{f^1}(x^2|y) \Longleftrightarrow \mu_{f^2}(x^1|y) \geq \mu_{f^2}(x^2|y)\,.$$

Theorem 37.7: The relation of order $N_\mu(f)$ for the maximal efficiency principle satisfies axiom 2 of independence of the principle of choice from the choice of elements $x \in X$, i.e. if for two multiobjective indicators f^1, $f^2 \in F$ and the elements x^1, $x^2 \in X$ the inequalities $f^1(x^1|y) \geq f^1(x^2|y)$, $f^2(x^1|y) \geq f^2(x^2|y)$ $\forall y \in Y$ are fulfilled, then there is $x^1 N_\mu(f^1) x^2 \Longleftrightarrow x^1 N_\mu(f^2) x^2$.

Theorem 37.8: The relation of order $N_\mu(f)$ for the maximal efficiency principle satisfies axiom 3 of universality of the principle of choice by $f \in F$.

Theorem 37.9: The relation of order $N_\mu(f)$ for the maximal efficiency principle may not satisfy axiom 4 of the "dictator" absence for the principle of choice, i.e. there are such multiobjective indicators $f^* \in F$ that $y^* \in Y$ is defined for which the fulfillment of $x^1 N_\mu(f^*) x^2$ is determined by the y^*th component of the multiobjective indicator f^*:

$$f^*(x^1|y^*) \geq f^*(x^2|y^*) \Longrightarrow x^1 N_\mu(f^*) x^2, \quad x^1, x^2 \in X\,.$$

Theorem 37.10: The relation of order $N_\mu(f)$ for the maximal efficiency principle satisfies axiom 5 of the weakened Pareto-optimality principle of choice.

Theorem 37.11: The relation of order $N_\mu(f)$ for the maximal efficiency principle satisfies axiom 7 of the principle of free choice.

Theorem 37.12: The relation of order $N_\mu(f)$ for the maximal efficiency principle satisfies axiom 16 of the "dictator" acceptability for the principle of choice.

Lemma 37.5: The element $x^0 \in X$ is improvable according to the multiobjective indicator $f \in F$ (there exists $x^* \in X$ such that $f(x^*|y) \geq f(x^0|y)\ \forall y \in Y$ if only one of the inequalities is strict) if and only if there exists $\lambda^0(y) \in \{Y \to R^1\}$ and the inequalities $\mu_f(x^0|y) \leq a_{\lambda^0}(y), \mu_f(x^0|y) < a_{\lambda^0}|y^*)$ are fulfilled for all $y \in Y$ and if only for one $y^* \in Y$, where $a_{\lambda^0}(y) = a_{\lambda^0} - \lambda^0(y)$,

$$a_{\lambda^0} = \max_{x \in X} \min_{y \in Y} [\,\mu_f(x|y) + \lambda^0(y)]\,.$$

Theorem 37.13: The maximal efficiency principle satisfies axiom 17 of unimprovability (Pareto-optimality) of the element $x^0 \in X$ optimal for the relation of order $N_\mu(f)$, i.e. if $x^0 \in X$ satisfies the conditions $x^0 N_\mu(f) x\ \forall x \in X \iff \min_{y \in Y} \mu_f(x^0|y) = \max_{x \in X} \min_{y \in Y} \mu_f(x|y)$ then there is no $x^* \in X \backslash X^0$ such that $f(x^*|y) \geq f(x^0|y)\ \forall y \in Y$ and if only one of the inequalities is strict.

The proofs for the above statements may be found in the author's work.[8]

The above statements enable us to establish that the relation of order $N_\mu(f)$ for the maximal efficiency principle satisfies axioms 1–3, and axiom 4 may not be fulfilled for some multiobjective indicators $f \in F$. The presence of Arrow's paradox and the result that with the strengthening of 6 of axiom 1 and the weakening of 7 of axiom 4 only the dominance principle satisfies axioms 2, 3, 5–7, and also the fulfillment of axioms 1–3, 5 for the maximal efficiency principle shows that when formulating a consistent system of axioms on a basis of 1–3, 5 one should modify axiom 4 by axiom 7 without strengthening axiom 1 and rejecting axiom 5. In this version the system of axioms 1–3, 5, 7 is satisfied by the relation of order $N_\mu(f)$ for the maximal efficiency principle.

It is to be noted that axiom 7 apparently is "excessive" in the sense of weakening axiom 4, and the assumption of axiom 7 leaves open the question of whether to give up completely the possibility of a "dictator" (to demand its absence as in axiom 4) or to assume its possibility in some cases. The answer to this question is partly contained in the statement of theorem (37.12).

Consider the properties of the maximal efficiency principle.

(A) Linearity of the principle of choice: the relation of order $N(f)$ may be presented as a linear combination of the multiobjective indicator components. The condition $\exists p(y) \in P : N(f) = \sum_{y\in Y} p(y)f(\cdot\,|\,y)$ is valid, and here P denotes a set $P = \{p(y) \in \{Y \to R^1\} : p(y) \geq 0, \sum_{y\in Y} p(y) = 1\}$. If Y is connected, then the sum by $y \in Y$ is taken as an integral.

(B) The membership of the region of the optimal compromise according to the principle of choosing an element. Thus we have $x^0 N(f)x \ \forall x \in X \Longrightarrow x^0 \in R$, where the compromise region R is defined as follows:

$$R = \{x^* \in X : \exists p^*(y) \in P\,,$$
$$\sum_{y\subset Y} p^*(y)f(x^*|y) = \max_{x\in X} \sum_{y\in Y} p^*(y)f(x|y)\,.$$

(C) Weak dominance of the optimal elements $x^0 \in X : x^0 N(f)x \ \forall x \in X$ according to the principle of choice, even though there is another principle of choice $N^*(f)$ such that for a particular $x^* \in X$, $x^* N^*(f)x \ \forall x \in X$. Here the conditions $x^0 N(f)x^*$, $x^* N^*(f)x^0$ are valid.

We have the following statements.

Theorem 37.14: The maximal efficiency principle satisfies the properties A, B, C, the optimal elements $x^0 \in X_{\mu^0}$ being dominant for $N_\mu(f)$ with respect to the optimal elements $x^* \in X$ according to any principle of choice $N(f)$ as distinct from $N_\mu(f)$ where

$$N_\mu(f) = \min_{y\in Y} \mu_f(\cdot\,|\,y)\,.$$

Lemma 37.6: If the components $f(x|y)$ of the multiobjective indicator are concave by $x \in X$ with $y \in Y$, then

$$\max_{x\in X} \min_{p(y)\in P} \sum_{y\in Y} p(y)\mu_f(x|y) = \min_{p(y)\in P} \max_{x\in X} \sum_{y\in Y} p(y)\mu_f(x|y)\,;$$

$$\max_{x\in X} \min_{y\in Y} \mu_f(x|y) = \min_{y\in Y} \max_{x\in X} \mu_f(x|y)\,,$$

i.e. there exists a saddle point $(x^0, y^0) \in X \times Y$ for $\mu_f(x|y)$, x^0 being an optimal element according to the maximal efficiency principle.

Using the maximal efficiency principle as the base, consider the characterization issues of the multiobjective optimization problem.

The characterization will be referred to as the parametric partition X_τ, Y_τ with $\tau \in T \subseteq R^1$ of the sets X, Y in the optimization problem of the multiobjective indicator $f(x) = \{f(x|y)\}$, $y \in Y$ when choosing the elements $x \in X$ so that with an increase of $\tau \in T$ there might occur a monotone increase (or, conversely, a decrease) in the fulfillment of a particular "property" of the elements $x \in X$ and multiobjective indicators $f_\tau(x) = \{f(x|y)\}$, $y \in Y_\tau$ with $x \in X_\tau$, or X_τ and $f_\tau(x)$ exhibit some "properties" differing from the properties $X_{\tau'}$, $f_{\tau'}(x)$ with $\tau \neq \tau'$, T now being unnecessarily from R^1.

The form of Bayesian sets represents one of the very first characterizations which has, comparatively speaking, long been utilized for decision making under uncertainty. In the adopted designations the form of Bayesian sets appears as $T = X$, $\tau = x$, $X_\tau = \{x\}$, $x \in X$, $Y_\tau = Y_\tau(x) = \{y' \in Y : f(x|y') \geq \max_{y \in Y} f(x|y)\}$, $f_\tau(x) = \{f(x|y)\}$, $y \in Y_\tau(x)$, where $Y_\tau(x)$ is a set of dominant objective terms at the point $\tau = x$ and $f_\tau(x)$ is a value of the multiobjective indicator $f(x)$ after excluding the dominated components at the point $\tau = x \in X$.

Based on the maximal efficiency principle, characterization of the multiobjective optimization problem is accomplished in the following way. Let $\tau = 0$, $\mu_{\tau=0} = 0$, $X_{\tau=0} = \Lambda$, $Y_{\tau=0} = \Lambda$, $\bar{X}_{\tau=0} = X$, $\bar{Y}_{\tau=0} = Y$, $f_{\tau=0} = f$.

(1) obtain μ_1^0 from the solution condition for the μ-problem in the form

$$\mu_1^0 = \max_{\substack{0 \leq \mu \leq 1 \\ x \in X}} [\mu | \mu_f(x|y) \geq \mu \quad \forall y \in Y] = \max_{x \in X} \left[\min_{y \in Y} \mu_f(x|y) \right],$$

suppose $\tau = 1$, $\mu_{\tau=1} = \mu_1^0$, thus we have

$$X_{\tau=1} = X_1 = \{x \in X : \mu_f(x|y) \geq \mu_1^0 \quad \forall y \in Y\}, \quad \bar{X}_1 = X \backslash X_1,$$

$$Y_{\tau=1} = Y_1 = \{y \in Y : \mu_f(x|y) = \mu_1^0 \quad \forall x \in X_1\}, \quad \bar{Y}_1 = Y \backslash Y_1,$$

$$f_{\tau=1}(x) = f_1(x) = \{f(x|y)\}, \quad y \in \bar{Y}_1, \quad x \in X;$$

(2) obtain μ_2^0 from the condition

$$\mu_2^0 = \max_{\substack{0 \leq \mu \leq 1 \\ x \in X}} [\mu | \mu_{f_1}(x|y) \geq \mu \quad \forall y \in \bar{Y}_1] = \max_{x \in X} \left[\min_{y \in \bar{Y}} \mu_{f_1}(x|y) \right],$$

suppose $\tau = 2$, $\mu_{\tau=2} = \mu_2^0$, thus we have

$$X_{\tau=2} = X_2 = \{x \in X : \mu_{f_1}(x|y) \geq \mu_2^0 \quad \forall y \in \bar{Y}_1\}, \quad \bar{X}_2 = \bar{X}_1 \backslash X_2,$$
$$Y_{\tau=2} = Y_2 = \{y \in Y : \mu_{f_1}(x|y) = \lambda_2^0 \quad \forall x \in X_2\}, \quad \bar{Y}_2 = \bar{Y}_1 \backslash Y_2,$$
$$f_{\tau=2}(x) = f_2(x) = \{f(x|y)\}, \quad y \in \bar{Y}_2, \quad x \in X.$$

$\mu_2^0 \geq \mu_1^0 \geq \mu_0^0 = 0$ is valid.

We continue this procedure in a similar way. Suppose that for some positive whole number $\tau \in R^1$ there have been derived μ_τ^0, X_τ, Y_τ, $\bar{X}_\tau = \bar{X}_{\tau-1} \backslash X_\tau$, $\bar{Y}_\tau = \bar{Y}_{\tau-1} \backslash Y_\tau$, $f_\tau(\cdot) = \{f(\cdot\,|\,y)\}$, $y \in \bar{Y}_\tau$, then obtain $\mu_{\tau+1}^0$ from the condition

$$\mu_{\tau+1}^0 = \max_{\substack{0 \leq \mu \leq 1 \\ x \in X}} \{\mu | \mu_{f_\tau}(x|y) \geq \mu \quad \forall y \in \bar{Y}_\tau\} = \max_{x \in X} \left[\min_{y \in \bar{Y}_\tau} \mu_{f_\tau}(x|y) \right],$$
$$\mu_{\tau+1} = \mu_{\tau+1}^0$$

assume that

$$X_{\tau+1} = \{x \in X : \mu_{f_\tau}(x|y) \geq \mu_{\tau+1}^0 \quad \forall y \in \bar{Y}_\tau\}, \quad \bar{X}_{\tau+1} = \bar{X}_\tau \backslash X_{\tau+1},$$
$$Y_{\tau+1} = \{y \in \bar{Y}_\tau : \mu_{f_\tau}(x|y) = \mu_{\tau+1}^0 \quad \forall x \in X_{\tau+1}\}, \quad \bar{Y}_{\tau+1} = \bar{Y}_\tau \backslash Y_{\tau+1},$$
$$f_{\tau+1}(x) = \{f(x|y)\}, \quad y \in \bar{Y}_{\tau+1}, \quad x \in X,$$

where $\mu_{\tau+1}^0 \geq \mu_\tau^0 \geq \mu_{\tau-1}^0 \geq \cdots \geq \mu_2^0 \geq \mu_1^0 \geq \mu_0^0 = 0$ etc.

As a result, with $\tau \in T = \{0, 1, 2, \ldots,\} \subset R^1$ we can obtain characterization of the optimization problem of the multiobjective indicator $f(x) = \{f(x|y)\}$, $y \in Y$ by $x \in X$ in the form of a sequence of characteristic numbers μ_τ^0 of the sets X_τ, $\bar{X}_\tau$ of elements from X, the sets Y_τ, $\bar{Y}_\tau$ of objective terms from Y, the multiobjective indicators $f_\tau \in F$. Thus we have

Theorem 37.15: The suggested procedure of characterizing the multiobjective optimization problem satisfies the following conditions:

(a) monotonocity of the characteristic numbers μ_τ^0 with $\tau \in T$ increasing (decreasing), i.e. $\mu_{\tau-1}^0 \leq \mu_\tau^0 \leq \mu_{\tau+1}^0$;
(b) monotonocity of the characteristic sets $\bar{Y}_\tau$ of objective terms from Y, i.e. $\bar{Y}_{\tau+1} \subseteq \bar{Y}_\tau \subseteq \bar{Y}_{\tau-1}$ with $\tau \in T$;

(c) monotonocity of the characteristic multiobjective indicators $f_\tau \in F$ with $\tau \in T$, i.e.

$$f_\tau(\cdot) = \{f(\cdot\,|\,y)\}\,, y \in \bar{Y}_\tau = \{f_{\tau-1}(\cdot)\}\backslash\{f(\cdot\,|\,y)\}\,, \quad y \in Y_\tau$$

therefore, f_τ is obtained from $f_{\tau-1}$ by excluding the components $y \in Y_\tau$;

(d) monotonocity of the characteristic sets X_τ of elements from X, i.e. we are "guaranteed" that the lowest degree of attaining the optimum of μ_τ^0 on the elements $x \in X_\tau$ increases (decreases) as $\tau \in T$ increases (decreases).

The proof of the theorem follows from the definition of the concepts utilized in the characterization procedure.

The characterization process in the case of the problem of vector optimization of $Y = I = \{1, 2, \ldots, m\}$ (the characterization technique based on the maximal efficiency principle), here the set T contains not more than $(m + 1)$ of the element and has a finite number of stages.

The studies of the natural conditions of the choice in the form of axioms 1–5 and their modifications show that relaxation of axiom 4 down to the possibilities of the "dictator" presence in the principle of choice results in solving Arrow's paradox, as a theorem of impossibility, using the proof of validity of these axioms for the maximal efficiency principle as the base.

§38. THE STUDY OF THE PARAMETRIC PRINCIPLE OF MAXIMAL EFFICIENCY IN MULTIOBJECTIVE OPTIMIZATION

In a sense, the maximal efficiency principle is "conservative": even a drastic increase in the majority of the multiobjective indicator components may not involve an increase of the degree, optimal by this principle, of $\mu^0 = \max\limits_{x \in X} \min\limits_{y \in Y} \mu_f(x|y)$ of achieving an optimum. In this respect, the critique of the guaranteed result principle in the theory of decision making under uncertainty gave rise to formulation of the Gurvitz parametric criterion. As far as the problem of optimization by the multiobjective indicator is concerned the parametric principle of maximal efficiency is formulated as follows: for a specified $p \in [0, 1]$ we need to obtain $x_p^0 \in X$ from a condition

$\mu_p^0 = p \min_{y \in Y} \mu_f(x_p^0|y) + (1-p) \max_{y \in Y} \mu_f(x_p^0|y) = \max_{x \in X} \Big[p \min_{y \in Y} \mu_f(x|y) + (1 - p) \max_{y \in Y} \mu_f(x|y) \Big]$, where p is a specified level of "weighing" the minimal and maximal possible degrees of achieving an optimum by the multiobjective indicator components.

In terms of the principle of choice the parametric principle of maximal average efficiency leads to the following relation of order: $N_\mu^p(f) = p \min_{y \in Y} \mu_f(\cdot \,|\, y) + (1-p) \max_{y \in Y} \mu_f(\cdot \,|\, y)$ moreover, for the optimal $x_p^0 \in X$ we have $x_p^0 N_\mu^p(f) x \; \forall x \in X$.

With $p = 1$ we have the maximal efficiency principle, whereas with $p = 0$ we have the dominant efficiency principle for which a set of optimal elements is defined, respectively, in the form

$$X_{p=1}^0 = \Big\{ x^0 \in X : \min_{y \in Y} \mu_f(x^0|y) = \max_{x \in X} \Big[\min_{y \in Y} \mu_f(x|y) \Big] \Big\},$$

$$X_{p=0}^0 = \Big\{ x^0 \in X : \max_{y \in Y} \mu_f(x^0|y) = \max_{x \in X} \Big[\max_{y \in Y} \mu_f(x|y) \Big] \Big\}.$$

For the dominant efficiency principle a set of optimal elements of $X_{p=0}^0$ coincides with a set of optimal elements for a dominant component of μ_f (or f) of the multiobjective indicator, thus we have

$$X_{p=0}^0 = \{ x^0 \in X : \exists y^0 \in Y | \, \mu_f(x^0|y^0) = \max_{x \in X} \mu_f(x|y^0) = 1 , \\ \mu_f(x^0|y^0) \geq \mu_f(x^0|y) \quad \forall y \in Y \} .$$

Investigation into concrete problems of multiobjective optimization, as a rule, starts with finding a region $X^0(y) = \{x^0 \in X : f(x^0|y) \geq f(x|y) \; \forall x \in X\}$ based on the solution of extremal problems for each of the components $f(x|y)$ of the multiobjective indicator $f(x)$ corresponding to the objective term $y \in Y$. In most cases the multiobjective indicator components are "inconsistent" (e.g. the minimal values of some components are achieved with the maximal values of the others), hence, it is natural that investigations of multiobjective optimization problems do not terminate in deriving $X^0(y)$, $y \in Y$. The values of the expression $\mu_f^p(x) = p \min_{y \in Y} \mu_f(x|y) + (1-p) \max_{y \in Y} \mu_f(x|y)$ for the fixed $x \in X$ with $p \in [0,1]$ represent the interval $\mathcal{J}_f^p(x) = [\mu_f^1(x), \mu_f^0(x)]$ of the mathematical expectation values of $\mu_f^p(x)$ of the form $\mu_f^p(x) = p\mu_f^1(x) + (1-p)\mu_f^0(x)$ for the chosen level of "weighing" $p \in [0,1]$.

Now we consider a relation of the maximal efficiency parametric principle to the axioms of choice. Thus we have

Theorem 38.1: The principle of choice $N_\mu^p(f)$ for the fixed $p \in [0,1]$ satisfies axioms 1–3, 16.

The proof of the theorem is carried out similarly to that of fulfilling axioms 1–3, 16 for $N_\mu^1(f) = N_\mu(f)$.

Instead of axiom 5 we formulate the following axiom 18. The axiom of linearity of the principle of choice, i.e. the relation of order may be presented as $N(f) = \sum_{y \in Y} p(y) f(\cdot \,|\, y)$,

$$p(y) \in S = \left\{ p(y) \in \{Y \to R^1\} : p(y) \geq 0 \quad \forall y \in Y\,, \quad \sum_{y \in Y} p(y) = 1 \right\}.$$

Theorem 38.2: The principle of choice $N_\mu^p(f)$ for the fixed $p \in [0,1]$ satisfies axiom 18.

Theorem 38.3: The optimal elements $x_p^0 \in X$ for the principle of choice $N_\mu^p(f)$ belong to the tradeoff domain

$$R = \left\{ \tilde{x} \in X : \exists \tilde{p}(y) \in S\,, \quad \sum_{y \in Y} \tilde{p}(y) f(\tilde{x}|y) = \max_{x \in X} \sum_{y \in Y} \tilde{p}(y) f(x|y) \right\}.$$

According to the statements the parametric principle of maximal efficiency satisfies axioms 1–3, 16, 18 and, therefore, along with the maximal efficiency principle, enables one to solve Arrow's paradox of the absence of the principle of choice satisfying axioms 1–5 and differing from the dominance principle of the form $N^{\text{dom}}(f) = \prod_{y \in Y} N_y$, where N_y is the relation of order on R^1 corresponding to $f(\cdot \,|\, y)$, i.e. with x^1, $x^2 \in X$, $x^1 N_\mu x^2 \iff f(x^1|y) \geq f(x^2|y)$.

The procedure of obtaining an optimal element $x_p^0 \in X$ for a specified $p \in [0,1]$ based on the maximal efficiency parametric principle is indeed the process of solving a game problem. At the same time, as in the case of the maximal efficiency principle, this problem may reduce to a maximin problem of the following form: derive δ^0, $q^0 \in [0,1]$, $x_p^0[\delta^0, q^0] \in X$ from the condition

$$[p\delta^0 + (1-p)q^0] = \max_{\substack{\delta \in [0,1], x \in X \\ \mu_f(x|y) \geq \delta \ \forall y \in Y}} \left\{ \min_{\substack{q \in [\delta, 1] \\ \mu_f(x|y) \leq q \ \forall y \in Y}} [p\delta + (1-p)q] \right\}.$$

Lemma 37.7: We have the inequality $p\delta^0 + (1-p)q^0 \geq \mu_p^0$.

Lemma 37.8: The inequality $[p\delta^0 + (1-p)q^0] \leq \mu_p^0$ is valid.

Theorem 38.4: A set of the elements $X_p^0 \subseteq X$ optimal by the maximal efficiency parametric principle coincides with the set of optimal elements $X_p^0(\delta^0, q^0)$ for a maximin problem.

The proof proceeds from the statements of Lemmas 37.7 and 37.8.

The set $X_p^0(\delta^0, q^0)$ of optimal elements may be presented as

$$\begin{aligned} X_p^0(\delta^0, q^0) &= \{x^0 \in X : \delta^0 \leq \mu_f(x^0|y) \leq q^0 \quad \forall y \in Y\} \\ &= \left\{ x^0 \in X : \min_{y \in Y} \mu_f(x^0|y) = \delta^0 \max_{y \in Y} \mu_f(x^0|y) = q^0 \right\}. \end{aligned}$$

Hence the set X_p^0 of the optimal elements by the maximal efficiency parametric principle appears as

$$X_p^0 = \{x \in X : \delta^0 \leq \mu_f(x|y) \leq q^0 \quad \forall y \in Y\}.$$

The procedure of reducing the maximin problem to the extremal enables one to present $[p\delta^0 + (1-p)q^0]$ in the following form

$$[p\delta^0 + (1-p)q^0] = \lambda_p^0 = \max_{\substack{\lambda \in [0,1], \delta \in [0,1] \\ x \in X, \mu_f(x|y) \geq \delta \ \forall y \in Y}} [\lambda | \lambda \leq p\delta + (1-p)q \ \forall q \in Q(x,\delta)]$$

where

$$\begin{aligned} Q(x,\delta) &= \{q \in R^1 : \delta \leq q \leq 1, \max_{y \in Y} \mu_f(x|y) \leq q\} \\ &= \{q \in R^1 : \delta \leq q \leq 1, \mu_f(x|y) \leq q \quad \forall y \in Y\}. \end{aligned}$$

In this case a set $Q(x, \delta)$ is nonempty because $q = 1$ belongs to $Q(x, \delta)$ for any $x \in X$, $\delta \in [0, 1]$.

Theorem 38.5: The maximal efficiency parametric principle is equivalent to an extremal problem, thus $\mu_p^0 = \lambda_p^0$ with a specified $p \in [0, 1]$.

The proof of the theorem follows from the fact that according to Lemmas 37.7 and 37.8 there is an inequality $\mu_p^0 = p\delta^0 + (1 - p)q^0$.

The maximal efficiency parametric principle makes it possible to implement parametric characterization of a set of optimal elements which might be termed a single-parameter tradeoff. Define the domain of a single-parameter tradeoff $\{R_p\} \subseteq X$ of the optimal elements $x_p^0 \in X$ in the following way: $\{R_p\} = \{X_p^0, 0 \leq p \leq 1\}$, where X_p^0 is a set of solutions for the original problem of obtaining μ_p^0 or an equivalent extremal problem.

Practical application of the domain of a single-parameter tradeoff $\{R_p\} \subseteq X$ is much more suitable and simpler than that of the tradeoff domain R in that $\{R_p\}$ has parametrization by scalar $p \in [0, 1]$, whereas R is parametrized on the m-dimensional unit simplex in the vector optimization problem (Y contains m elements) or on a space of pseudoprobabilistic distributions on Y. The characterization method for the multiobjective optimization problem on the basis of the maximal efficiency principle which coincides with the parametric principle with $p = 1$ is formulated above. The characterization process based on the maximal efficiency parametric principle is carried out either for the fixed $p \in [0, 1]$ or for all $p \in [0, 1]$ in the way somewhat analogous to the foregoing. In this case it is possible to formulate and prove[8] a theorem analogous to theorem (37.15) for the maximal efficiency principle.

Parametric characterization supplements characterization based on the maximal efficiency principle in that an increase in the ("guaranteed" from below) degree of $\delta_\tau^0(\mu_\tau^0)$ is naturally related to a decrease in the ("determined" from above) degree of q_τ^0 of achieving an optimum with an increase of $\tau \in T$ for the fixed $p \in [0, 1]$, here for different $p \in [0, 1]$ we obtain a characterization "tube" of values of δ_τ^0, q_τ^0 narrowing as $\tau \in T$ increases.

Chapter 10

THE STUDY OF IMPROVABILITY AND PRIORITY ISSUES IN MULTIOBJECTIVE OPTIMIZATION PROBLEMS

§39. THE STUDY OF THE PROBLEM OF FEASIBLE SOLUTION IMPROVABILITY

Assume that X is the set from a space $\tilde{X}$ of the arbitrary elements x identified as solutions and a vector of estimation criteria $f(x) = \big(f_1(x), f_2(x), \ldots, f_m(x)\big)$ is specified on this set, moreover, each component $f_i(x)$, $i \subset I = \{1, \ldots, m\}$, $x \in X$ is a unique representation of the space X in the space R^1 and expressed in one scale. Assume that all estimation criteria are specified in a positive ingredient, i.e. when choosing solutions of $x \in X$ one should try to increase their values. To study the concept of comparing solutions on the set X, define the relation of order $\underset{m}{\geq}$ on a space of m-dimensional vectors.

Definition 39.1: The solution $x^0 \in X$ is better than the solution $x^1 \in X$ according to the vector indicator f and the relation of order $\underset{m}{\geq}$ on R^m if the conditions $f(x^0) \underset{m}{\geq} f(x^1) \Longleftrightarrow f_i(x^0) \geq f_i(x^1)\ \forall i \in I$ are fulfilled.

Define one more approach to the preference relation concept.

Definition 39.2: The solution $x^0 \in X$ is assumed to be better than the solution $x^1 \in X$ according to the vector indicator f if the inequalities

$$f_i(x^0) \geq f_i(x^1), \qquad f_j(x^0) > f_j(x^1) \tag{39.1}$$

are fulfilled for all i and if only for one j from I.

We call this relation a simple order of preference and denote it by $\underset{m}{\succ}$. It may be assumed, without restricting generality, that the condition should be fulfilled if $x^0 \underset{m}{\succ} x^1$ ($x^0 \in X$ is better than $x^1 \in X$ in the sense of the definition (39.2). Thus $x^0 \underset{m}{\geq} x^1$ (x^0 is better than x^1 in the sense of the definition 39.1), or in other words, $\left(\underset{m}{\succ}\right) \subset \left(\underset{m}{\geq}\right)$.

Hence of interest is the question whether for an arbitrary solution $x^1 \in X$ there is or there isn't x^0 in X such that $x^0 \underset{m}{\succ} x^1$, the inequalities (39.1) are fulfilled. Thus we have:

Definition 39.3: The solution $x^0 \in X$ is called improvable on the set X by the vector indicator f, if there exists a solution $x^1 \in X$ such that $x^1 \underset{m}{\succ} x^0$.

Assume that each of the estimation criteria f_i approaches extreme values on the set X.

Theorem 39.1: The solution $x^0 \in X$ is improvable by the vector indicator f on the set X if and only if there exists an m-dimensional vector $\lambda = (\lambda_1, \ldots, \lambda_m) \in R^m$ such that the inequalities

$$f_i(x^0) \leq a_i(\lambda), \qquad f_j(x^0) < a_j(\lambda) \tag{39.2}$$

hold for all i and if only for one j from I, where the values $a_i(\lambda)$ have the form $a_i(\lambda) = a(\lambda) - \lambda_i$,

$$a(\lambda) = \max_{x \in X} \min_{i \in I} [\, f_i(x) + \lambda_i] \,.$$

We set forth the statements allowing a restriction of a set of coefficients λ without violating validity of theorem (39.1).

Theorem 39.2: For $\lambda \in R^m$ to satisfy the inequalities (39.2) with a certain $x^0 \in X$ it should satisfy the inequalities

$$\max_{x \in X} f_i(x) - \min_{x \in X} f_j(x) \geq \lambda_j - \lambda_i \tag{39.3}$$

for all i, $j \in I$.

Define a set of the form $\Lambda = \{\lambda \in R^m : \max_{x \in X} f_i(x) - \min_{x \in X} f_i(x) \geq \lambda_j - \lambda_i, \forall i, j \in I$, here $\lambda_1 = 0$.

Theorem 39.3: If for a certain λ from R^m and a certain $x^0 \in X$ the inequalities (39.2) are true, then there exists $\lambda' \in \Lambda$ such that the inequalities (39.2) hold for x^0 and λ'.

Theorem 39.4: In the case of two components of the indicator $f(x)$, a set Λ under conditions of the preceding theorem may be defined as follows: $\Lambda = \{(\lambda_1, \lambda_2) \in R^2 : f_1(x^2) - f_2(x^2) \leq \lambda_2 - \lambda_1 \leq f_1(x') - f_2(x')\}$, here x^1, $x^2 \in X$ are obtained from the condition $(i = 1, 2)$ $f_i(x^i) = \max_{x \in X} f_i(x)$.

Definition 39.4: The solution $x^0 \in X$ is called unimprovable (Pareto-optimal) on the set by the vector indicator $f(x)$ if there is no solution $x^1 \in X$ which is better than the solution x^0 by f in the sense of the definition 39.2, i.e. there is no $x^1 : f(x^1) \underset{m}{\succ} f(x^0)$.

Theorem 39.5: The solution $x^0 \in X$ is unimprovable (Pareto-optimal) by the vector indicator f if and only if the inequalities (39.2) of theorem (39.1) are inconsistent for any m-dimensional vector $\lambda \in R^m$.

Corollary 39.1: The statement of theorem (39.5) remains valid if the m-dimensional vector $\lambda = (\lambda_1, \ldots, \lambda_m)$ is chosen from the region Λ.

Theorem 39.6: The solution $x^0 \in X$ is unimprovable (Pareto-optimal) by the vector indicator $f(x)$ if and only if there is an m-dimensional vector $\lambda \in \Lambda \subseteq R^m$ such that the equality

$$\min_{i \in I}[f_i(x^0) + \lambda_i] = \max_{x \in X} \left\{ \min_{i \in I}[f_i(x) + \lambda_i] \right\} \tag{39.4}$$

exists.

Theorem 39.7: If for $x^0 \in X$, $\lambda^0 \in \Lambda \subseteq R^m$ the equality (39.4) is fulfilled, then x^0 is unimprovable (Pareto-optimal) and the inequalities $f_i(x^0) \geq a_i(\lambda^0)$, $i = 1, \ldots, m$ exist.

Theorem 39.8: There are no feasible solutions of $x^0 \in X$ and an m-dimensional vector $\lambda^0 \in \Lambda \subseteq R^m$ for which the following inequalities are fulfilled

$$f_i(x^0) > a_i(\lambda^0)\,, \qquad i = 1, \dots, m\,. \tag{39.5}$$

The statements of the theorems (39.1), (39.5), (39.7), (39.8) set a boundary (from above or from below, respectively) for the values of components of the vector indicator f by the values of the magnitudes $a_i(\lambda)$ with $\lambda \in \Lambda \subseteq R^m$.

Theorem 39.9: For the feasible solution $x^0 \in X$ to be improvable (Pareto-optimal) by the vector indicator f, it is essential and sufficient that the following inequalities ($i \in I$) might be fulfilled (inconsistent):

$$f_i(x^0) \leq \max_{\lambda \in \Lambda} \left\{ \max_{x \in X} \left[\min_{j \in I} (f_j(x) + \lambda_j) \right] - \lambda_i \right\}.$$

The latter inequalities are not only of paramount importance for "constructively" testing the possibility of improvability or Pareto-optimality of the solutions $x^0 \in X$, but also enable one to set limits on the possibility to improve the values of the multiobjective indicator components because the right-hand sides of these inequalities do not depend on x^0 and are defined only for a concrete specification of an index i of the vector indicator component f_i the set X and the indicator f itself here Λ is itself defined merely by f and X. The proofs of the foregoing statements are given in the author's work.[8]

Suppose that $\Lambda(X^1) = \{\lambda(x) = (\lambda_1(x), \dots, \lambda_m(x))\ \forall x \in X^1 : \lambda_i(x) = f_1(x) - f_i(x)\ \forall i \in I\}$ then we have $\Lambda(X^1) \subseteq \Lambda$ for any $X^1 \subseteq X$.

Theorem 39.10: For any $\lambda \in R^m$ there exists $\lambda^1 \in \Lambda(X^1)$ such that $a_i(\lambda^1) = a_i(\lambda)$ for all $i \in I$.

The statement of the latter theorem enables one to substitute a set Λ by $\Lambda(X)$ without violating validity of the theorem (39.3). Denote $[-f_i(x)]$ through $f_i^-(x)$ for all $x \in X$ and all $i \in I : a_j^-(\lambda) = \max_{x \in X} \min_{i \in I}[\lambda_i - f_i(x)] - \lambda_j$.

Definition 39.5: The solution x^* is termed the dual solution on the set X by the vector indicator $f(x)$ if there exists $\lambda \in R^m$ such that

$$f_i(x^*) \geq a_i^-(\lambda)\,, \qquad f_j(x^*) > a_j^-(\lambda) \tag{39.6}$$

for all i and if only for one j from I.

If we denote by X^0 a set of improvable solutions from $X : X^*$ is a set of dual solutions from X, it is evident that $X^* \cap X^0 = \varnothing$, $X^* \cup X^0 \subseteq X$, $X^e = X \backslash X^0$, $X^* \subseteq X^e$ here X^e is a set of efficient (Pareto-optimal) solutions of the form $X^e = \{x \in X | \bar{\exists} x^1 \in X : f(x^1) \underset{m}{\succsim} f(x)\}$. Assume that for a given vector indicator $f(x)$ and a set $X \subseteq \tilde{X}$ the sets X^0 and X^* are nonempty. Thus we have

Theorem 39.11: The inequalities (39.2) are fulfilled for x^0 and $\lambda(x^*)$, moreover, $x^0 \in X$, $x^* \in X^*$ if and only if for x^* and $\lambda(x^0)$ the inequalities (39.6) are completed.

Theorem 39.12: For any $x^0 \in X$ there exists $\lambda \in \Lambda(X^*)$ such that the inequalities (39.2) are fulfilled.

Theorem 39.13: The set X^* is empty if and only if the set X^0 is empty.

Theorem 39.14: For the solution $x^0 \in X$ to be improvable by the vector indicator $f(x)$ it is essential and sufficient that there might exist the dual element $x^* \in X$ such that the inequalities (39.2) would be fulfilled with $\lambda = \lambda(x^*)$.

Consider other approaches to the issue of solution improvability.

Define the sets $I_1, \ldots, I_R$ so that $\bigcup_{r=1}^{R} I_r = I$. Denote by $f^{(r)}(x)$ the vector indicator ordered according to an increase in indices which is obtained from the components corresponding to the indices from a set I_r.

Definition 39.6: The solution $x' \in X$ is better than the solution $x^0 \in X$ by the vector indicator $f(x)$ if the inequalities

$$f^{(r)}(x') \underset{I_r}{\gtrsim} f^{(r)}(x^0), \qquad f^{(c)}(x') \underset{I_c}{>} f^{(c)}(x^0) \tag{39.7}$$

are fulfilled for all r and if only for one c from $\{1, 2, \ldots, R\}$. The definition 39.6 is equivalent to the definition 39.2 with $R = 1$, $I_1 = 1$.

Definition 39.7: The solution $x^0 \in X$ is called improvable by the vector indicator $f(x)$ if there exists the solution $x' \in X$ such that the conditions of (39.7) are fulfilled.

Theorem 39.15: For the solution $x^0 \in X$ to be improvable by the vector indicator $f(x)$ on the set X, it is essential that it might be improvable by $f^{(r)}(x)$ for a certain $r \in \{1, 2, \ldots, R\}$.

Definition 39.8: The solution $x' \in X$ is assumed to be better than the solution $x^0 \in X$ by the vector indicator $f(x)$ with the priority relation $\langle I, \triangleright \rangle$ if there exists

$$i \in I : f_j(x') > f_j(x^0) \quad \forall j \in I_i \,.$$

Here, on a set of estimation criteria, we specify the lexicographic relation of priority $\{I, \triangleright\}$, suppose $I_i = \{j \in I : j \triangleright i\}$.

Definition 39.9: The solution $x^0 \in X$ is called improvable on the set X by the vector indicator $f(x)$ with the priority relation $\langle I, \triangleright \rangle$ if there exist the solutions $x' \in X$ and $i \in I$ such that $f_j(x') > f_j(x^0)$ $\forall j \in I_i$.

Theorem 39.16: The solution $x^0 \in X$ is improvable by $f(x)$ with the priority relation $\langle I, \triangleright \rangle$ if there exists such $I_i \subseteq I$ that it is improvable by the vector indicator $f^{(i)}(x)$.

Consider the problem of solution improvability with respect to the multiobjective conceptual indicator.

As stated above, we assume that there is the set X from set $\tilde{X}$ of the arbitrary elements x termed solutions, and also there is the set Y from the set $\tilde{Y}$, and a multiobjective indicator is specified which is the functional defined on the set $X \times Y$ with the values $f(x|y)$. Let $f(x) = \{f(x|y)\}$, $y \in Y$, $\mathcal{F} = \{f(x) : x \in X\}$. Employ the order of preference $\langle \mathcal{F}, \triangleright \rangle$ for defining the concept of a better solution on the set X.

Definition 39.10: The solution x^0 is assumed to be better than the solution x' by the multiobjective indicator f and the order of preference $\langle \mathcal{F}, \triangleright \rangle$, if $f(x^0) \triangleright f(x')$. The following approach is also conceivable in defining the order of preference.

Definition 39.11: The solution x^0 is assumed to be better than the solution x' by f if for all y and if only for one y' from Y the inequalities

$$f(x^0|y) \geq f(x'|y) \,, \qquad f(x^0|y') > f(x'|y') \tag{39.8}$$

are completed.

This relation is termed the simple order of preference and it is denoted by $\langle \mathcal{F}, \underset{Y}{\succsim} \rangle$. It may be assumed, without limiting generality, that for any order of preference the relation $(\underset{Y}{\succ}) \subseteq (\triangleright)$ has to be fulfilled.

Of vital importance for an arbitrary solution is the issue of testing whether it can or cannot be improved by a simple relation of preference.

Definition 39.12: The solution x' is assumed to be improvable on the set X by f if there exists x^0 from X such that inequalities (39.8) are fulfilled.

Definition 39.13: The solution x' is assumed to be unimprovable (Pareto-optimal) on the set X by the multiobjective indicator f if there is no $x^0 \in X$ such that inequalities (39.8) are fulfilled.

Let $f(x|y)$ be restricted by x and y, the sets X and Y being compact.

Theorem 39.17: The solution x' is improvable by f on X if and only if there exists the functional $\lambda = \{\lambda(y)|y \in Y\}$ defined on the set Y such that the inequalities are fulfilled for all y and if only for one y' from Y:

$$\begin{aligned} f(x'|y) &\le a(y|\lambda), \quad f(x'|y') < a(y'|\lambda)\,, \\ a(y|\lambda) &= a(\lambda) - \lambda(y)\,, \\ a(\lambda) &= \max_{x\in X} \min_{y \in Y} [\, f(x|y) + \lambda(y)]\,. \end{aligned} \tag{39.9}$$

We set forth the statements refining the set of choice of functionals λ and retaining the validity of the latter theorem. Thus we have:

Theorem 39.18: If for a particular $\lambda' \in \{\tilde{Y} \to R^1\}$ and a certain $x' \in X$ in equalities (39.9) are fulfilled, then the following inequalities are fulfilled for all y and y' from Y:

$$\max_{x\in X} f(x|y) - \min_{x \in X} f(x|y') \ge \lambda(y') - \lambda(y)\,. \tag{39.10}$$

Define the following set

$$\begin{aligned} \Lambda = \{\lambda \in \{\tilde{Y} \to R^1\} : &\max_{x\in X} f(x|y) - \min_{x\in X} f(x|y^0) \\ &\ge \lambda(y^0) - \lambda(y)\ \forall y, y^0 \in Y\}\,. \end{aligned}$$

Theorem 39.19: If for a particular $\lambda \in \Lambda \subseteq \{\tilde{Y} \to R^1\}$ and $x' \in X$ inequalities (39.9) are fulfilled, then there exists $\lambda' \in \Lambda'$ such that inequalities (39.9) are fulfilled for x' and λ' where $\Lambda' = \{\lambda \in \{\tilde{Y} \to R^1\} : \lambda(y') = 0\}$, $\max_{x \in X} f(x|y) - f(x'|y^0) \geq \lambda(y^0) - \lambda(y)\ \forall y, y^0 \in Y$.

Theorem 39.20: The solution x' is unimprovable by the multiobjective indicator f on the set X only when there exists the functional $\lambda(y)$ defined on the set Y such that

$$\min_{y \in Y}[f(x'|y) + \lambda(y)] = \max_{x \in X} \min_{y \in Y}[f(x|y) + \lambda(y)]\,. \tag{39.11}$$

The conditions of the theorem (39.20) are sufficient if $f(x|y)$ is concave by x for all y.

Theorem 39.21: The solution $x^0 \in X$ is improvable (unimprovable) by the multiobjective indicator f on the set X if and only if the following inequalities prevail (are inconsistent) with $y \in Y$:

$$f(x^0|y) \leq \max_{\lambda \in \Lambda} \max_{x \in X} \left\{ \min_{y^0 \in Y} [f(x|y^0) + \lambda(y^0)] - \lambda(y) \right\},$$

which are equivalent to the following equality:

$$\max_{y \in Y} \left\{ f(x^0|y) - \max_{\lambda \in \Lambda} \max_{x \in X} \left[\min_{y^0 \in Y} [f(x|y^0) + \lambda(y^0)] - \lambda(y) \right] \right\} = 0\,.$$

Definition 39.14: The solution $x' \in X$ is identified as active if there exists x^0 from X such that for all y and if only for one y' from Y the inequalities $f(x'|y) \geq f(x^0|y)$, $f(x'|y') > f(x^0|y')$ are fulfilled.

Theorem 39.22: The solution $x' \in X$ is the active solution on the set X by f if and only if there exists the functional $\lambda = \{\lambda(y)|y \in Y\} \in \{Y \to R^1\}$ defined on the set Y such that for all y and if only for one y' from Y the following conditions are fulfilled $f(x'|y) \geq a^-(y)$,

$$\begin{aligned} f(x'|y') > a^-(y')\,, \quad a^-(y) &= a^- + \lambda(y)\,, \\ a^- = \max_{x \in X} \min_{y \in Y}&[-\lambda(y) - f(x|y)]\,. \end{aligned} \tag{39.12}$$

The active solutions are improvable by the multiobjective indicator with the negative ingredient of the criterion. All the above statements can be reformulated for the case of the negative ingredient of the multiobjective indicator components.

Suppose we have the following subsets of the set X: improvable X_Y, unimprovable X_H, those of the active X_a solutions on X by the multiobjective indicator f. Thus we have

Theorem 39.23: Among the sets, the following relations are fulfilled

$$X_a = \varnothing \Longleftrightarrow X_Y = \varnothing\,; \qquad X_H \cap X_Y = \varnothing\,,$$
$$X_H \cup X_Y = X\,, \qquad X_a \subseteq X_H\,.$$

Introduce the representation of X in G where G is the totality of all subsets $\{\tilde{Y} \to R^1\}$ placing in correspondence for each $x' \in X$ the set $\Lambda(x')$ of the functionals $\lambda(y)$ for which the inequalities (39.9) are fulfilled.

Theorem 39.24: The condition $f(x) \underset{Y}{\succsim} f(x')$ is fulfilled only when $\Lambda(x) \subseteq \Lambda(x')$.

Denote by $A(x')$, $B^+(x')$, $B^-(x')$ the representations of X in G placing in correspondence for each x' from X the sets of the functionals $\lambda(y)$ for which inequalities (39.12), equalities (39.11) and the following equalities are fulfilled

$$\max_{y\in Y}[f(x'|y) + \lambda(y)] = \min_{x\in X}\max_{y\in Y}[\,f(x|y) + \lambda(y)]\,.$$

Let $B(x) = B^+(x) \cap B^-(x)$, denote for $X' \subseteq X$ by $\Lambda(X')$ the image of the set X' with the representation $\Lambda(x)$, $x \in X'$ similar to $A(X')$, $B(X')$.

Definition 39.15: The set $\Lambda \subseteq G$ is termed sufficient provided $\Lambda(X) \cap \Lambda = \varnothing$, if and only if $\Lambda(X) = \varnothing$.

Theorem 39.25: The set $\Lambda = B(X_a)$ is sufficient.

Definition 39.16: The set $A \subseteq G$ is called a-sufficient provided $A(x) \cap A = \varnothing$, if and only if $A(x) = \varnothing$ with $x \in X$.

Theorem 39.26: The set $A = B(X_Y)$ is a-sufficient.

Now consider the case of arbitrary X and Y, having rejected the compactness assumption.

Definition 39.17: The solution $x' \in X$ is identified as ε-improvable on the set X by f if there exist $x^0 \in X$ and $\varepsilon \geq 0$ such that the inequalities are fulfilled for all y and if only for one y' from Y:

$$f(x'|y) \leq f(x^0|y) + \varepsilon\,, \quad f(x'|y') < f(x^0|y') + \varepsilon\,.$$

Definition 39.18: The solution $x' \in X$ is referred to as improvable on the set X by f if there exists $\varepsilon > 0$ such that x' is ε-improvable.

Theorem 39.27: The solution $x' \in X$ is improvable by f on the set X (in the sense of the definition 39.18) if and only if there exists the functional of $\lambda = \{\lambda(y)\}$, $y \in Y$ such that for all y and if only for one y' from Y

$$f(x'|y) \leq a(y|\lambda)\,, \qquad f(x'|y') < a(y'|\lambda)\,,$$
$$a(y|\lambda) = a(\lambda) - \lambda(y)\,, \qquad a(\lambda) = \max_{x \in X}\Big[\min_{y \in Y} f(x|y) + \lambda(y)\Big]\,.$$

The latter relations may be written in the following, sometimes more suitable, form for all y, y' and if only for single $\bar{y}$, $\bar{y}'$ from Y:

$$\max_{x \in X} f(x|y) - f(x'|y') \geq \lambda(y') - \lambda(y)\,,$$
$$\max_{x \in X} f(x|\bar{y}) - f(x'|\bar{y}') > \lambda(\bar{y}') - \lambda(y)\,.$$

Now suppose on the set $\mathcal{F} = \{f(x) : x \in X\}$ several orders of preference $\langle \mathcal{F}, \trianglerighteq_1 \rangle$, $\langle \mathcal{F}, \trianglerighteq_2 \rangle$, $\ldots$, $\langle \mathcal{F}, \trianglerighteq_z \rangle$ are specified.

Definition 39.19: The solution $x' \in X$ is called improvable on the set X by the multiobjective indicator f and a set of the preference orders $\{\trianglerighteq_1, \trianglerighteq_2, \ldots, \trianglerighteq_z\}$ specified on $\mathcal{F}$ if there exists x from X such that for all l and if only for one l' from $\{1, 2, \ldots, Z\}$ there exists

$$f(x) \trianglerighteq_l f(x')\,, \qquad f(x) \triangleright_{l'} f(x')\,.$$

Now suppose on the set Y, the lexicographical relation of order $\langle Y, \triangleright\rangle$, called the priority relation, is specified. Denote $Y_y = \{y' \in Y : y' \triangleright y\}$, where $y \in Y$.

Definition 39.20: The solution $x' \in X$ is better than the solution $x^0 \in X$ by the multiobjective indicator f with the specified priority relation $\langle Y, \triangleright\rangle$ if there exists y^0 from Y such that x' is better than x^0 by $\mathcal{F}(x|y^0)$ and a simple order of preference on Y_{y^0} where

$$\mathcal{F}(x|y^0) = \{f(x|y')\}, \quad y' \in Y^0 .$$

It may be shown that the introduced relation of preference is asymmetric, transitive and antireflexive.

Definition 39.21: The solution $x' \in X$ is better than the solution $x^0 \in X$ by the multiobjective indicator f if there exists a complimentary unique representation $p = \{p(y)\}$, $y \in Y$ of the set Y in Y such that for all y and if only for one y' from Y there exists

$$f(x'|y) \geq f(x|\, p(y)) , \quad f(x'|y') > f(x|\, p(y')) .$$

The above relation of preference is asymmetric, transitive and antireflexive.

Definition 39.22: The solution $x^0 \in X$ is improvable by the preference relation introduced in the last-mentioned definition on the components of the multiobjective indicator f on the set X if there exist $y' \in Y$ and the solution $x' \in X$ which is better than $x^0 \in X$ according to this preference relation.

Denote for the fixed $x' \in X$; y, $y' \in Y$; $X_{yy'} = \{x \in X : f(x|y) \geq f(x'|y')\}$ and define the content of these sets as $|X_{y'}| = \bigcup_{u \leq U} \bigcap_{y \in U} X_{yy'}$, where U is the set of all subsets of Y.

Theorem 39.28: For the solution $x' \in X$ to be improvable in the sense of definition 39.22, it is essential that there might exist $y' \in Y$ such that $|X_{y'}| \neq \varnothing$.

Suppose we have the following representation of Y in Y:

$$q(x, x'|y) \in \{y' \in Y^*(y) : f(x|y') = \min_{y'' \in Y^*(y)} f(x|y'')\} ,$$

here the set $Y^*(y)$ is defined as:

$$Y^*(y) = \{y' \in Y : f(x'|y) \geq f(x|y')\} \backslash q(x, x'|\, Y'(y)),$$
$$Y'(y) = \{y' \in Y : f(x|y') \leq f(x|y)\}, q(x, x'|\, Y')$$
$$= \bigcup_{y \in Y'} q(x, x'|\, y).$$

Theorem 39.29: The solution $x' \in X$ is better than the solution $x \in X$ if and only if for all y and for at least one y' from Y

$$f(x'|y) \geq f(x|q(x, x'|y)), \quad f(x'|y') > f(x|q(x, x'|y')).$$

The foregoing definitions and statements form merely part of the diversity that occurs owing to employment of the simple preference relation on the components of the multiobjective indicator f on the set $\mathcal{F}$ for obtaining the preference relations of one solution as compared with the other as well as for formulating different approaches to defining the concepts of improvability and generalized Pareto-optimality (unimprovability) and the necessary and sufficient conditions of improvability or optimality according to the preference relations introduced.

Consider the problem of improvability depending on the optimum achievement degree.

Instead of the multiobjective indicator f in the definitions of improvability of the solution $x^0 \in X$ we shall utilize $\mu_f(x) = \{\mu_f(x|y)\}$, $y \in Y$ the components of which are the magnitudes $\mu_f(x|y)$ is the degree of achieving an optimum at the point $x \in X$ by the yth component $f(x|y)$, thus we get:

$$\Lambda_\mu = \{\lambda \in \{\tilde{Y} \to R^1\} : |\,\lambda(y^0) - \lambda(y)| \leq 1, \forall y, y^0 \in Y\}. \tag{39.13}$$

Definition 39.23: The solution $x^0 \in X$ is improvable according to the multiobjective indicator μ_f on the set X if there exists a solution $x' \in X$ which is better than the solution $x^0 \in X$ in that for all $y \in Y$ and if only for one $y' \in Y$ the following inequalities are completed

$$\mu_f(x'|y) \geq \mu_f(x^0|y), \quad \mu_f(x'|y') > \mu_f(x^0|y'). \tag{39.14}$$

Definition 39.24: The solution $x^0 \in X$ cannot be improved (Pareto-optimal) according to the multiobjective indicator μ_f on the set X if there exists a solution $x' \in X$ which is better than $x^0 \in X$ in the sense of these inequalities.

Theorem 39.30: The solution $x^0 \in X$ is improvable (unimprovable) according to the multiobjective indicator μ_f if and only if there prevails (are inconsistent) the following inequalities with $y \in Y$:

$$\mu_f(x^0|y) \le \max_{\lambda \in \Lambda_\mu} \max_{x \in X} \Big\{ \min_{y^0 \in Y} [\mu_f(x|y^0) + \lambda(y^0)] - \lambda(y) \Big\},$$

which are equivalent to the following equality:

$$\max_{y \in Y} \Big\{ \mu_f(x^0|y) - \max_{\lambda \in \Lambda_\mu} \max_{x \in X} \Big[\min_{y^0 \in Y} [\mu_f(x|y^0) + \lambda(y^0)] - \lambda(y) \Big] \Big\} = 0 .$$

In the same way we may also reduce the conditions for improvability of the solutions $x^0 \in X$ or their Pareto-optimality by using μ_f instead of f.

Definition 39.25: The element $x^* \in X$ is referred to as superoptimal according to the generalized principle of maximal efficiency if the conditions are fulfilled for all $y \in Y$:

$$\begin{aligned} \mu_f(x^*|y) &= \max_{\lambda \in \Lambda_\mu} \Big\{ \min_{y^0 \in Y} [\mu_f(x^*|y^0) + \lambda(y^0)] - \lambda(y) \Big\} \\ &= \max_{\lambda \in \Lambda_\mu} \max_{x \in X} \Big\{ \min_{y^0 \in Y} [\mu_f(x|y^0) + \lambda(y^0)] - \lambda(y) \Big\} \end{aligned}$$

The superoptimal solution set X^*_μ is the contraction limiting by $\lambda \in \Lambda_\mu$ not only of the Pareto-optimal solution set X_H but also of the maximal efficiency criterion-optimal element set X^0_μ, i.e. $X^*_\mu \subseteq X^0_\mu \subseteq X^0_H \subseteq X$.

On the one hand, it enables one to carry out characterization of the problem of the f multiobjective optimum on the set X, using the generalized principle of maximal efficiency as the base, and formulate the parametric generalized principle of maximal efficiency, on the other.

In this case the values $\mu_f(x^*|y)$ with $y \in Y$ constitute the limiting bounds for possibilities to improve the solution of $x^0 \in X$ (in that $x^0 \in X$ can be improved if and only if $\mu_f(x^0|y) \le \mu_f(x^*|y)$ for all $y \in Y$) as well as the limiting bounds for the best choice of $x^0 \in X^0_H$ (Pareto-optimal) by specifying conditions for $x^0 \in X^*_\mu$.

§40. THE STUDY OF THE PRIORITY PROBLEM IN MULTIOBJECTIVE OPTIMIZATION

Consider the problem of priority in multiobjective optimization.

Priority in the problem of optimal choice of an element $x^0 \in X$ by the multiobjective indicator $f(x) = \{f(x|y)\}$, $y \in Y$, where Y is a set of objective terms with the fixed $X \subseteq \tilde{X}$, $Y \subseteq \tilde{Y}$, $f \in F$ is taken as specification of the priority relation on the set Y according to importance, on the components $f(\cdot\,|\,y)$ of the multiobjective indicator f with $y \in Y$ according to significance (preference), and also on the values of $f(x) \in \mathcal{F}$ of the multiobjective indicator f with $x \in X$ according to efficiency (a degree of increase in the values of f). The priority relation may also be specified on the elements $x \in X$ themselves, but in this case it is understood as specification of the principle of choice. In multiobjective optimization problems, the priority on X (i.e. the principle of choice) generally is not specified, because, if it is not specified, then there is no problem of deriving the principle of choice, and there remains the calculative problem of deriving optimal elements. If there is no specification of priority on Y, $f(x)$ with $x \in X$, $\mathcal{F}$ and priority is specified on X as the principle of choice, then it remains to obtain a solution of an appropriate extremal or maximin problem. Hence the problem of taking into account a specified priority aggravates solving the problem of deriving the principle of choice which satisfies a specified or unspecified totality of axioms and properties for which this priority should be fulfilled on optimal elements.

The theory of multiobjective optimization is still in the making, hence it is clear that the problem of priority is complex and differently understood by various authors. A common approach to the problem of definition, specification and incorporation of priority is still lacking. Keeping this in mind, one should not anticipate an exhaustive solution of the problem. We focus on the following three aspects of the concept of priority in multiobjective optimization: priority on the multiobjective terms of $y \in Y$, on the multiobjective indicator components $f(\cdot\,|\,y)$ with $y \in Y$ and on the values $f(x) \in \mathcal{F}$ of the multiobjective indicator f. Definition of the priority relation is based on the comparison of the objective terms $y \in Y$ in the scale C_Y of the importance of $y \in Y$, the components of $f(x|y)$ with $x \in X$ for the fixed $y \in Y$ of the multiobjective indicator f in the scale $C_{f(\cdot\,|\,y)}$ of significances of the quantitative values of $f(x|y)$ of the y quality being obtained when specifying $x \in X$, and the values of $f(x) \in \mathcal{F}$ in the efficiency scale $C_{\mathcal{F}}$ for a totality of the quantitative values of $\{f(x|y)\}$, $y \in Y$ of all qualities of $y \in Y$ obtained when choosing $x \in X$. The scale C_Y is identified as a particular order or a list of indices on a specified set Y of the objective terms (expressing different qualities, characteristics, properties, etc.) generated by specifying the elements $x \in X$. This order determines the importance either among objective terms or objective terms and subsets from Y or subsets from Y. When comparing objective terms or subsets from Y the importance is taken as one or several estimates of utility (values, etc.) of the qualities under consideration corresponding to objective terms y from Y. Sometimes this order indicating the importance of objective terms in the scale C_Y is not given and, in compliance with it, with the index enumeration of objective terms y from Y the partitions of a set of indices by the importance (which may not be specified, though related to its quality and every quality is related to its index. In a number of instances that kind of order is specified as mixed, i.e. when comparing some subsets by the importance and giving indices or their subsets. But here we need to correlate specification of the importance of subsets of objective terms as qualities and specification of the importance of these subsets for terms as indices.

The scale $C_{f(\cdot\,|y)}$ is taken as a total of the obtained various degrees (gradations) of the y quality in the quantity $f(x|y)$ of different elements $x \in X$, the ordering on which is termed the significance (preference) of obtaining some quantitative values of the quality of $f(\cdot\,|\,y)$ as against the others. If Y is a totality of quality indices as objective terms and there is no indication in the scale C_y as to the importance when comparing different objective terms y from Y, then the significance in the scale $C_{f(\cdot\,|y)}$ can be presented as the importance of obtaining some quantitative values of the y quality as against the others. Moreover, here the importance of obtaining quantitative values of $f(\cdot\,|y)$ from a particular set can be specified when comparing them with a set of quantitative values of $f(\cdot\,|y)$ of another quality $y' \neq y$. Let $C_f = \{C_{f(\cdot\,|y)}\}$, $y \in Y$.

The scale $C_{\mathcal{F}}$ is taken as specification of a natural order N^{dom} on the multiobjective indicator values of $f(x)$ in the following form for $x^1, x^2 \in X$:

$$f(x^1)N^{\mathrm{dom}}f(x^2) \iff f(x^1) \triangleright f(x^2) \iff f(x^1|y) \geq f(x^2|y)\ \forall y \in Y\,,$$

or the relation of order $\hat{N}^{\text{dom}}$ of the form

$$f(x^1)\hat{N}^{\text{dom}} f(x^2) \Longleftrightarrow \{[f(x^1)N^{\text{dom}} f(x^2)] \\ \wedge [\exists y^0 \in Y : f(x^1|y^0) > f(x^2|y^0)]\}\,.$$

Specification of the set X of admissible elements is supplemented, in a number of problems with constraints on the multiobjective indicator values, though, basically, it leads to substitution of X by its particular subset $X' \subseteq X$.

When comparing the objective terms of $y \in Y$, the components $f(\cdot\,|\,y)$ of the multiobjective indicator $f(\cdot) = \{f(\cdot\,|\,y)\}$, the values of $f(x)$ with $x \in X$ of the multiobjective indicator $f \in F$ from the subset $\mathcal{F} = \{f(x)\}$, $x \in X$ there may prevail equivalence of the objective terms of y, the components $f(\cdot\,|\,y)$ and the values of $f(x)$ in their scales C_Y, $C_{f(\cdot\,|\,y)}$, $C_{\mathcal{F}}$ respectively. In this case, the equivalence $y^1 \underset{C_Y}{\backsim} y^2$ of the objective terms y^1, $y^2 \in Y$ in compliance with the scale C_Y is understood as the situation when the importance of the objective terms is the same or each of them enters into subsets of Y_1 and $Y_2 \subseteq Y$. The equivalence of Y_1 and $Y_2 \subseteq Y$ in this case signifies the same importance of a generalized quality of Y_1, Y_2 as a totality of qualities expressed by the objective terms $y \in Y_1$, $y \in Y_2$. By the equivalence of $f(x|y^1) \underset{C_f}{\backsim} f(x|y^2)$ with $y^1 \neq y^2 \in Y$ for a particular element $x \in X$ we mean fulfillment of the following relations:

$$f(x|y^1) \underset{C_f}{\backsim} f(x|y^2) \Longleftrightarrow \{[y^1 \underset{C_Y}{\backsim} y^2) \vee (y^1 \underset{C_Y}{\square} y^2)] \\ \wedge (C_{f(\cdot\,|\,y^1)} \underset{C_f}{\backsim} C_{f(\cdot\,|\,y^2)}) \vee [(C_{f(\cdot\,|\,y^1)} \underset{C_f}{\square} C_{f(\cdot\,|\,y^2)}) \\ \wedge [(f(x|y^1) \in K_1) \wedge (f(x|y^2) \in K_2) : (K_1 \underset{C_f}{\backsim} K_2), (K_1 \wedge K_2) \subseteq C_f]]\}\,,$$

i.e. $f(x|y^1)$ is equivalent to $f(x|y^2)$ with $y^1 \neq y^2$, y^1, $y^2 \in Y$ if and only if y^1 and y^2 are equivalent and incomparable in the scale C_Y, the scales $C_{f(\cdot\,|\,y^1)}$ and $C_{f(\cdot\,|\,y^2)}$ are equivalent and incomparable in the scale C_f and $f(x|y^1)$, $f(x|y^2)$ belong respectively to the sets $k_1, k_2 \in C_f$ equivalent by the scale C_f, where by $\square$ we denote the binary relation of incomparatibility. The equivalence $f(x^1) \underset{C_f}{\backsim} f(x^2)$ corresponds to fulfillment of the following relations:

$$f(x^1) \underset{C_{\mathcal{F}}}{\backsim} f(x^2) \iff \{[f(x^1)N^{\text{dom}}f(x^2) \wedge f(x^2)N^{\text{dom}}f(x^1)] \vee [f(x^1)\hat{N}^{\text{dom}}f(x^2) \wedge f(x^2)\hat{N}^{\text{dom}}f(x^1)]\},$$

in other words, the values $f(x^1)$, $f(x^2) \in \mathcal{F}$ are equivalent in the scale C_f if and only if these values are equivalent either by the relation of order N^{dom} or by $\hat{N}^{\text{dom}}$. An example of equivalence is provided by $y^1 = y^2$, $f(x^1|y^1) = f(x^2|y^2)$, although $x^1 \neq x^2$ is possible.

It is possible to secure another form of comparison for the objective terms $y \in Y$, the components $f(\cdot\,|\,y)$ of the multiobjective indicator f and the values of $f(x) \in \mathcal{F}$ with $x \in X$ where unequivalence prevails, by which we mean the situation when at least one of the forms of equivalence is absent in the scales C_Y, C_f, $C_{\mathcal{F}}$. In this case (of unequivalence), when specifying $y^1 \neq y^2 \in Y$, $x^1 \neq x^2 \in X$ and $f \in F$ we define relations $\underset{C_Y}{\triangleright}$, $\underset{C_f}{\triangleright}$ and $\underset{C_{\mathcal{F}}}{\triangleright}$ in the following way: $y^1 \underset{C_Y}{\triangleright} y^2$ if y^1 is more important than y^2 in the scale C_Y; $f(\cdot\,|\,y^1) \underset{C_{\mathcal{F}}}{\triangleright} f(\cdot\,|\,y^2)$ if the y^1th component of $f(\cdot\,|\,y^1)$ of the multiobjective indicator f has more significance (preference) in the scale C_f as against the y^2th component of $f(\cdot\,|\,y^2)$; $f(x^1) \underset{C_{\mathcal{F}}}{\triangleright} f(x^2)$ if the value of $f(x^1)$ of the multiobjective indicator f on the element $x^1 \in X$ exhibits more efficiency as against the value of $f(x^2)$ on the element $x^2 \in X$.

The binary relations of preference $\triangleright$ on the scales C_Y, C_f, $C_{\mathcal{F}}$ enable one to introduce similar relations between an element and a combination of elements from Y, $\{f(\cdot\,|\,y)\}, y \in Y$, $\mathcal{F}$ combinations of elements of these sets both separately by every scale and jointly. Thus we have

$$Y \underset{C_Y}{\triangleright} y_2, y_1 \underset{C_Y}{\triangleright} Y_2, \{f(x)\}_{x \in X_1} \underset{C_{\mathcal{F}}}{\triangleright} f(x^2) \ldots.$$

We consider now the definition of priority and briefly discuss description of methods for its specification. At the outset we have the case of specifying the two-component multiobjective indicator and the two-element set X, i.e. $Y_1 = \{y^1, y^2\}$, $X_1 = \{x^1, x^2\}$ for some $y^1 \neq y^2 \in Y$ and $x^1 \neq x^2 \in X$. Suppose the scales C_Y, C_f, $C_{\mathcal{F}}$ are given with some specified relations of importance $\underset{C_Y}{\triangleright}$, significance $\underset{C_f}{\triangleright}$ and efficiency $\underset{C_{\mathcal{F}}}{\triangleright}$, and if it is clear which scale is utilized, then it will be omitted under the sign $\triangleright$. For the elements $y^1 \neq y^2 \in Y$ in the scale C_Y by the binary relation C_Y one and only one of the following relations can exist: $y^1 \triangleright y^2$, $y^2 \triangleright y^1$, $y^1 \backsim y^2$, $\overline{y^1 \triangleright y^2}$, $\overline{y^2 \triangleright y^1}$, $\overline{y^1 \backsim y^2}$, $y^1 \square y^2$.

For components $f(\cdot\,|\,y^1)$, $f(\cdot\,|\,y^2)$ of the multiobjective indicator $f \in F$ in the scale C_f by the binary relation $\underset{C_{\mathcal{F}}}{\triangleright}$ it is possible to fulfill only one of the following relations:

$$f(\cdot\,|\,y^1) \triangleright f(\cdot\,|\,y^2), f(\cdot\,|\,y^2) \triangleright f(\cdot\,|\,y^1), f(\cdot\,|\,y^1) \backsim f(\cdot\,|\,y^2),$$
$$\overline{f(\cdot\,|\,y^1) \triangleright f(\cdot\,|\,y^2)}, \overline{f(\cdot\,|\,y^2) \triangleright f(\cdot\,|\,y^1)}, \overline{f(\cdot\,|\,y^1) \backsim f(\cdot\,|\,y^2)}, f(\cdot\,|\,y^1) \square f(\cdot\,|\,y^2).$$

In the case with two values of $\mathcal{F}_1 = \{f(x^1|y^1), f(x^1|y^2)\}$, $\mathcal{F}_2 = \{f(x^2|y^1), f(x^2|y^2)\}$ of the two-component objective indicator $\{f(\cdot\,|\,y^1), f(\cdot\,|\,y^2)\}$ in the scale $C_{\mathcal{F}}$ by the relation $\underset{C_{\mathcal{F}}}{\triangleright}$ only one of the following relations can be satisfied: $\mathcal{F}_1 \triangleright \mathcal{F}_2$; $\mathcal{F}_2 \triangleright \mathcal{F}_1$; $\mathcal{F}_1 \backsim \mathcal{F}_2$; $\overline{\mathcal{F}_1 \triangleright \mathcal{F}_2}$; $\overline{\mathcal{F}_2 \triangleright \mathcal{F}_1}$; $\overline{\mathcal{F}_1 \backsim \mathcal{F}_2}$; $\mathcal{F}_1 \square \mathcal{F}_2$. In this case the priority for the objective terms $y^1, y^2 \in Y$ is taken as a requirement to fulfill only one of the relations $y^1 \underset{C_Y}{\triangleright} y^2$; $y^2 \underset{C_Y}{\triangleright} y^1$ we have $P_{r_i}(y^1, y^2) \iff y^1 \underset{C_Y}{\triangleright} y^2$, $P_{r_i}(y^2, y^1) \iff y^2 \underset{C_Y}{\triangleright} y^1$. Priority for the components $f(\cdot\,|\,y^1)$, $f(\cdot\,|\,y^2)$ of the multiobjective indicator from these two components is understood as a requirement to fulfill one and only one of the relations:

$$f(\cdot\,|\,y^1) \underset{C_f}{\triangleright} f(\cdot\,|\,y^2), \quad f(\cdot\,|\,y^2) \underset{C_f}{\triangleright} f(\cdot\,|\,y^1),$$

i.e.

$$P_{r_i}(f(\cdot\,|\,y^1), f(\cdot\,|\,y^2)) \iff f(\cdot\,|\,y^1) \underset{C_f}{\triangleright} f(\cdot\,|\,y^2),$$
$$P_{r_i}(f(\cdot\,|\,y^2), f(\cdot\,|\,y^1)) \iff f(\cdot\,|\,y^2) \underset{C_f}{\triangleright} f(\cdot\,|\,y^1).$$

Priority for the values of $\mathcal{F}_1(x^1)$ and $\mathcal{F}_2(x^2)$ of the multiobjective indicator f is assessed as a requirement to fulfill one of the relations

$$\mathcal{F}_1 \underset{C_{\mathcal{F}}}{\triangleright} \mathcal{F}_2, \qquad \mathcal{F}_2 \underset{C_{\mathcal{F}}}{\triangleright} \mathcal{F}_1,$$

i.e.

$$P_{r_i}(\mathcal{F}_1(x^1), \mathcal{F}_2(x^2)) \Longleftrightarrow \mathcal{F}_1 \underset{C_{\mathcal{F}}}{\triangleright} \mathcal{F}_2\,,$$

$$P_{r_i}(\mathcal{F}_2(x^2), \mathcal{F}_1(x^1)) \Longleftrightarrow \mathcal{F}_2 \underset{C_{\mathcal{F}}}{\triangleright} \mathcal{F}_1\,,$$

and the meaning of $P_{r_i}(\mathcal{F}_1(x^1), \mathcal{F}_2(x^2))$ resides in the following. For the element x^1 the value $\mathcal{F}_1(x^1)$ of the two-component multiobjective indicator displays more efficiency in the scale $C_{\mathcal{F}}$ as against $\mathcal{F}_2(x^2)$, for the element x^2, respectively, the opposite is true.

When on the scales C_Y, C_f, $C_{\mathcal{F}}$ the relations of importance $\underset{C_Y}{\triangleright}$, significance $\underset{C_f}{\triangleright}$ and efficiency $\underset{C_{\mathcal{F}}}{\triangleright}$ are specified as perfect orders, the priority on the sets of objective terms Y, components of the multiobjective indicator f and values of $\mathcal{F}$ is understood as specifying on them the subsets of chains (series) completely ordered by appropriate relations $\underset{C}{\triangleright}$, and the concept of priority specification $P_{r_i}(\mathcal{K}_1, \mathcal{K}_2) \Longleftrightarrow \mathcal{K}_1 \underset{C}{\triangleright} \mathcal{K}_2$ for two subsets or $P_{r_i}(\mathcal{K}_1, \mathcal{K}_2, \ldots, \mathcal{K}_m) \Longleftrightarrow \{\mathcal{K}_1 \underset{C}{\triangleright} \mathcal{K}_2 \underset{C}{\triangleright} \ldots \underset{C}{\triangleright} \mathcal{K}_m\}$ will be defined by using generalization of the binary relation of priority in the requisite scale C as the base. Here specifying the relations of importance, significance and efficiency enables one to establish priority not only separately in each scale but also in a common scale $C_0 = C_Y \cup C_f \cup C_{\mathcal{F}}$ on a direct product $O = Y \times f \times \mathcal{F}$, where f is taken as a set of the multiobjective indicator components, in other words, $f = \{f(\cdot\,|\,y)\}, y \in Y$.

Therefore, when specifying the scale C_0 and the relation $\underset{C}{\triangleright}$ the priority $P_{r_i}\{O,\, C_0\}$ is taken as a chain (one or several) of subsets from O ordered by the relation $\underset{C_0}{\triangleright}$. In the general case we can also place in $P_{r_i}\{O, C_0\}$ specification of a set of relations $\{\underset{C_0^l}{\triangleright}\}_{l\in\mathcal{L}}$, or the scales $\{C_0^l\}_{l\in\mathcal{L}}$ themselves, or their part, or the part that has been specified before formulating priority in the statement of the problem.

The multiobjective optimization problem is assumed to be specified if we determine the sets $X \subseteq \tilde{X}$ of feasible elements x, $Y \subseteq \tilde{Y}$ of objective terms, where $f \in F$ is the objective indicator, $\mathcal{F}$ is the value $f(x)$ of the multiobjective indicator with $x \in X$ and priority is $P_{r_i}\{O, C_0\}$.

After the multiobjective optimization problem has been formulated the problem arises with respect to determining the principle of choice and the elements optimal by this principle. This is according to the specified priority or the incorporation of the specified priority with the utilization of a principle of choice depending on which of the following assumed set of possible cases is realized:

Is there the scale C_Y on Y?

If so, then is the order $\underset{C_Y}{\triangleright}$ specified by importance?

Is the priority on objective terms and subsets from Y specified with the use of that order and scale?

Is there the scale C_f on the multiobjective indicator components $f \in F$?

If so, then is the order $\underset{C_f}{\triangleright}$ specified by importance?

Did we specify with the use of that order and scale the priority on the components of the multiobjective indicator f and the multiobjective indicators from F determined by various combinations of these components (special indicators for f)?

Is there a dominant element $x^0 \in X$ by the relation of order N^{dom} according to the specified X, Y, $f \in F$ and the priority $P_{r_i}\{O, C_0\}$?

If so, then x^0 is the multiobjective optimum.

If it is not, then is there the Pareto-optimal element $x^0 \in X$ by the relation of order $\bar{\tilde{N}}^{\text{dom}}$ according to the specified X, Y, f and $P_{r_i}\{O, C_0\}$?

If there are no Pareto-optimal elements, then the stated problem of multiobjective optimization has no solution.

If the Pareto-optimal element is unique, then it represents a solution of the multiobjective optimization problem.

If the set of Pareto-optimal elements contains a considerable number of elements, then it is here that the problem arises with respect to restriction of this set on the basis of incorporating priority, constraints, etc., formulating and selecting the principle of choice (as the relation of order on X by the values $f(x) \in \mathcal{F}$ with $x \in X$), obtaining optimal and superoptimal elements, characterization, etc.

An important direction of solving multiobjective optimization problems, when specifying priority in the form of scales, relations of order according to importance, significance and efficiency and the principles of choice as an order relation, represents utilization of the concept of comparing orders (here we employ the above idea that one order is stronger than another) and finding a stronger order in a linear closure or linear space with the basis given by the specified relations of order. As noted above, realization of that principle is based on generalization of the principles of choice as a linear convolution with pseudoprobabilistic distributions as well as generalization of the maximal efficiency principle. We shall set forth a few illustrative examples of utilization of the possibilities of these approaches as applied to vector optimization problems for incorporating priority and obtaining a multiobjective optimum for several special cases. We start with incorporating priority in the maximal efficiency principle.

We have the set X of feasible elements, the set $Y = I = \{1, \ldots, m\}$ of the objective term indices and the multiobjective vector indicator $f(x) = \{f_1(x), \ldots, f_m(x)\}$ as a totality of quantitative values of the objective terms (qualities) corresponding to the feasible elements $x \in X$. Assume that the components f_i of the vector indicator f have a positive ingredient (when selecting elements $x \in X$ one should seek an increase in their values). Suppose we do not specify the scales of importance and significance (preference) on the set Y and components of the vector indicator f, and there is no priority specification. There exists the possibility of introducing the efficiency scale on the basis of utilizing natural normalization of the vector indicator components as $(i \in I)$, $\mu_i(x) = [f_i(x) - f_i^0]/[f_i^* - f_i^0]$. Here $\mu_i(x)$ is the degree of achieving an optimum by the ith component of the vector indicator f for the element $x \in X$, $f_i^0 = \min_{x \in X} f_i(x)$, $f_i^* = \max_{x \in X} f_i(x)$ and the scales of efficiency by each component μ_i in a quantitative form are dimensionless and represent intervals $[0, 1]$ (for any $i \in I$ the inequalities $0 \le \mu_i(x) \le 1$ are fulfilled). For the maximal efficiency principle in the conceptual form of specification, verifications of the fulfillment of axioms as well as properties and characterization of the sets of elements optimal by this principle are given above. This enables one to transfer properly the axioms, properties and characterization onto the multiobjective vector indicator case. Moreover, the above properties of improvability, Pareto-optimality and many others for the feasible elements $x \in X$ can also be properly extended to the case of the vector indicator f.

Following the maximal efficiency principle, the optimal element $x^0 \in X$ is obtained from solving the following maximin problem: $\max_{x \in X} \min_{i \in I} \mu_i(x) = \min_{i \in I} \mu_i(x^0)$, the latter problem may be reduced to the extremal problem:

$$\mu^0(x^0) = \min_{i \in I} \mu_i(x^0) = \max_{\substack{0 \le \mu \le 1 \\ x \in X, \mu_i(x) \ge \mu \\ i \in I}} \mu = \max_{\substack{0 \le \mu \le 1 \\ x \in X}} [\mu | \, \mu_i(x) \ge \mu, \forall i \in I] .$$

Let $\mu(x) = \big(\mu_1(x), \ldots, \mu_m(x)\big)$, the maximal efficiency principle is termed the maximal efficiency μ-criterion or simply the μ-criterion, and the corresponding maximal or extremal problem is identified as the μ-problem. Consider several specific cases of priority specification in the vector optimization problem.

(1) Assume that on the vector indicator components $f(x) = \big(f_1(x), \ldots, f_m(x)\big)$ we specify a simple relation of priority (lexicographic order) of the form $1 \triangleright 2 \triangleright \cdots \triangleright i \triangleright i+1 \triangleright \cdots \triangleright m$ which does not change when utilizing the natural normalization $\mu_i(x), i \in I$. For that case the μ-criterion is formulated as follows: $\mu^0_{\text{lex}} = \max\limits_{\substack{0 \le \mu \le 1 \\ X_\mu}} \mu_{\text{lex}} \ne \varnothing$, where via X^{lex}_μ denotes the set of elements $x \in X$ as $X^{\text{lex}}_\mu = \{x \in X : \mu(x) \ge (m-i+1)\mu, i = \overline{1,m}\}$. In so doing, the magnitude μ^0_{lex} has the form $\mu^0_{\text{lex}} = \max\limits_{x \in X}\{\min\limits_{i \in I}[\mu_i(x)/(m-i+1)]\}$, here the μ-criterion optimal element $x^0 \in X$ satisfies the inequalities $\mu_i(x^0) \ge (m-i+1)\mu^0_{\text{lex}}$.

(2) Assume that on the vector criterion components the priority binary relation $\triangleright$ is specified such that $i \triangleright j$, $j \in I_i$, $(i = \overline{1,m})$, here we designate $I_i = \{j \in \{1, \ldots, m\} : i \triangleright j\}$. In this case the μ-criterion is formulated as $\mu^0(\triangleright) = \max\limits_{\substack{0 \le \mu \le 1 \\ X_\mu(\triangleright) \ne \varnothing}} \mu$, here $X_\mu(\triangleright)$ is the set of elements $x \in X$ of the form

$$X_\mu(\triangleright) = \{x \in X : \mu_i(x) \ge (m_i + 1)\mu \quad (i = \overline{1,m})\},$$

m_i is a number of elements of the set $I_i \subseteq I = \{1, \ldots, m\}$. Similarly it may be shown that the magnitude $\mu^0(\triangleright)$ takes the form

$$\mu^0(\triangleright) = \max_{x \in X}\Big\{\min_{i \in I}[\mu_i(x)/(m_i + 1)]\Big\},$$

where the μ-criterion optimal element $x^0 \in X$ satisfies the inequalities $\mu_i(x^0) \ge (m_i + 1)\mu^0(\triangleright)$, $i \in I$.

(3) If priority is specified as $i \triangleright I^i_s$ $(s = 1, \ldots, m_i, i \in I)$, where $I^i_s \subseteq I$, the μ-criterion is stated in the following way:

$$\mu^0(\triangleright) = \max_{\substack{0 \le \mu \le 1 \\ X_\mu(\triangleright) \ne \varnothing}} \mu\,,$$

here

$$X_\mu(\triangleright) = \left\{ x \in X : \mu_i(x) \geq \left[\sum_{s=1}^{m_i} m(I_s^i) + 1 \right] \mu, \quad i \in I \right\},$$

$m(I_s^i)$ is a number of elements of the set $I_s^i \subseteq I$, and the magnitude $\mu^0(\triangleright)$ is equal to

$$\mu^0(\triangleright) = \max_{x \in X} \left\{ \min_{i \in I} \left\{ \mu_i(x) \Big/ \left[\sum_{s=1}^{m} m(I_s^i) + 1 \right] \right\} \right\}$$

here the μ-criterion optimal element $x^0 \in X$ satisfies the inequality

$$\mu_i(x^0) \geq \left[\sum_{s=1}^{m_i} m(I_s^i) + 1 \right] \mu^0(\triangleright).$$

Consider the possibility of applying the μ-criterion to derive the weight coefficients of linear convolution. In the vector optimization theory one of the most essential principles of optimality is the linear convolution criterion of the form

$$g_\mu(x^0|\alpha) = \sum_{i=1}^{m} \alpha_i \mu_i(x),$$

where the weight coefficients $\alpha_1, \ldots, \alpha_m$ satisfy the conditions $0 \leq \alpha_i \leq 1, (i = \overline{1,m}), \sum\limits_{i=1}^{m} \alpha_i = 1$. In those cases when the weight coefficients $\alpha_1, \ldots, \alpha_m$ are specified, the optimal element $x^0 \in X$ is derived from the condition $g_\mu(x^0|\alpha) = \max\limits_{x \in X} g_\mu(x|\alpha)$.

The problem of formulating the methods for obtaining weight coefficients still remains unsolved. Consider the application of the μ-criterion when solving this problem. In the above priority specification cases, the μ-criterion for obtaining weight coefficients is stated in the following way.

(1) Specifying a simple relation of priority (the case of lexicographic order) of the form $1 \triangleright 2 \triangleright \cdots \triangleright i \triangleright i+1 \triangleright \cdots \triangleright m$, obtain

$$\mu_{\text{lex}}^0 = \max_{\substack{0 \leq \mu \leq 1 \\ A_\mu^{\text{lex}} \neq \varnothing}} \mu,$$

where $A_\mu^{\text{lex}} = \{(\alpha_1, \ldots, \alpha_m) \in A : \alpha_i \geq (m - i + 1)\mu \ (i = \overline{1,m})\}$, $A = \left\{ (\alpha_1, \ldots, \alpha_m) \in R^m,\ 0 \leq \alpha_i \leq 1 \ (i = \overline{1,m}),\ \sum\limits_{i=1}^{m} \alpha_i = 1 \right\}$, and the μ_{lex}^0 satisfies the condition

$$\mu^0_{\text{lex}} = \max_{(\alpha_1,\ldots,\alpha_m)\in A} \left\{ \min_{i\in I}[\alpha_i/(m-i+1)] \right\}.$$

Moreover, $\mu^0_{\text{lex}} = 2/[m(m+1)]$.

In this case the μ-criterion optimal weight coefficients $\alpha^0_1, \ldots, \alpha^0_m$ satisfy the conditions ($i \in I$)

$$0 \leq \alpha^0_i \leq 1\,, \quad \alpha^0_i \geq 2(m-i+1)/[m(m+1)]\,, \quad \sum_{j=1}^{m} \alpha^0_j = 1\,.$$

These conditions, particularly, are satisfied by the weight coefficients of the form $\hat{\alpha}^0_i = 2(m-i+1)/[m(m+1)]$, $(i = 1, \ldots, m)$, which represent the point estimates for a simple relation of priority and implement the maximin problem solution.

(2) Specifying the binary relation of the priority $i \triangleright j$ for $j \in I_i = \{l \in \{\overline{1,m}\} : l \triangleright i\}$, $i \in I$, obtain

$$\mu^0(\triangleright) = \max_{\substack{0\leq\mu\leq 1 \\ A_\mu(\triangleright)\neq\varnothing}} \mu\,,$$

where $A_\mu(\triangleright) = \{(\alpha_1, \ldots, \alpha_m) \in A : \alpha_i \geq (m_i+1)\mu,\ i \in I\}$ and the magnitude $\mu^0(\triangleright)$ is represented as

$$\mu^0(\triangleright) = \max_{(\alpha_1,\ldots,\alpha_m)\in A} \left\{ \min_{i\in I}[\alpha_i/(m_i+1)] \right\}.$$

It may be shown that

$$\mu^0(\triangleright) = 1 \Bigg/ \left[\sum_{j=1}^{m} (m_j+1) \right],$$

and the μ-criterion optimal weight coefficients $\alpha^0_1, \ldots, \alpha^0_m$ satisfy the conditions ($i \in I$), $0 \leq \alpha^0_i \leq 1$,

$$\alpha^0_i \geq (m_i+1) \Bigg/ \left[\sum_{j=1}^{m} (m_j+1) \right], \quad \sum_{j=1}^{m} \alpha^0_j = 1\,.$$

These conditions are satisfied, in particular, by the weight coefficients $(i \in I)$, $\hat{\alpha}_i^0 = (m_i + 1)\Big/\Big[\sum_{j=1}^{m}(m_i + 1)\Big]$ which complete the maximin problem solution.

(3) Specifying priority of the form $i \rhd I_s^i$ $(s = 1, \ldots, m_i; i = 1, \ldots, m)$, where $I_s^i \subseteq I$, obtain

$$\mu^0(\rhd) = \max_{\substack{0 \le \mu \le 1 \\ A_\mu(\rhd) \ne \varnothing}} \mu\,,$$

$$A_\mu(\rhd) = \left\{(\alpha_1, \ldots, \alpha_m) \in A : \alpha_i \ge \left[\sum_{s=1}^{m_i} m(I_s^i) + 1\right]\mu, \quad \forall i \in I\right\},$$

$$\mu^0(\rhd) = \max_{(\alpha_1,\ldots,\alpha_m)\in A}\left\{\min_{i\in I}\left\{\alpha_i\Big/\left[\sum_{s=1}^{m_i} m(I_s^i) + 1\right]\right\}\right\};$$

it may be shown that

$$\mu^0(\rhd) = 1\Bigg/\left\{\sum_{j=1}^{m}\left[\sum_{s=1}^{m_j} m(I_s^i) + 1\right]\right\},$$

the μ-criterion optimal weight coefficients satisfy the conditions

$$0 \le \alpha_i^0 \le 1,\ \sum_{j=1}^{m}\alpha_j^0 = 1,\ \alpha_i^0 \ge \left[\sum_{s=1}^{m_i} m(I_s^i) + 1\right]\Bigg/\left[\sum_{j=1}^{m}\left\{\sum_{s=1}^{m_j} m(I_s^i) + 1\right\}\right].$$

These conditions, in particular, are satisfied by the weight coefficients of the form

$$\hat{\alpha}_i^0 = \left[\sum_{s=1}^{m_i} m(I_s^i) + 1\right]\Bigg/\left\{\sum_{j=1}^{m}\left[\sum_{s=1}^{m_j} m(I_s^i) + 1\right]\right\}$$

which realize the solution of the appropriate maximin problem.

The above specific cases of specifying and incorporating priority enable one to formulate general approaches to employing the maximal efficiency principle and obtaining the weight coefficients of linear convolution to solve the problem of incorporating more complex priority specification forms in multiobjective optimization problems.

Chapter 11

PROBLEMS OF MULTIOBJECTIVE OPTIMIZATION UNDER INFORMATION DEFICIENCY

§41. PROBLEMS OF DECISION MULTIOBJECTIVE OPTIMIZATION UNDER UNCERTAINTY

The situation of decision making in multiobjective optimization is referred to as the pair $\{X, f\}$, here $X = (x_1, \ldots, x_{m_X})$ is a set of decisions of the governing unit $M; f(x) = \{f_l(x)\}_{l=1}^{L}$ is a vector of the estimate functionals defined on X and assuming their values from R^1. In the thus defined multiobjective situation of decision making $\{X, f\}$, the decision making problem resides in the fact that the governing unit M must select one decision optimal by the principle of choice adopted by this unit. The multiobjective decision making problem is associated with three factors $\{u, v, w\}$, here u is a normalization procedure; v is a convolution criterion; w is a priority relation. In the given case the normalization procedure u is identified as a unique representation of R^I in R^L. Normalization is used for substitution of an ingredient and transition to the comparable scales in the estimate functionals. The priority relation w is referred to as the vector of estimates $(w_1, \ldots, w_L)$, in particular cases priority may be specified by the relation of order. The convolution criterion v is identified as the rule determining the decision for a given situation of multiobjective decision making which may be adopted as optimal. The convolution criterion represents, as a rule, the functional mapping R^L in R^1.

The main directions of research into the multiobjective decision making problems should involve development of the choice of the main elements of convolution, the methods for normalization and priority incorporation, and development of approaches to the problems of analyzing and substantiating the choice of these elements.

Suppose $\{X, f\}$ is a decision making situation, where X is the set of decisions of the management unit M; $f(x)$ is the vector estimate functional on the solution $x \in X$. Assume that we have two sets: $U = \{u_q, q \in Q\}$, the set of convolution criteria and $V = \{v_\omega, \omega \in \Omega\}$, the set of normalization procedures.

An unambiguous definition of the normalization procedure and convolution criterion is complex and even impractical for many vector optimization problems.

In our case we have several versions corresponding to various conditions for selecting an element from the set:

the choice of a unique element from the set (A) is determined;
none of the elements of the set is satisfactory (B);
all the elements of the set are acceptable (C).

In this case for the set U we have the cases A_I, B_I, C_I, for $V - A_2$, B_2, C_2 and for the multiobjective decision problem we have nine versions: A_IA_2; A_IB_2; A_IC_2; B_IA_2; B_IB_2; B_IC_2; C_IA_2; C_IB_2; C_IC_2. Let $\mathcal{F}_q(x)$ denote the vector estimate functional for $x \in X$ obtained after employing q normalization ($q \in Q$), similarly (with $\omega \in \Omega$) let $v_\omega^q(x) = v_\omega(\mathcal{F}_q(\omega))$ for $x \in X$, $v_\omega^q(x^{q\omega}) = \operatorname*{opt}_{x_k \in X} v_\omega^q(x_k)$.

For B_I we assume that the convolution criterion is determined on the non-normalized components of the estimate functional and, therefore, the situation coincides with A_I. In the cases A_IA_2 the unchanging index q or ω is omitted. Consider several approaches to formulating the problems of choosing u, v based on the guaranteed result criterion and employment of the uncertainty functions. Approaches to formulation of these problems may be divided into two groups: the first includes convolution procedures with regard to all convolution criteria (or all normalization procedures, or both according to a version); the second comprises the methods of determining one convolution criterion (a normalization procedure). Denote:

$$\delta_\omega^q(x_k) = [v_\omega^q(x^{q\omega}) - v_\omega^q(x_k)]/v_\omega^q(x^{q\omega}),$$

$$\gamma_\omega^q(x_k) = \max_{l=1,\dots,L}\{[\mathcal{F}_q^l(x^{q\omega}) - \mathcal{F}_q^l(x_k)]/\mathcal{F}_q^l(x^{q\omega})\},$$

$$\varepsilon_\omega^q(x_k) = \|x^{q\omega} - x_k\|,$$

here $\|\cdot\|$ denotes a number of nonzero components. Consider the formulation procedures for the problem of choosing u, v for each version. Thus we have:

In the version $A_I A_2$ a unique criterion of v convolution is used for a specified unique normalization of u when deriving $x_{k^0} \in X$ as an optimal solution from the condition $v\{u[f(x_{k^0})]\} = \max_{x_k \in X} v\{u[f(x_k)]\}$.

In the version $B_I A_2$ a unique criterion for v convolution is used when obtaining x_{k^0} as an optimal solution from the condition (the estimate functional components are not normalized)

$$v[f(x_{k^0})] = \max_{x_k \in X} v[f(x_k)].$$

In the version $C_I A_2$ we have a set of normalized vector estimate functionals $\mathcal{F}_q(x_k)$ and, respectively, $\delta^q(x_k)$, $\gamma^q(x_k)$, $\varepsilon^q(x_k)$ with $q \in Q$, $x_k \in X$, and the unique criterion for v convolution is utilized when obtaining a solution to one of the following problems:

(1) $H(x_{k^0}) = \max_{x_k \in X} H(x_k)$, $H(x_k) = -\sum_{q \in Q} \bar{v}^q(x_k) \ln \bar{v}^q(x_k)$,

$\bar{v}^q(x_k) = v\Big[\mathcal{F}_q(x_k)\Big] \Big/ \Big\{ \sum_{q \in Q} v[\mathcal{F}_q(x_k)] \Big\}$;

(2) $\Delta^{q_0}(x_{k^0}) = \min_{x_k \in X} \max_{q \in Q} \Delta^q(x_k)$, here as $\Delta^q(x_k)$ we can use $\delta^q(x_k)$, $\gamma^q(x_k)$, $\varepsilon^q(x_k)$ with the fixed $\omega \in \Omega$ corresponding to the specified convolution of v;

(3) $H(q^0) = \max_{q \in Q} H(q)$, $H(q) = -\sum_{q_1 \in Q} \bar{v}^q(x^{q_1 \omega}) \ln \bar{v}^q(x^{q_1 \omega})$;

(4) $\Delta^{q_0}(x^{q_1 \omega}) = \min_{q \in Q} \max_{q_1 \in Q} \Delta^q(x^{q_1 \omega})$.

In the versions $A_I B_2$; $B_I B_2$; $C_I B_2$ we have respectively a unique normalization of $u[f(x)]$, the non-normalized $f(x)$ and a set of normalized $\mathcal{F}_q(x)$ with $q \in Q$, $x \in X$, and convolution criteria are unacceptable. In this case, in the above problems (1)–(4) the choice of normalization and an optimal solution are formulated in the same way, if formally as a criterion for v convolution we employ transformation of summing by the estimate functional components $v[\mathcal{F}_q(x_k)] = \sum_{l=1}^{L} \mathcal{F}_q^l(x_k)$.

In the versions $A_I C_2$; $B_I C_2$; $C_I C_2$ we have, respectively, the unique normalization of $u[f(x)]$, the non-normalized $f(x)$, a set of normalized

$\mathcal{F}_q(x)$ with $q \in Q$, $x \in X$ and the set $v_\omega^q(x) = v_\omega[\mathcal{F}_q(x)]$ of convolution criteria with $\omega \in \Omega$.

In the version $C_I C_2$ the problems of the optimal choice of the optimal solution according to the multiobjective estimate functional are formulated as follows:

(1) $H(x_{k^0}) = \max\limits_{x_k \in X} H(x_k)$, $H(x_k) = -\sum\limits_{q \in Q} \bar{v}_\omega^q(x_k) \ln \bar{v}_\omega^q(x_k)$,

$$\bar{v}_\omega^q(x_k) = v_\omega^q(x_k) \Bigg/ \left[\sum_{\substack{q_1 \in Q \\ \omega_1 \in \Omega}} v_{\omega_1}^{q_1}(x_k) \right];$$

(2) $\Delta_{\omega^0}^{q_0}(x_{k^0}) \min\limits_{x_k \in X} \max\limits_{\substack{q \in Q \\ \omega \in \Omega}} \Delta_\omega^q(x_k)$, here as $\Delta_\omega^q(x)$ we may use $\delta_\omega^q(x)$, $\gamma_\omega^q(x)$, $\varepsilon_\omega^q(x)$;

(3) $H(q^0, \omega^0) = \max\limits_{\substack{q \in Q \\ \omega \in \Omega}} H(q, \omega)$, $H(q, \omega) = -\sum\limits_{\substack{q_1 \in Q \\ \omega_1 \in \Omega}} \hat{v}_\omega^q(x^{q_1 \omega_1}) \ln \hat{v}_\omega^q(x^{q_1 \omega_1})$,

$$\hat{v}_\omega^q(x^{q_1 \omega_1}) = \bar{v}_\omega^q(x^{q_1 \omega_1}) \Bigg/ \left[\sum_{\substack{q_2 \in Q \\ \omega_2 \in \Omega}} \bar{v}_\omega^q(x^{q_2 \omega_2}) \right];$$

(4) $\Delta_{\omega_0}^{q_0}(x^{q_1^0 \omega_1^0}) = \min\limits_{\substack{q \in Q \\ \omega \in \Omega}} \max\limits_{\substack{q_1 \in Q \\ \omega_1 \in \Omega}} \Delta_\omega^q(x^{q_1 \omega_1})$.

In the version $A_I C_2$ these problems are formulated provided that the set Q consists of the unique element q corresponding to the specified normalization $u[f(x)]$ and for the version $B_I C_2$ utilization of the non-normalized $f(x)$ means that Q also consists of the unique element q corresponding here to the identical normalization $u[f(x)] \equiv f(x)$ with $x \in X$.

In the general case the convolution criteria are dependent on the priority incorporation procedure w, in other words, $v_w[\mathcal{F}_q(x)] = v[\mathcal{F}_q(x)|w]$ with $x \in X$. Suppose we have the specified set W of the w priority incorporation procedures, then it is possible to determine the cases A_3, B_3, C_3 as before and formulate the problems of choice for each of the twenty-seven versions, at the same time we can consider $v_w[\mathcal{F}_q(x)]$ for the fixed normalizations of u, the priority incorporation of w and the convolution of v just as different convolution procedures with different $w \in W$, and then the cases A_3, B_3, C_3 go over into A_2, B_2, C_2, respectively.

Consider the multiobjective decision problems under uncertainty.

In the case of the unique estimate functional in decision making under uncertainty there arises the multiobjective decision problems determined by specifying the set of criteria $k_s = \{\nu_s^1, \ldots, \nu_s^{k_s}\}$ of decision making in various informational situations I_s of opening the uncertainty. Within the limits of the fixed informational situation I_s, we can specify several estimate functionals $\{f_s^l\}_{l=1}^L$ which can form a multiobjective indicator both for the specified criterion $\nu_s \in k_s$ and for a set of criteria from k_s. The problems constituting special cases of possible versions of formulating multiobjective decision problems under uncertainty are given below.

Assume that the management unit M has $L(L > 1)$ situations of decision making $\{X, Q, f^1\}$, $\{X, Q, f^2\}$, $\ldots$, $\{X, Q, L\}$ differing from one another by the estimate functional, and it is required to determine one multiobjective decision for all L situations of decision making under uncertainty of selecting the state $Q_j \in Q$ by the environment. The above first multiobjective decision problem under uncertainty arises when it is necessary to incorporate several indicators say, expenditures and efficiency. The multiobjective situation $\{X, f\}$ of decision making in the present problem will have the estimate functionals $f(x|Q)$ depending on Q, thus we have $f(x_k|Q) = \{f^l(x_k|Q_j)\}_{j,l=1}^{n,L}$ with $x_k \in X$. Employing the principal elements $\{u, v, w\}$ of multiobjective decision making enables one to consider the situation of decision making under uncertainty with one scalar estimate functional. In the obtained situation of decision making we determine the informational situation I_s, $s \in \{1, \ldots, S\}$ and choose a decision criterion. The subsequent process of decision making is carried out according to the criterion for the informational situation obtained.

Describe the second multiobjective decision problem under uncertainty. Assume that the unit M has L situations of decision making $\{X, Q, f^1\}$, $\{X, Q, f^2\}, \ldots, \{X, Q, f^L\}$ differing from one another by the estimate functional. Assume that for all L decision situations the same information situation I_s prevails and the unit M has chosen a decision criterion. Having applied the decision criterion in each of the L situations, we obtain the multiobjective decision situation. We next choose the principal factors of multiobjective decision making $\{u, v, w\}$, obtain the optimal solution $x^0 \in X$ or using now in the multiobjective decision situation the maximal efficiency principle we get the optimal solution $x^0 \in X$.

Formulate the third decision problem under uncertainty. Assume that the unit M in the decision situation $\{X, Q, f\}$ has the information situation

I_s. The given information situation I_s is correlated with the set of decision criteria $k_s = \{\nu_s^1, \ldots, \nu_s^{k_s}\}$ from which the management unit M determines not one criterion but the subset $k_s^0 \subseteq k_s$. Applying each of the decision criteria $\nu_s \in k_s^0$ to the present situation of decision making we obtain for each decision x_s^0 the vector of the estimate functional values and, consequently, the multiobjective decision situation in which a decision is chosen in the same way as had been done in the above problems.

Assume that in the fourth multiobjective decision problem the unit M has the decision situation $\{X, Q, f\}$ and let the set of the information situations I be specified. For each of the information situations $I_s \subseteq I$ the management unit determines the decision criterion $\nu_s^0 \in k_s^0$. This situation of decision making prevails when the medium changes its state after the management unit has made a decision. Applying each of the chosen criteria to the decision situation $\{X, Q, f\}$ we get the multiobjective decision situation.

The last example of the decision problem under uncertainty constitutes a combination of several problems given above. Suppose we have a combination of problems of the second and third types where the management unit M has L situations of making decisions $\{X, Q, f^l\}_{l=1}^L$ pertaining to one information situation I_s identified with the set of decision criteria k_s from which the management unit M determines a subset of criteria. Solving, for each of the L situations, the multiobjective decision problem of the third type we obtain the multiobjective decision problem of the second type.

The problems of optimality and improvability in the multiobjective indicator components have been considered above for the case of Pareto-optimality. Here we consider the necessary and sufficient conditions of improvability in the multiobjective decision problems. Suppose we have the multiobjective decision situation $\{X, f\}$. We call the solution $x \in X = \{x_1, \ldots, x_{m_X}\}$ unimprovable (Pareto-optimal) if there is no solution $x' \in X$ such that $f^l(x') \geq f^l(x)$ for any $l = 1, \ldots, L$, and if only one of the inequalities is strict. Hence, the solution $x \in X$ is termed improvable (on the set X by the multiobjective indicator) if there exists the solution $x' \in X$ such that $f^l(x') \geq f^l(x)$ for any $l = 1, \ldots, L$, and if only one of the inequalities is strict.

The Pareto set is independent of normalization and monotone transformations when incorporating priority and choosing convolution that is why employment of the Pareto-optimal multiobjective solutions, particularly under uncertainty, is the most promising. Moreover, the Pareto-optimality is one of the axioms (of natural conditions) of determining the principle of choice by the multiobjective indicator.

Let us formulate the necessary and sufficient conditions for improvability of multiobjective decisions in the situations $\{X, f\}$.

Theorem 41.1: The solution $x^0 \in X$ is improvable in the multiobjective decision situation $\{X, f\}$ if and only if there exists the L-dimensional vector $\gamma = (\gamma_1, \ldots, \gamma_L)$ such that the inequalities

$$f^l(x^0) \leq c_l(\gamma), \quad f^{l_0}(x^0) < c_{l_0}(\gamma) \tag{41.1}$$

hold for all $l = 1, \ldots, L$ and if only for one $l_0 \in \{1, \ldots, L\}$ where there is

$$c_l(\gamma) = c(\gamma) - \gamma_l, \quad c(\gamma) = \max_{x \in X} \min_{l=1,\ldots,L} [f^l(x) + \gamma_l]. \tag{41.2}$$

Theorem 41.2: The solution $x^0 \in X$ is improvable in the multiobjective decision situation $\{X, f\}$ if the inequalities

$$f^l(x^0) \leq c_l(\bar{f}), \; l = 1, \ldots, L; \; f^{l_0}(x^0) < c_{l_0}(\bar{f}), \; l_0 \in \{1, \ldots, L\} \tag{41.3}$$

hold where the constants $c_l(\bar{f})$ are defined as $(l = 1, \ldots, L)$

$$c_l(\bar{f}) = \min_{x \in X} f^l(x) + c(\bar{f}) \Big[\max_{x \in X} f^l(x) - \min_{x \in X} f^l(x) \Big], \tag{41.4}$$

$$c(\bar{f}) = \max_{x \in X} \Big[\min_{l=1,\ldots,L} \bar{f}^l(x) \Big], \tag{41.5}$$

here $\bar{f}^l(x)$ denotes the expression of the lth component of the multiobjective indicator $f(x)$ normalized as follows:

$$\bar{f}^l(x) = \Big[f^l(x) - \min_{y \in X} f^l(y) \Big] \Big/ \Big[\max_{y \in X} f^l(y) - \min_{y \in X} f^l(y) \Big].$$

In this case, note that the magnitudes $c(\gamma)$ and $c(\bar{f})$ of the form (41.2) and (41.5) can be obtained (for the given L-dimensional vector γ) by solving the extremal problems:

$$c(\gamma) = \lambda_0(\gamma) = \max_{\substack{\lambda \in R^1 \\ x \in X}} \{\lambda | [f^l(x) + \gamma_l] \geq \lambda, \quad l = 1, \ldots, L\}, \tag{41.6}$$

$$c(\bar{f}) = \lambda_0 = \max_{\substack{0 \leq \lambda \leq 1 \\ x \in X}} \{\lambda | \bar{f}^l(x) \geq \lambda, \quad l = 1, \ldots, L\}. \tag{41.7}$$

Theorem 41.3: There is no solution $x^0 \in X \setminus \bar{X}$ such that for all $l = 1, \ldots, L$ the inequalities

$$f^l(x^0) \geq c_l(\bar{f}) \tag{41.8}$$

are legitimate where the constants $c_l(\bar{f})$ take the form (41.4), and via $\bar{X}$ we denote the set of solutions $\bar{x} \in X$ of the maximin problem (41.5), i.e.,

$$\bar{X} = \left\{\bar{x} \in X : \min_{l=1,\ldots,L} f^l(\bar{x}) = \max_{x\in X}\left[\min_{l=1,\ldots,L} \bar{f}^l(x)\right]\right\}.$$

The result of theorem (41.3) supplements the sufficient conditions of (41.3) for improvability of the solution $x^0 \in X$ in the situation $\{X, f\}$.

Corollary 41.1: The set $\bar{X} \subseteq X$ of the solutions $\bar{x} \in X$ of the maximin problem (41.5) or problem (41.7) coincides with the set $X_{\lambda_0} = \{x \in X : f^l(x) \geq \lambda_0, l = 1, \ldots, L\}$ of solutions from X satisfying the maximal efficiency principle in the multiobjective decision situation $\{X, f\}$.

Theorem 41.4: Any solution $\bar{x} \in \bar{X} \subseteq X$ cannot be improved in the multiobjective decision situation $\{X, f\}$.

Theorem 41.5: The vectors $\gamma = (\gamma_1, \ldots, \gamma_L)$, for which the inequalities

$$f^l(x) \leq c_l(\gamma), \quad f^{l_0} < c^{l_0}(\gamma) \tag{41.9}$$

hold for all $l \in \{1, \ldots, L\}$ and a particular $l_0 \in \{1, \ldots, L\}$ with some solution $x \in X$, satisfy the inequalities

$$\gamma_{l_2} - \gamma_{l_1} \leq \max_{x\in X} f^{l_1}(x) - \min_{x\in X} f^{l_2}(x) \tag{41.10}$$

for all $l_1, l_2 \in \{1, \ldots, L\}$.

Corollary 41.2: If we assume that components of the multiobjective estimate functional f are normalized in the natural way, then the L-dimensional vectors $\gamma = (\gamma_1, \ldots, \gamma_L) \in G$ satisfying inequalities (41.10) belong to the set $G(\bar{f})$ of the form $G(\bar{f}) = \{\gamma \in R^L : |\gamma_{l_1} - \gamma_{l_2}| < 1\}, l_1, l_2 \in \{1, \ldots, L\}$.

Theorem 41.6: The solution $x^0 \in X$ can be improved in the situation $\{X, f\}$ if and only if there exists the L-dimensional vector $\gamma \in G$ for which inequalities (41.1) are fulfilled.

Theorem 41.7: The solution $x^0 \in X$ cannot be improved (Pareto-optimal) in the multiobjective decision situation only when there exists the vector

$\gamma \in G$ such that

$$\min_{l=1,\dots,L}[f^l(x^0)+\gamma_l] = \max_{x\in X}\min_{l=1,\dots,L}[f^l(x)+\gamma_l]\,. \tag{41.11}$$

Corollary 41.3: Condition (41.11) of theorem (41.7) is sufficient for concave $f^l(x)$ with $x \in X$, $l = 1,\dots,\mathcal{L}$, i.e. the set of Pareto-optimals solutions in the situation $\{X, f\}$ coincides in this case with the set $X^0 = \{x \in X \,|\, \exists\gamma \in G : (41.11)\}$, which determines characterization of the set of Pareto-optimal solutions differing from the tradeoff domain.

Theorem 41.8: For the solution $x^0 \in X$ to be improvable (Pareto-optimal) in the situation $\{X, f\}$, it is necessary and sufficient that the following inequalities are fulfilled (inconsistent)

$$f^l(x_0) \le \max_{\gamma\in G}\Big\{\max_{x\in X}\Big[\min_{l'=1,\dots,L}(f^{l'}(x)+\gamma_{l'})\Big] - \gamma_l\Big\}\,.$$

The proof of the above and subsequent statements is given in the author's work.[8]

In the multiobjective decision situation $\{X, f\}$ under uncertainty it is possible to obtain the necessary and sufficient conditions for improvability of the solutions $x^0 \in X$, which is achieved in the same way via concretization of multiobjective information situations.

Consider some problems of multiobjective decision stability.

Suppose we have the multiobjective decision situation $\{X, f\}$ in which $X = \{x_1, \dots, x_{m_X}\}$, $f(x) = \{f^l(x)\}_{l=1}^{L}$, here the components of $f(x)$ are specified in a positive ingredient. We say that the solution $x^0 \in X$ is Pareto-optimal on X in that there is no $x^* \in X$ for which the inequalities $f^l(x^*) \ge f^l(x^0)$, $\forall l \in \{1,\dots,L\}$ are fulfilled and if only one inequality is strict. Let the set of Pareto-optimal solutions for the multiobjective indicator f be denoted via $M(f)$. The stability of multiobjective decisions is understood as the property to remain optimal with small changes in the specification of the relation of natural order $\hat{N}^{\text{dom}}(f)$ used in the Pareto-optimality principle. Define the following estimation of stability of the solution x^0 on the set $X : \varepsilon(x^0) = \min_{x\in X}\|f(x^0) - f(x)\|_{N^{m_X}}$, here $\|\cdot\|_{N^{m_X}}$

can be defined, say, with $n \in N^{m_X}$ in the following way:

$$\|n\| = \max_{1 \le i \le m_X} \max[0, n_i],$$

$$\|n\| = \sum_{i=1} \max[0, n_i],$$

$$\delta(x^0) = \min_{x \in X \setminus M(f)} \|x^0 - x\|_{N^{m_X}}.$$

Denote by $\succ$ the relation of order $\hat{N}^{\text{dom}}(f)$. For arbitrary numbers $\varepsilon > 0$, $\delta > 0$ define the sets

$$M_\varepsilon(f) = \{x \in X \,|\, \bar{\exists} x' \in X : f(x') \underset{\varepsilon}{\succ} f(x)\},$$

$$M_\delta(f) = \{x \in X \,|\, \bar{\exists} x' \in X : f(x') \underset{\delta}{\succ} f(x)\},$$

here by $\underset{\varepsilon}{\succ}$, $\underset{\delta}{\succ}$ we denote the relations of order $N(f) \subset N^L$ determined by the following conditions $f(x') \underset{\varepsilon}{\succ} f(x)$ if and only if there exists $\alpha \in N^L$, $\alpha \ge 0$, $\|\alpha\|_{N^L} \le \varepsilon$ such that $f(x') \succ f(x) - \alpha$; $f(x') \underset{\delta}{\succ} f(x)$ if and only if $\|x' - x\|_{N^{m_X}} \ge \delta$, $f(x') \succ f(x)$. We call the sets $M_\varepsilon(f)$, $M_\delta(f)$ respectively the regions of ε and δ stability, and the magnitudes ε and δ the radii of multiobjective decision stability. Thus we have:

Theorem 41.9: The conditions $M_\varepsilon(f) \subseteq M(f) \subseteq M_\delta(f)$ are fulfilled.

The absolute stability domain is identified as $M_\alpha(f) = \{x \in X \,|\, \bar{\exists} x' \in X : f(x') \underset{\alpha}{\succ} f(x)\}$, here $f(x') \underset{\alpha}{\succ} f(x)$ with $\alpha = (\varepsilon, \delta) \in R^2$ if and only if $[f(x') \underset{\varepsilon}{\succ} f(x)] \wedge \{[\|x' - x\|_{N^{m_X}} \le \delta] \vee [f(x') \underset{\delta}{\succ} f(x)]\}$ and $M_\alpha(f) \subseteq M(f)$

Suppose we have the following sets with $\varepsilon > 0$, $\delta > 0$:

$$M^*_\varepsilon(f) = \{x \in X \,|\, \bar{\exists} x' \in X : f(x') \overset{*}{\underset{\varepsilon}{\succ}} f(x)\},$$

$$M^*_\delta(f) = \{x \in X \,|\, \bar{\exists} x' \in X : f(x') \overset{*}{\underset{\delta}{\succ}} f(x)\},$$

here $f(x') \overset{*}{\underset{\varepsilon}{\succ}} f(x)$ if and only if for any $\alpha \in N^L$, $\alpha \ge 0$, $\|\alpha\|_{N^L} = 1$ the conditions $f(x') \underset{\varepsilon}{\succ} f(x) - \alpha$ are fulfilled, similarly $f(x') \overset{*}{\underset{\delta}{\succ}} f(x)$ if and only if $f(x') \underset{\delta}{\succ} f(x'')$ for any $x'' \in X$ such that $\|x - x''\|_{N^{m_X}} \le \delta$.

Theorem 41.10: Fulfill the following relations:

$$M^*_\varepsilon(f) \subseteq M(f) \subseteq M^*_\delta(f).$$

Define with $\alpha = (\varepsilon, \delta)$ the set

$$M_\alpha^*(f) = \{x \in X \mid \bar{\exists} x' \in X : f(x') \underset{\alpha}{\overset{*}{\succ}} f(x)\},$$

here $f(x') \underset{\alpha}{\overset{*}{\succ}} f(x)$ if and only if for any $x'' \in X$ such that $\|x - x''\|_{N^m X} \leq \delta$ the condition $f(x'') \underset{\varepsilon}{\overset{*}{\succ}} f(x)$ holds.

Suppose we have the set $S(f) = \{x \in X \mid \bar{\exists}\gamma \in N^L : f(x) \prec \bar{c}(\gamma)\}$, here $\bar{c}(\gamma) = (c_1(\gamma), \ldots, c_L(\gamma))$, $c_l(\gamma) = c(\gamma) - \gamma_l$, $c(\gamma) = \max_{x \in X} \min_{1 \leq l \leq L} [f^l(x) + \gamma_l]$.

Theorem 41.11: There exists $M(f) = S(f)$.

Define

$$S_\delta(f) = \{x \in X \mid \bar{\exists}\gamma \in N^L : f(x) \prec \bar{c}_\delta(\gamma, x)\},$$

$$S_\varepsilon(f) = \{x \in X \mid \bar{\exists}\gamma \in N^L : f(x) \underset{\varepsilon}{\prec} \bar{c}(\gamma)\},$$

$$S_\delta^*(f) = \{x \in X \mid \bar{\exists}\gamma \in N^L : f_\delta^*(x) \underset{\varepsilon}{\prec} \bar{c}(\gamma)\},$$

$$S_\varepsilon^*(f) = \{x \in X \mid \bar{\exists}\gamma \in N^L : f(x) \prec \bar{c}_\varepsilon^*(\gamma)\},$$

where

$$c_\delta(\gamma) = \max_{\substack{x' \in X \\ \|x - x'\|_{N^m X} \geq \delta}} \min_{1 \leq l \leq L} [f^l(x') + \gamma_l], \quad c_\varepsilon^*(\gamma) = c(\gamma) + \varepsilon,$$

$$f_\delta^{*l}(x) = \max_{\substack{x' \in X \\ \|x - x'\|_{N^m X} \leq \delta}} f^l(x').$$

Theorem 41.12: The following conditions are satisfied: $S_\delta(f) = M_\delta(f)$, $S_\delta^*(f) = M_\delta^*(f)$, $S_\varepsilon(f) = M_\varepsilon(f)$, $S_\varepsilon^*(f) = M_\varepsilon^*(f)$.

Consider the set $B(f)$ of the form $B(f) = \{x \in X \mid \bar{\exists}\gamma \in N^L : f(x) \underset{\sim}{\succ} \bar{c}(\gamma)\}$. We have $S(f) \subseteq B(f)$. Similarly, it is possible to define the appropriate sets $B_\varepsilon(f)$, $B_\varepsilon^*(f)$, etc., and also the set $(\beta > 0)$

$$S_\beta(f) = \{x \in X \mid \bar{\exists}\gamma \in N^L : f(x) \prec \bar{c}^\beta(\gamma)\},$$

$$c^\beta(\gamma) = \max_{\substack{\gamma' \in N^L \\ \|\gamma - \gamma'\|_{N^m X} \leq \beta}} \max_{x' \in X} \min_{1 \leq l \leq L} [f^l(x') + \gamma_l'].$$

The above concepts of stability make it possible to extend the formulation of the solutions optimal by the principle of choice on the basis of generalizing the natural order $\hat{N}^{\rm dom}$ and the concepts of improvability of solutions by the multiobjective indicator so that they reflect different properties of sensibility, stability invariance, tolerance, vagueness, etc. of the relations of order and appropriate optimal multiobjective solutions.

§42. MULTIOBJECTIVE OPTIMIZATION PROBLEMS FOR DYNAMIC CONTROL SYSTEMS

The vector optimization problem for dynamic control systems with flexible extremities is referred to as the problem of the form: obtain the controls $\varphi^0(x) = (\varphi_1^0(x), \ldots, \varphi_z^0(x))$ of numbers x_0^0, x_1^0, vectors y_0^0, y_1^0 and trajectory $y(x|\varphi^0)$ of changes in the state of a dynamic system $dy/dx = f(y, \varphi^0(x), x)$, $x_0^0 \le x \le x_1^0$, $y = y_0^0$ with $x = x_0^0$, $y = y_1^0$ with $x = x_1^0$ from the condition for achieving the vector criterion minimum $V_0(x_0, x_1, y_0, y_1, \varphi) = \{V_0^1(x_0, x_1, y_0, y_1, \varphi), \ldots, V_0^N(x_0, x_1, y_0, y_1, \varphi)\}$ by choosing elements $\{x_0, x_1, y_0, y_1, \varphi\} \in M_1 \cap M_2$, where x is a scalar independent variable which is most commonly time; x_0, x_1 are the values of x determining the beginning and the end of the control action φ; the control φ is an r-dimensional vector-function $\varphi(x) = (\varphi_1(x), \ldots, \varphi_r(x)) \in L_r^2[x_0, x_1]$; y_0 and y_1 are the n-dimensional vectors of the origin and the extremity of the trajectory $y(x|y)$ of solving the differential equation

$$\frac{dy}{dx} = f(y, \varphi(x), x)\,, \quad x_0 \le x \le x_1 \tag{41.12}$$

otherwise $y = y_0$ with $x = x_0$, $y = y_1$ with $x = x_1$; M_1 is a set of elements $\{x_0, x_1, y_0, y_1, \varphi\} \in R^1 \times R^1 \times R^n \times R^n \times L_r^2[x_0, x_1]$, such that $a_0 \le x_0 \le b_0$, $a_1 \le x \le b_1$, $v_l(x_0, x_1, y_0, y_1) = 0$ $(l = 1, \ldots, k)$, $\varphi_j^1(x) \le \varphi_j(x) \le \varphi_j^2(x)$ with $x \in [x_0, x_1]$ $(j = 1, \ldots, r)$ and M_2 is a set of elements $\{x_0, x_1, y_0, y_1, \varphi\} \in M_1$ such that $V_i(x_0, x_1, y_0, y_1, \varphi) \le 0$ $(i = 1, \ldots, m)$ the functionals V_0^s, V_i $(s = \overline{1, N})$, $(i = \overline{1, m})$ take the form

$$V_0^s(x_0, x_1, y_0, y_1, \varphi) = g_0^s(x_0, x_1, y_0, y_1) + \int_{x_0}^{x_1} f_0^s(y(x|\varphi), \varphi(x), x)dx\,,$$

$$V_i(x_0, x_1, y_0, y_1, \varphi) = g_i(x_0, x_1, y_0, y_1) + \int_{x_0}^{x_1} f_i(y(x|\varphi), \varphi(x), x)dx\,.$$

Let there be given the numbers of a_0, b_0, a_1, b_1 ($b_0 \leq a_1$) r-dimensional vector-functions $\varphi^1(x)$, $\varphi^2(x) \in L_r^2[a_0, b_1]$; the n-dimensional vector-function $f(y, \varphi, x)$; the scalar functions $f_0^s(y, \varphi, x)$, $f_i(y, \varphi, x)$ ($s = 1, \ldots, N$; $i = 1, \ldots, m$) continuously differentiable with respect to $y \in R^n$, $\varphi \in R^r$ and piecewise-continuous in $x \in [a_0, b_1]$, the scalar functions $g_0^s(x_0, x_1, y_0, y_1)$, $g_i(x_0, x_1, y_0, y_1)$ continuously differentiable with respect to $x_0 \in [a_0, b_0]$, $x_1 \in [a_1, b_1]$, $y_0 \in R^n$, $y_1 \in R^n$. From the mathematical point of view it might be good to improve the vector optimization problem formulation, indicate the rule in the sense of how one should understand the solution of the multiobjective optimization problem for the dynamic control system with flexible extremities when the constraints have the form of equalities and inequalities (the principle of choosing an optimal control according to the vector criterion V_0). Consider natural conditions or axioms for the principle of choice in the dynamic control system with flexible extremities.

We call the order (or preference) the binary relation N between two elements $\{x_0^1, x_1^1, y_0^1, y_1^1, \varphi^1\}$, $\{x_0^2, x_1^2, y_0^2, y_1^2, \varphi^2\} \in M_1 \cap M_2$ if the following conditions are fulfilled:

(1) $\{x_0, x_1, y_0, y_1, \varphi\} N \{x_0, x_1, y_0, y_1, \varphi\}$ for any $\{x_0, x_1, y_0, y_1, \varphi\} \in M_1 \cap M_2$ (reflexivity);

(2) if $\{x_0^1, x_1^1, y_0^1, y_1^1, \varphi^1\} N \{x_0^2, x_1^2, y_0^2, y_1^2, \varphi^2\}$, $\{x_0^2, x_1^2, y_0^2, y_1^2, \varphi^2\} N \{x_0^3, x_1^3, y_0^3, y_1^3, \varphi^3\}$, then $\{x_0^1, x_1^1, y_0^1, y_1^1, \varphi^1\} N \{x_0^3, x_1^3, y_0^3, y_1^3, \varphi^3\}$ holds for any $\{x_0^1, x_1^1, y_0^1, y_1^1, \varphi^1\}$, $\{x_0^2, x_1^2, y_0^2, y_1^2, \varphi^2\}$, $\{x_0^3, x_1^3, y_0^3, y_1^3, \varphi^3\} \in M_1 \cap M_2$ (transitivity);

(3) either $\{x_0^1, x_1^1, y_0^1, y_1^1, \varphi^1\} N \{x_0^2, x_1^2, y_0^2, y_1^2, \varphi^2\}$, or $\{x_0^2, x_1^2, y_0^2, y_1^2, \varphi^2\} \times N \{x_0^1, x_1^1, y_0^1, y_1^1, \varphi^1\}$ for any $\{x_0^1, x_1^1, y_0^1, y_1^1, \varphi^1\}$, $\{x_0^2, x_1^2, y_0^2, y_1^2, \varphi^2\} \in M_1 \cap M_2$ (asymmetry);

(4) if $\{x_0^1, x_1^1, y_0^1, y_1^1, \varphi^1\} N \{x_0^2, x_1^2, y_0^2, y_1^2, \varphi^2\}$, $\{x_0^2, x_1^2, y_0^2, y_1^2, \varphi^2\} N \{x_0^1, x_1^1, y_0^1, y_1^1, \varphi^1\}$, then the elements $\{x_0^1, x_1^1, y_0^1, y_1^1, \varphi^1\}$, $\{x_0^2, x_1^2, y_0^2, y_1^2, \varphi^2\} \in M_1 \cap M_2$ are equivalent.

The order N is identified as the natural order if it is the order (preference) determined by the scalar real criterion values $V_0(x_0, x_1, y_0, y_1, \varphi)$, in other words, $\{x_0^1, x_1^1, y_0^1, y_1^1, \varphi^1\} N \{x_0^2, x_1^2, y_0^2, y_1^2, \varphi^2\}$, if and only if $V_0(x_0^1, x_1^1,$

$y_0^1, y_1^1, \varphi^1) \leq V_0(x_0^2, x_1^2, y_0^2, y_1^2, \varphi^2)$ for any $\{x_0^1, x_1^1, y_0^1, y_1^1, \varphi^1\}$, $\{x_0^2, x_1^2, y_0^2, y_1^2, \varphi^2\} \in M_1 \cap M_2$.

Each of the components V_0^s $(s = 1, \ldots, N)$ of the vector criterion establishes the natural order on the elements $\{x_0, x_1, y_0, y_1, \varphi\} \in M_1 \cap M_2$ we denote it by N_s. Assume that for the elements $\{x_0, x_1, y_0, y_1, \varphi\} \in M_1 \cap M_2$ and the vector criterion $V_0 = \{V_0^1, \ldots, V_0^N\}$ the relation of order $N(V_0^1, \ldots, V_0^N)$ is defined, then the vector optimization problem for a dynamic system is the problem of deriving such $\{x_0^0, x_1^0, y_0^0, y_1^0, \varphi^0\} \in M_1 \cap M_2$ that there exists $\{x_0^0, x_1^0, y_0^0, y_1^0, \varphi^0\} N(V_0^1, \ldots, V_0^N)\{x_0, x_1, y_0, y_1, \varphi\}$ for any $\{x_0, x_1, y_0, y_1, \varphi\} \in M_1 \cap M_2$.

It is the rule of specifying that kind of relation of order $N(V_0^1, \ldots, V_0^N)$ for the specified vector criterion $V_0 = \{V_0^1, \ldots, V_0^N\}$ that represents the principle of choice.

We will use the axioms derived above and formulate the axioms as applied to the dynamic control system with flexible extremities, and obtain the natural conditions to be satisfied with the principle of choice as the relation of order $N(V_0^1, \ldots, V_0^N)$:

(1) independence from positive linear transformation of special criteria, thus we have $N(V_0^1, \ldots, V_0^N) \Longleftrightarrow N(k_1 V_0^1 + l_1, \ldots, k_N V_0^N + l_N)\, \forall k_1, \ldots, k_N > 0, l_1, \ldots, l_N$;

(2) the condition of independence when choosing the elements $\{x_0, x_1, y_0, y_1, \varphi\} \in M_1 \cap M_2$, in other words, if for any two vector criteria there exists ${}^*V_0 = \{{}^*V_0^1, \ldots, {}^*V_0^N\}$, ${}^0V_0 = \{{}^0V_0^1, \ldots, {}^0V_0^N\}$ the elements $\{x_0^1, x_1^1, y_0^1, y_1^1, \varphi^1\}$, $\{x_0^2, x_1^2, y_0^2, y_1^2, \varphi^2\} \in M_1 \cap M_2$ are such that ${}^*V_0(x_0^1, x_1^1, y_0^1, y_1^1, \varphi^1) = {}^0V_0(x_0^1, x_1^1, y_0^1, y_1^1, \varphi^1)$; ${}^*V_0(x_0^2, x_1^2, y_0^2, y_1^2, \varphi^2) = {}^0V_0(x_0^2, x_1^2, y_0^2, y_1^2, \varphi^2)$ then we get that $\{x_0^1, x_1^1, y_0^1, y_1^1, \varphi^1\} N({}^*V_0^1, \ldots, {}^*V_0^N)\{x_0^2, x_1^2, y_0^2, y_1^2, \varphi^2\}$ if and only if $\{x_0^1, x_1^1, y_0^1, y_1^1, \varphi^1\} N({}^0V_0^1, \ldots, {}^0V_0^N)\{x_0^2, x_1^2, y_0^2, y_1^2, \varphi^2\}$;

(3) the universality condition, i.e. the principle of choosing $N(V_0^1, \ldots, V_0^N)$ is defined for any vector criterion $V_0 = (V_0^1, \ldots, V_0^N)$ of the control system optimization;

(4) the "dictator" absence condition, i.e. there is no $s \in \{1, \ldots, N\}$ such that fulfillment of the N_s order relation for the elements $\{x_0^1, x_1^1, y_0^1, y_1^1, \varphi^1\}$, $\{x_0^2, x_1^2, y_0^2, y_1^2, \varphi^2\} \in M_1 \cap M_2$ implies fulfillment of the relation of

order $N(V_0^1, \ldots, V_0^N)$ for these elements, i.e.

$$\exists s \in \{1, \ldots, N\} : \{x_0^1, x_1^1, y_0^1, y_1^1, \varphi^1\} N_s \{x_0^2, x_1^2, y_0^2, y_1^2, \varphi^2\}$$
$$\Rightarrow \{x_0^1, x_1^1, y_0^1, y_1^1, \varphi^1\} N(V_0^1, \ldots, V_0^N) \{x_0^2, x_1^2, y_0^2, y_1^2, \varphi^2\};$$

(5) the weakened "Pareto principle," thus we have $(N_1 \times N_2 \times \cdots \times N_N) \subseteq N(V_0^1, \ldots, V_0^N)$, i.e. if the element $\{x_0^0, x_1^0, y_0^0, y_1^0, \varphi^0\} \in M_1 \cap M_2$ is optimal according to each of the special criteria $V_0^s (s = 1, \ldots, N)$, then $\{x_0^0, x_1^0, y_0^0, y_1^0, \varphi^0\}$ is optimal according to the principle of choosing $N(V_0^1, \ldots, V_0^N)$ defined for the vector criterion $V_0 = (V_0^1, \ldots, V_0^N)$.

As applied to the formulated natural conditions we have the problem of the existence of the relations of order $N(V_0^1, \ldots, V_0^N)$ satisfying these conditions (the investigations into this problem resulted in the Arrow paradox of nonexistence of the order relation satisfying these axioms). Thus the efforts to change the formulation of these conditions so that there might exist the relation of order $N(V_0^1, \ldots, V_0^N)$ satisfying even new requirements are quite legitimate.

Assume that $V_0 = (V_0^1, \ldots, V_0^N)$, $'V_0 = ('V_0^1, \ldots, 'V_0^N)$ are two vector optimization criteria for the dynamic control system, and $N_1, \ldots, N_N$; $N_1', \ldots, N_N'$ are the natural order relations determined by the special criteria $V_0^1, \ldots, V_0^N$; $'V_0^1, \ldots, 'V_0^N$ respectively. In this case $'V_0 = \{'V_0^1, \ldots, 'V_0^N\}$ are referred to as the improvement of $V_0 = \{V_0^1, \ldots, V_0^N\}$ (and V_0 the deterioration of $'V_0$ respectively), if from fulfillment of the condition $\{x_0^1, x_1^1, y_0^1, y_1^1, \varphi^1\}(N_1 \times \cdots \times N_N)\{x_0^2, x_1^2, y_0^2, y_1^2, \varphi^2\}$ for the elements $\{x_0^1, x_1^1, y_0^1, y_1^1, \varphi^1\}, \{x_0^2, x_1^2, y_0^2, y_1^2, \varphi^2\} \in M_1 \cap M_2$ it follows that $\{x_0^1, x_1^1, y_0^1, y_1^1, \varphi^1\}(N_1' \times \cdots \times N_N')\{x_0^2, x_1^2, y_0^2, y_1^2, \varphi^2\}$.

Modify Conditions 1–5 so as to augment 1, relax 4 and reject 5 (the consequence of the preceding four conditions). In other words, instead of Conditions 1–5 one should consider Conditions 1′, 2, 3 and 4′, where Conditions 1′ and 4′ take the form:

(1′) the monotonicity condition, i.e. if for the elements $\{x_0^1, x_1^1, y_0^1, y_1^1, \varphi^1\}$, $\{x_0^2, x_1^2, y_0^2, y_1^2, \varphi^2\}$ we fulfill the conditions

(a) $\{x_0^1, x_1^1, y_0^1, y_1^1, \varphi^1\} N(V_0^1, \ldots, V_0^N) \{x_0^2, x_1^2, y_0^2, y_1^2, \varphi^2\}$, $'V_0 = ('V_0^1, \ldots, 'V_0^N)$ for improvement $V_0 = \{V_0^1, \ldots, V_0^N\}$, then $\{x_0^1, x_1^1, y_0^1, y_1^1, \varphi^1\} N('V_0^1, \ldots, 'V_0^N) \{x_0^2, x_1^2, y_0^2, y_1^2, \varphi^2\}$;

(b) $\{x_0^1, x_1^1, y_0^1, y_1^1, \varphi^1\}\bar{N}(V_0^1,\dots,V_0^N)\{x_0^2, x_1^2, y_0^2, y_1^2, \varphi^2\}$, $'V_0 = \{'V_0^1,\dots,'V_0^N\}$ with deterioration $V_0 = \{V_0^1,\dots,V_0^N\}$, then $\{x_0^1, x_1^1, y_0^1, y_1^1,\varphi^1\}\bar{N}('V_0^1,\dots,'V_0^N)\{x_0^2, x_1^2, y_0^2, y_1^2, \varphi^2\}$ the line above the order relation denotes negation of the relation;

(4′) the condition of the free principle of choice which resides in the fact that for any elements $\{x_0^1,x_1^1,y_0^1,y_1^1,\varphi^1\}$, $\{x_0^2,x_1^2,y_0^2,y_1^2,\varphi^2\} \in M_1 \cap M_2$ differing from one another there exist the vector criteria $V_0 = \{V_0^1,\dots,V_0^N\}$, $'V_0 = \{'V_0^1,\dots,'V_0^N\}$ such that $\{x_0^1, x_1^1, y_0^1, y_1^1, \varphi^1\} N(V_0^1, \dots, V_0^N)\{x_0^2, x_1^2, y_0^2, y_1^2, \varphi^2\}$; $\{x_0^1, x_1^1, y_0^1, y_1^1, \varphi^1\}\bar{N}('V_0^1,\dots, 'V_0^N)\{x_0^2, x_1^2, y_0^2, y_1^2,\varphi^2\}$.

It may be shown that the relaxed "Pareto principle" is the consequence of Conditions 1′, 2, 3 and 4′ and Conditions 1′, 2, 3 and 4′ are fulfilled for the principle of choosing $N(V_0^1,\dots,V_0^N)$ determined according to the vector criterion $V_0 = \{V_0^1,\dots,V_0^N\}$ if and only if it is trivial, i.e. if it is the dominance principle, thus we have: $N(V_0^1,\dots,V_0^N) = N_1 \times \cdots \times N_N$.

In essence, we have obtained a negative answer to the question of whether there exists in the class of order relations the principle of choice different from the trivial satisfying the natural conditions stated above. The need arises for developing the axiomatic approach to the study of vector optimization problems and formulating the system of axioms on the class of binary relations (of the type of order or quasi-order) and the system of "natural" conditions so that there might exist the principles of choice differing from the trivial. It is to be noted that employing the principles of maximal and parametric maximal efficiency as the principles of choice solves this paradox in the sense that for them Conditions 1–3 are fulfilled, Condition 4 as dictator admissibility is relaxed, and Condition 5 as Pareto optimality is augmented.

For the purposes of investigation and solution of vector optimization concrete problems, attempts are currently made to define different principles of choice in intuitive and subjective ways. We formulate the main definitions of efficient (optimal) elements in the optimization problem of the dynamic control system for the vector indicator $V_0 = \{V_0^1,\dots,V_0^N\}$ similar to the definitions of effective (optimal) points in the vector optimization problems for the functions of many variables with a negative ingredient of the components V_0.

Definition 42.1: The element $\{x_0^0, x_1^0, y_0^0, y_1^0, \varphi^0\} \in M_1 \cap M_2$ is called optimal (dominant) if the conditions $V_0^s\{x_0^0, x_1^0, y_0^0, y_1^0, \varphi^0\} \leq V_0^s(x_0, x_1, y_0, y_1, \varphi)$ $(s = 1, \ldots, N)$ are fulfilled for any $\{x_0, x_1, y_0, y_1, \varphi\} \in M_1 \cap M_2$.

Definition 42.2: (a) The element $\{x_0^0, x_1^0, y_0^0, y_1^0, \varphi^0\} \in M_1 \cap M_2$ is called Pareto-efficient in reference to the vector criterion $V_0(x_0, x_1, y_0, y_1, \varphi) = \{V_0^1(x_0, x_1, y_0, y_1, \varphi), \ldots, V_0^N(x_0, x_1, y_0, y_1, \varphi)\}$ on the set $M_1 \cap M_2$ if there is no element $\{x_0, x_1, y_0, y_1, \varphi\} \in M_1 \cap M_2$ such that $V_0^s(x_0, x_1, y_0, y_1, \varphi) \leq V_0^s(x_0^0, x_1^0, y_0^0, y_1^0, \varphi^0)$ $(s = 1, \ldots, N)$ and if only one inequality is strict.

(b) The element $\{x_0^0, x_1^0, y_0^0, y_1^0, \varphi^0\} \in M_1 \cap M_2$ is Pareto-efficient if for any element $\{x_0, x_1, y_0, y_1, \varphi\} \in M_1 \cap M_2$ either there exist the indices $s_0 \in \{1, \ldots, N\}$ such that $V_0^{s_0}(x_0^0, x_1^0, y_0^0, y_1^0, \varphi^0) < V_0^{s_0}(x_0, x_1, y_0, y_1, \varphi)$ or the equalities with $s \in \{1, \ldots, N\} V_0^s(x_0^0, x_1^0, y_0^0, y_1^0, \varphi^0) = V_0^s(x_0, x_1, y_0, y_1, \varphi)$ are fulfilled.

(c) The element $\{x_0^0, x_1^0, y_0^0, y_1^0, \varphi^0\} \in M_1 \cap M_2$ is Pareto-efficient if the condition (with $s = 1, \ldots, N$) $V_0^s(x_0^0, x_1^0, y_0^0, y_1^0, \varphi^0) \geq V_0^s(x_0, x_1, y_0, y_1, \varphi)$ for a certain element $\{x_0, x_1, y_0, y_1, \varphi\} \in M_1 \cap M_2$ implies that $V_0^s(x_0^0, x_1^0, y_0^0, y_1^0, \varphi^0) = V_0^s(x_0, x_1, y_0, y_1, \varphi)$.

It might be shown that the three definitions of the Pareto-efficient element $\{x_0^0, x_1^0, y_0^0, y_1^0, \varphi^0\} \in M_1 \cap M_2$ are equivalent, and sometimes other names are used: the non-dominant, unimprovable, Pareto-optimal or minimal element.

Definition 42.3: The element $\{x_0^0, x_1^0, y_0^0, y_1^0, \varphi^0\} \in M_1 \cap M_2$ is Sleighter-optimal in reference to the vector criterion $V_0 = \{V_0^1, \ldots, V_0^N\}$ on the set $M_1 \cap M_2$ if there is no element $\{x_0, x_1, y_0, y_1, \varphi\} \in M_1 \cap M_2$ such that

$$V_0^s(x_0, x_1, y_0, y_1, \varphi) < V_0^s(x_0^0, y_0^0, x_1^0, y_1^0, \varphi^0) \quad (s = 1, \ldots, N).$$

Definition 42.4: The element $\{x_0^0, x_1^0, y_0^0, y_1^0, \varphi^0\} \in M_1 \cap M_2$ is Kuhn–Tucker efficient if there is no element $\{x_0, x_1, y_0, y_1, \varphi\} \in M(x_0^0, x_1^0, y_0^0, y_1^0, \varphi^0)$ such that for all $s = 1, \ldots, N$ the inequalities

$$\Big\{[\mathrm{grad}_{x_0^0} V_0^s(x_0^0, x_1^0, y_0^0, y_1^0, \varphi^0)]x_0 + [\mathrm{grad}_{x_1^0} V_0^s(x_0^0, x_1^0, y_0^0, y_1^0, \varphi^0)]x_1$$

$$+ [\mathrm{grad}_{y_0^0} V_0^s(x_0^0, x_1^0, y_0^0, y_1^0, \varphi^0)]^* y_0 + [\mathrm{grad}_{y_1^0} V_0^s(x_0^0, x_1^0, y_0^0, y_1^0, \varphi^0)]^* y_1$$

$$+ \int_{x_0^0}^{x_1^0} [\mathrm{grad}_{\varphi^0} V_0^s(x_0^0, x_1^0, y_0^0, y_1^0, \varphi^0)]^* \varphi(x)dx\Big\} \leq 0\,,$$

are fulfilled, and at least one of the inequalities is strict.

Definition 42.5: The element $\{x_0^0, x_1^0, y_0^0, y_1^0, \varphi^0\} \in M_1 \cap M_2$ is called equilibrium if

$$\max_{s=1,\dots,N} V_0^s(x_0^0, x_1^0, y_0^0, y_1^0, \varphi^0) = \min_{s=1,\dots,N} V_0^s(x_0^0, x_1^0, y_0^0, y_1^0, \varphi^0)\,.$$

Definition 42.6: The element $\{x_0^0, x_1^0, y_0^0, y_1^0, \varphi^0\} \in M_1 \cap M_2$ is called the tradeoff one (or belonging to the tradeoff domain) if there exists $\alpha^0 = (\alpha_1^0, \dots, \alpha_N^0) \in N^N$ such that $\alpha_s^0 \geq 0$ $(s = 1, \dots, N)$, $\sum_{s=1}^{N} \alpha_s^0 = 1$,

$$\sum_{s=1}^{N} \alpha_s^0 V_0^s(x_0^0, x_1^0, y_0^0, y_1^0, \varphi^0) = \min_{M_1 \cap M_2} \sum_{s=1}^{N} \alpha_s^0 V_0^s(x_0, x_1, y_0, y_1, \varphi)\,.$$

Definition 42.7: The element $\{x_0^0, x_1^0, y_0^0, y_1^0, \varphi^0\} \in M_1 \cap M_2$ is called minimal if

$$\min_{s=1,\dots,N} V_0^s(x_0^0, x_1^0, y_0^0, y_1^0, \varphi^0) = \min_{M_1 \cap M_2} \Big[\min_{s=1,\dots,N} V_0^s(x_0, x_1, y_0, y_1, \varphi)\Big]\,.$$

Definition 42.8: The element $\{x_0^0, x_1^0, y_0^0, y_1^0, \varphi^0\} \in M_1 \cap M_2$ is called maximal if

$$\max_{s=1,\dots,N} V_0^s(x_0^0, x_1^0, y_0^0, y_1^0, \varphi^0) = \max_{M_1 \cap M_2} \Big[\max_{s=1,\dots,N} V_0^s(x_0, x_1, y_0, y_1, \varphi)\Big]\,.$$

Definition 42.9: The element $\{x_0^0, x_1^0, y_0^0, y_1^0, \varphi^0\} \in M_1 \cap M_2$ is called efficient from below (or minimax) if

$$\max_{s=1,\dots,N} V_0^s(x_0^0, x_1^0, y_0^0, y_1^0, \varphi^0) = \min_{M_1 \cap M_2} \Big[\max_{s=1,\dots,N} V_0^s(x_0, x_1, y_0, y_1, \varphi)\Big]\,.$$

Definition 42.10: The element $\{x_0^0, x_1^0, y_0^0, y_1^0, \varphi^0\} \in M_1 \cap M_2$ is called efficient from above (or maximin) if

$$\min_{s=1,\ldots,N} V_0^s(x_0^0, x_1^0, y_0^0, y_1^0, \varphi^0) = \max_{M_1 \cap M_2} \left[\min_{s=1,\ldots,N} V_0^s(x_0, x_1, y_0, y_1, \varphi) \right].$$

Definition 42.11: The element $\{x_0^0, x_1^0, y_0^0, y_1^0, \varphi^0\} \in M_1 \cap M_2$ is called α-efficient from below (or Gurvitz-efficient from below with the α indicator) if there exists

$$\left[\alpha \min_{s=1,\ldots,N} V_0^s(x_0^0, x_1^0, y_0^0, y_1^0, \varphi^0) + (1-\alpha) \max_{s=1,\ldots,N} V_0^s(x_0^0, x_1^0, y_0^0, y_1^0, \varphi^0)\right]$$
$$= \min_{M_1 \cap M_2} \left[\alpha \min_{s=1,\ldots,N} V_0^s(x_0, x_1, y_0, y_1, \varphi) + (1-\alpha) \max_{s=1,\ldots,N} V_0^s(x_0, x_1, y_0, y_1, \varphi)\right].$$

Here the value $\alpha (0 \le \alpha \le 1)$ is assumed to be specified.

Definition 42.12: The element $\{x_0^0, x_1^0, y_0^0, y_1^0, \varphi^0\} \in M_1 \cap M_2$ is called α-efficient from above (or Gurvitz-efficient from above with the α indicator) if there exists

$$\left[\alpha \max_{s=1,\ldots,N} V_0^s(x_0^0, x_1^0, y_0^0, y_1^0, \varphi^0) + (1-\alpha) \min_{s=1,\ldots,N} V_0^s(x_0^0, x_1^0, y_0^0, y_1^0, \varphi^0)\right]$$
$$= \max_{M_1 \cap M_2} \left[\alpha \max_{s=1,\ldots,N} V_0^s(x_0, x_1, y_0, y_1, \varphi) + (1-\alpha) \min_{s=1,\ldots,N} V_0^s(x_0, x_1, y_0, y_1, \varphi)\right].$$

Here the value $\alpha (0 \le \alpha \le 1)$ is specified.

Definition 42.13: The element $\{x_0^0, x_1^0, y_0^0, y_1^0, \varphi^0\} \in M_1 \cap M_2$ is called Trukhayev-efficient if

$$H[V_0(x_0^0, x_1^0, y_0^0, y_1^0, \varphi^0)] = \max_{M_1 \cap M_2} H[V_0(x_0, x_1, y_0, y_1, \varphi)],$$

where $H[V_0]$ is the uncertainty function of the values of the vector criterion V_0 determined, say, when H is taken to be the Shannon entropy

$$H[V_0(x_0, x_1, y_0, y_1, \varphi)] = H[V_0^1(x_0, x_1, y_0, y_1, \varphi), \ldots, V_0^N(x_0, x_1, y_0, y_1, \varphi)]$$
$$= -\sum_{s=1}^{N} \tilde{V}_0^s(x_0, x_1, y_0, y_1, \varphi) \lg[\tilde{V}_0^s(x_0, x_1, y_0, y_1, \varphi)],$$

here $(s = 1, \dots, N)$

$$\tilde{V}_0^s(x_0, x_1, y_0, y_1, \varphi) = \bar{V}_0^s(x_0, x_1, y_0, y_1, \varphi) \Big/ \sum_{\sigma=1}^{N} \tilde{V}_0^{\sigma}(x_0, x_1, y_0, y_1, \varphi),$$

$$\bar{V}_0^s(x_0, x_1, y_0, y_1, \varphi) = \Big[V_0^s(x_0, x_1, y_0, y_1, \varphi) - \min_{M_1 \cap M_2} V_0^s(x_0, x_1, y_0, y_1, \varphi)\Big] \times \Big[\max_{M_1 \cap M_2} V_0^s(x_0, x_1, y_0, y_1, \varphi) - \min_{M_1 \cap M_2} V_0^s(x_0, x_1, y_0, y_1, \varphi)\Big]^{-1}.$$

Definition 42.14: The element $\{x_0^0, x_1^0, y_0^0, y_1^0, \varphi^0\} \in M_1 \cap M_2$ is called optimal from below by the λ-criterion if we fulfill the condition $\{x_0^0, x_1^0, y_0^0, y_1^0, \varphi^0\} \in M_{\lambda^0}$, where M_λ denotes a set of such elements $\{x_0, x_1, y_0, y_1, \varphi\} \in M_1 \cap M_2$ that $\bar{V}_0^s(x_0, x_1, y_0, y_1, \varphi) \leq \lambda$ $(s = 1, \dots, N)$, and the value of λ^0 is the solution of the extremal problem

$$\lambda_0 = \min_{\substack{0 \leq \lambda \leq 1, \{x_0, x_1, y_0, y_1, \varphi\} \in M_1 \cap M_2 \\ \bar{V}_0^s(x_0, x_1, y_0, y_1, \varphi) \leq \lambda, s=1,\dots,N}} \lambda,$$

where the functionals $\bar{V}_0^s$ take the form as given above.

Definition 42.15: The element $\{x_0^0, x_1^0, y_0^0, y_1^0, \varphi^0\} \in M_1 \cap M_2$ is called optimal from above by the λ-criterion if we fulfill the condition $\{x_0^0, x_1^0, y_0^0, y_1^0, \varphi^0\} \in M_{\lambda^0}$ where M_λ denotes a set of such elements $\{x_0, x_1, y_0, y_1, \varphi\} \in M_1 \cap M_2$ that $\bar{V}_0^s(x_0, x_1, y_0, y_1, \varphi) \geq \lambda$, $s = 1, \dots, N$ and the magnitude λ^0 is the solution of the extremal problem

$$\lambda^0 = \max_{\substack{0 \leq \lambda \leq 1, \{x_0, x_1, y_0, y_1, \varphi\} \in M_1 \cap M_2 \\ \bar{V}_0^s(x_0, x_1, y_0, y_1, \varphi) \geq \lambda, s=1,\dots,N}} \lambda.$$

Consider the conditions for the dynamic control system improvability according to the vector criterion.

Assume that the feasible $\{x_0^1, x_1^1, y_0^1, y_1^1, \varphi^1\} \in M_1 \cap M_2$ and the vector criterion $V_0 = \{V_0^1, \dots, V_0^N\}$ which determines the trivial relation of order $N(V_0^1, \dots, V_0^N)$, are specified. The question arises as to whether there exists another feasible set $\{x_0^2, x_1^2, y_0^2, y_1^2, \varphi^2\} \in M_1 \cap M_2$ which is more preferable than $\{x_0^1, x_1^1, y_0^1, y_1^1, \varphi^1\} \in M_1 \cap M_2$ in the sense of the trivial relation of order $N(V_0^1, \ldots, V_0^N) = N_1 \times \cdots \times N_N$, i.e. for any $s = 1, \ldots, N$ the inequality

$$V_0^s(x_0^2, x_1^2, y_0^2, y_1^2, \varphi^2) \geq V_0^s(x_0^1, x_1^1, y_0^1, y_1^1, \varphi^1) \tag{41.13}$$

holds and at least one inequality is strict, what the conditions for the existence of such an element $\{x_0^2, x_1^2, y_0^2, y_1^2, \varphi^2\} \in M_1 \cap M_2$ are. Here we suppose that approaching an optimum of the vector indicator resides in increasing each of the components of the vector criterion $V_0 = \{V_0^1, \ldots, V_0^N\}$.

The following theorem formulates a sufficient condition for improvability of the element $\{x_0^1, x_1^1, y_0^1, y_1^1, \varphi^1\} \in M_1 \cap M_2$ according to the vector criterion.

Theorem 41.13: The feasible element $\{x_0^1, x_1^1, y_0^1, y_1^1, \varphi^1\} \in M_1 \cap M_2$ is improvable according to the vector criterion $V_0 = \{V_0^1, \ldots, V_0^N\}$, i.e. there exists the element $\{x_0^2, x_1^2, y_0^2, y_1^2, \varphi^2\} \in M_1 \cap M_2$ for which inequality (41.14) holds with any $s = 1, \ldots, N$ if for the element $\{x_0^1, x_1^1, y_0^1, y_1^1, \varphi^1\} \in M_1 \cap M_2$ the condition $(s = 1, \ldots, N)$

$$V_0^s(x_0^1, x_1^1, y_0^1, y_1^1, \varphi^1) < C_s(V_0) \tag{41.14}$$

is fulfilled, here the constants $C_s(V_0)$ are defined as follows

$$\begin{aligned}(s = 1, \ldots, N) : C_s(V_0) = & \min_{M_1 \cap M_2} V_0^s(x_0, x_1, y_0, y_1, \varphi) \\ & + C(\bar{V}_0)\Big[\max_{M_1 \cap M_2} V_0^s(x_0, x_1, y_0, y_1, \varphi) \\ & - \min_{M_1 \cap M_2} V_0^s(x_0, x_1, y_0, y_1, \varphi)\Big]\end{aligned} \tag{41.15}$$

where the constant $C(\bar{V}_0)$ takes the form

$$C(\bar{V}_0) = \max_{M_1 \cap M_2}\Big[\min_{s=1,\ldots,N} \bar{V}_0^s(x_0, x_1, y_0, y_1, \varphi)\Big] \tag{41.16}$$

and $\bar{V}_0^s$ is a normalized value of the sth special criterion V_0^s.

Condition (41.15) is not only the sufficient condition for improvement but also is close to the necessary condition. Thus we have

Theorem 41.14: There is no feasible element $\{x_0^1, x_1^1, y_0^1, y_1^1, \varphi^1\} \in M_1 \cap M_2$ for which the inequality $(s = 1, \ldots, N)$

$$V_0^s(x_0^1, x_1^1, y_0^1, y_1^1, \varphi^1) > C_s(V_0), \tag{41.17}$$

holds if among special criteria of the vector criterion $V_0 = \{V_0^1, \dots, V_0^N\}$ there are no degenerate ones.

Degeneracy of the special criterion V_0^s of the vector criterion $V_0 = \{V_0^1, \dots, V_0^N\}$ is taken as fulfilment of the equality

$$V_0^s(x_0, x_1, y_0, y_1, \varphi) = \min_{M_1 \cap M_2} V_0^s = \max_{M_1 \cap M_2} V_0^s = \text{const}\,.$$

We set forth a generalization of the improvability conditions.

Theorem 41.15: The feasible element $\{x_0^1, x_1^1, y_0^1, y_1^1, \varphi^1\} \in M_1 \cap M_2$ can be improved according to the vector criterion $V_0 = \{V_0^1, \dots, V_0^N\}$ if and only if there exists the vector $\gamma = (\gamma_1, \dots, \gamma_N) \in R^N$ such that the inequalities

$$\bar{V}_0^s(x_0^1, x_1^1, y_0^1, y_1^1, \varphi^1) \le C_s(\gamma)\,, \quad \bar{V}_0^\sigma(x_0^1, x_1^1, y_0^1, y_1^1, \varphi^1) < C_\sigma(\gamma) \quad (41.18)$$

hold for any $s \in \{1, \dots, N\}$ and for at least one $\sigma \in \{1, \dots, N\}$, where

$$C_s(\gamma) = C(\gamma) - \gamma_s \; (s = 1, \dots, N)\,,$$
$$C(\gamma) = \max_{M_1 \cap M_2} \Big\{ \min_{s=1,\dots,N} [\bar{V}_0^s(x_0, x_1, y_0, y_1, \varphi) + \gamma_s] \Big\}.$$

If the feasible element from $M_1 \cap M_2$ is specified, the problem arises with respect to determining in R^N the range of values of the vector γ for which inequalities (41.18) are fulfilled.

Theorem 41.16: Any vector $\gamma \in R^N$ for which (41.18) are fulfilled with some feasible elements $\{x_0^1, x_1^1, y_0^1, y_1^1, \varphi^1\} \in M_1 \cap M_2$ satisfies inequalities for all s, $\rho \in \{1, \dots, N\}$:

$$\max_{M_1 \cap M_2} \bar{V}_0^s(x_0, x_1, y_0, y_1, \varphi) - \min_{M_1 \cap M_2} \bar{V}_0^\rho(x_0, x_1, y_0, y_1, \varphi) \ge \gamma_\rho - \gamma_s\,. \quad (41.19)$$

We have the following equalities from the definition of the normalized value $\bar{V}_0^s$ with $s = 1, \dots, N$

$$\min_{M_1 \cap M_2} \bar{V}_0^s(x_0, x_1, y_0, y_1, \varphi) = 0\,; \quad \max_{M_1 \cap M_2} \bar{V}_0^s(x_0, x_1, y_0, y_1, \varphi) = 1$$

then the range of values of the vector $\gamma \in N^N$ determined by inequalities (41.19) coincides with the set G of the form

$$G = \{\gamma \in N^N : \gamma_s - \gamma_\rho \leq 1, \quad \forall S, \rho \in \{1, \ldots, N\} \tag{41.20}$$

with $N = 2$ we have $G = \{(\gamma_1, \gamma_2) \in N^2\colon |\gamma_1 - \gamma_2| < 1\}$.

Theorem 41.17: If inequalities (41.18) hold for $\{x_0^1, x_1^1, y_0^1, y_1^1, \varphi^1\} \in M_1 \cap M_2$ and a particular vector $\gamma^1 \in N^N$, then they also hold for $\gamma^\delta = \gamma^1 - \delta = (\gamma_1^1 - \delta, \gamma_2^1 - \delta, \ldots, \gamma_N^1 - \delta)$ with any finite $\delta \in R^1$.

The set G (41.20) can be restricted, say, to the set $G' = \{\gamma \in N^N : \gamma_1 = 0, \gamma_s - \gamma_\rho \leq 1, \forall s, \rho \in \{1, \ldots, N\}$, in particular, with $N = 2$, $G' = \{(\gamma_1, \gamma_2) \in R^2 : \gamma_1 = 0, |\gamma_2| \leq 1\}$.

Corollary 42.1: (to theorem (41.15)). The feasible element $\{x_0^1, x_1^1, y_0^1, y_1^1, \varphi^1\} \in M_1 \cap M_2$ can be improved according to the vector criterion $V_0 = \{V_0^1, \ldots, V_0^N\}$ if and only if such that there is a vector $\gamma \in G$.

Theorem 41.18: The feasible element $\{x_0^1, y_0^1, x_1^1, y_1^1, \varphi^1\} \in M_1 \cap M_2$ cannot be improved (Pareto-optimal) according to the vector criterion $V_0 = \{V_0^1, \ldots, V_0^N\}$ if and only if there is no vector $\gamma = (\gamma_1, \ldots, \gamma_n) \in G$ such that inequalities (41.18) are consistent.

Note, that application of the maximal efficiency principle to solve the multiobjective optimization problems for dynamic control systems enables one to formulate the relation of order N_λ satisfying the axioms of choice and leads to an extremal problem (λ-problem) of the form for which the apparatus has been adequately developed for obtaining optimality conditions and numerical techniques of solution, in other words, the maximal efficiency principle looks rather promising for applications when solving the multiobjective optimization problems of dynamic control systems.

Chapter 12

METHODOLOGY OF VECTOR OPTIMIZATION

§43. OPTIMIZATION METHODOLOGY FOR HIERARCHICAL SEQUENCE OF QUALITY CRITERIA

We formulate, in a simplified form, the vector optimization problem as follows: assume that system behavior is characterized by the n-dimensional vector $x = \{x_1, \ldots, x_n\} \in X \subset E^n$ and estimated by a k-dimensional vector function $I(x) = \{I_1(x), \ldots, I_k(x)\}$ the components of which constitute specified real functions of a variable x, where it is required to find a point of $x^0 \in X$ optimizing, in a sense, values of functions $I_1(x), \ldots, I_k(x)$. Many of the studies on vector optimization methodology can be grouped according to the following directions:

- optimization methodology of hierarchical sequence of quality criteria;
- methodology of determining the sets of unimprovable points;
- methodology of determining a tradeoff decision.

We consider all these approaches in succession. Optimization methodology for hierarchical sequence of quality criteria is based on introducing an order relation on a set of criteria. Assume that an order is set such that $I_1(x) > I_2(x) > \cdots > I_k(x)$, the solution of $x^0 \in X$ is defined as satisfying the relations:

$$I_1(x^0) = \min_{x \in X_0 \subset X} I_1(x) \, ,$$

$$I_2(x^0) = \min_{x \in X_1 \subset X_0} I_2(x),$$

$$I_3(x^0) = \min_{x \in X_2 \subset X_1} I_3(x),$$

$$I_k(x^0) = \min_{x \in X_{k-1} \subset X_{k-2}} I_k(x),$$

here the sets $X_i \subset X_{i-1}$ $(i = 1, 2, \ldots, k-1)$ are defined as $X_i = \{x : I_i(x) = \min_{x \in X_{i-1}} I_i(x)\}$.

When solving most of the practical problems, employment of that kind of approach proves to be ineffective since optimizing even by the first criterion leads to a unique optimal solution (if any) and everything results in optimizing only by the first criterion. The problem arises with respect to determining relative importance of quality criteria in vector problems of managerial decision optimization. We consider the problem in its general statement.

In vector optimization problems, "utility," "quality," "efficiency," "value" and other features of objects are estimated by the criteria $k_1, k_2, \ldots, k_m$ $(m \geq 2)$ generally identified as the function mapping a set of objects Q in a subset X_i (containing no less than two points) of a numerical line R_l which is known as the ith criterion scale, and its elements are scale estimates. A set of criteria k_i forms the vector criterion $k = (k_1, \ldots, k_m)$ mapping the set Q in the set $\chi = X_1 \times X_2 \times \cdots \times X_m$ of vectors, the components of which are scale estimates. For analyzing a particular problem and obtaining the necessary information to solve it, it is convenient to manipulate vectors from χ which do not necessarily correspond to real objects from Q but which can be regarded as characteristics of some hypothetical objects; such vectors are commonly identified as vector estimates. In this case, some combinations of scale estimates may prove to be unfeasible because the vectors involving these estimates will be the meaningless sets of numbers. Hence, in general, inclusion of $X \subseteq \chi$ is strict. Based on the information obtained on the problem, in the set of vector estimates X we construct the binary relations of preference and indifference which are then, along with supplementary hypotheses and considerations (in particular, the maximal guaranteed result principle employed when indeterminate factors are available), used for constructing preference relations on a set of strategies U and isolating an optimal strategy. Furthermore the relations of preference, indifference, nonstrict preference and incomparability will be respectively labelled by P, I, R, N which, if need be, will be provided with appropriate super- and subscripts. According to what has been previously said in our previous discussion, for any two vector estimates x and y only one of the following cases is possible: (a) xPy x is preferable to y; (b) yPx; (c) xIy x and y are the same by preference, i.e. it does not matter which of the two vector estimates is selected as the best; (d) xNy x and y cannot be compared by preference. xRy should imply that either xPy or xIy is true, in other words, there exists $R = P \cup I$. Relations P, I and N are recovered by R (we can say that the relation R generates P, I and N): xIy when simultaneously xRy and yRx; when xPy, xRy but yRx is not true; xNy when neither xRy nor yRx is true.

Assume that criteria are independent by preference and, for each of them, the large values are preferred over the small, then the vector estimate is preferred over the vector estimate $(x\|y_i)$ obtained from x by substituting x_i by the scale estimate y_i with any $i \in M = \{1, \ldots, m\}$, $y_i < x_i$. It is natural to introduce into X, and formally also into R_{l^m}, the relation of nonstrict preference R^0 defined as follows: xR^0y when there exist the inequalities

$$x_i \geq y_i\,, \qquad i = 1, \ldots, m\,. \tag{43.1}$$

Evidently, the relation R^0 generates the relations I^0, P^0 and N^0. It is clear that xP^0y when if only one of inequalities (43.1) is strict, and I^0 is the relation of vector equality.

In the analysis of vector optimization problems the relations introduced are rather useful, but they do not allow one to solve concrete problems in full since many vector estimates prove to be incomparable. A need arises for extending these relations, and it can be done only when obtaining additional information. We will consider constructing the relations of preference and indifference using information on the importance of criteria as the base.

Information on the relative importance of criteria is given by a totality of letters, figures and other signs, i.e. it can be represented by the set Ω, we will also employ the set $\Omega^0 = \Omega \cup \{0\}$. Each message $\omega \in \Omega$ enables one to compare the vector estimates of special types by preference, in other words, to establish in X the relations of strict preference P^ω and indifference I^ω, or else only one of such relations, R^ω is taken as a combination of these relations. Since, for comparing two vector estimates it is possible to employ the relation R^ω for any $\omega \in \Omega$, as well as R^0, then the information Ω determines in X the binary relation of nonstrict preference R_*^Ω, strict preference P_*^Ω and indifference I_*^Ω, thus

$$P_*^\Omega = \bigcup_{\omega \in \Omega^0} P^\omega; \quad I_*^\Omega = \bigcup_{\omega \in \Omega^0} I^\omega; \quad R_*^\Omega = P_*^\Omega \cup I_*^\Omega. \tag{43.2}$$

The relations introduced are beneficial for solving vector optimization problems. The relations P_*^Ω, I_*^Ω, R_*^Ω, generally speaking, are not transitive. If it is desired to have the transitive relations of preference and indifference P^Ω, I^Ω, R^Ω in the problem, then they can be obtained by extending the relations.

Assume that for the vector estimates x and y there are such s of vector estimates z^k and $s+1$ of symbols $\omega^k \in \Omega^0$ that the statements $xR^{\omega^1}z^1$, $z^1R^{\omega^2}z^2, \dots, z^sR^{\omega^{s+1}}y \Longleftrightarrow xR^{\omega^1}z^1R^{\omega^2}\dots z^sR^{\omega^{s+1}}y$ are valid (43.2). If R_*^Ω is not transitive, then $xR_*^\Omega y$ can be untrue.

Supplement R_*^Ω with the pairs of estimates (x, y) for which there exist chains of the type (43.2). Extension of binary relations in this fashion is termed a transitive closure (denote it by T_rCl) and the extended relations the transitive closures of initials. Using information Ω on the importance of criteria as the base, we then construct a reflexive and transitive relation of nonstrict preference — a quasi-order $R^\Omega = T_rClR_*^\Omega$.

Since for vector estimates x and y there exists a chain of the type (43.2) in which all R are I, then these estimates are to be taken as the same by preference. Thus, the relation of indifference (Ω-equivalence) $I^\Omega = T_rClI_*^\Omega$ is defined in X. Since for vector estimates x and y there exists a chain of the type (43.2) in which even one R is P, then x is assumed to be preferable to y. By this means the relation of strict preference P^Ω is determined. Hence $R^\Omega = I^\Omega \cup P^\Omega$ and P^Ω offers the property of generalized transitivity by I^Ω from $xP^\Omega y$ and $yI^\Omega z$, as well as from $xI^\Omega y$ and $yP^\Omega z$ we have $xP^\Omega z$. We consider the definition of ordering by the importance of homogeneous criteria.

Criteria $k_{i_1}, \dots, k_{i_n}$ are termed homogeneous if they have a common scale $X_{i_1} = \dots = X_{i_n}$ and a set $X \subseteq R_{l^m}$ is symmetrical in reference to coordinate axes $X_{i_1}, \dots, X_{i_n}$, in other words, if for any vector estimate $x = (x_1, \dots, x_m)$ it can be stated that the vector obtained from x by any rearrangement of coordinates $x_{i_1}, \dots, x_{i_n}$ proves to be a vector estimate, i.e. belongs to X. In general, the idea of symmetry is used and underlies any concept of equality, equivalence, etc. and conversely any idea of preference, superiority, etc. is based on negation of symmetry defined in one form or other.

Suppose that we have two homogeneous criteria k_r and k_t and the vector estimates x and $\tilde{x}^{rt}$ are symmetrical and obtained from x by rearranging its components x_r and x_t. If two vector estimates x and $\tilde{x}^{rt}$ are regarded as equally preferable, then the criteria k_r and k_t should be regarded as equivalent, equally important.

Definition 43.1: Criteria k_r and k_t are equivalent (denoted by rSt) when any two vector estimates x and $\tilde{x}^{rt}$ are equal by preference.

Assume that the stated circumstance represents the message ω_1, then we say that the message ω_1 establishes the relation of indifference I^{ω_1} between vector estimates x and $y = \tilde{x}^{rt} : xI^{\omega_1}y$.[11] The best of any two estimates is thought to be that (x or $\tilde{x}^{rt}$) in which the rth coordinate is larger, in such an event, the symmetry of the rth and tth criteria is disturbed, the criterion k_2 being more important here.

Definition 43.2: We say that the criterion k_r is more important than the criterion k_t ($\omega_2 = rBt$) when any vector estimate is x in which $x_r > x_t$ is preferable to $y = \tilde{x}^{rt}(xP^{\omega_2}y)$.

Consider the problem of comparing a combination of criteria by importance.

Assume that we have two non-intersecting sets of criteria ($n \geq 1$, $l \geq 1$)

$$\{k_{r_1}, \ldots, k_{r_n}\}, \quad \{k_{t_1}, \ldots, k_{t_l}\}. \tag{43.3}$$

Suppose that all criteria are homogeneous, we have a homogeneous set $n + l$ of criteria. In homogeneous sets for ordering criteria by importance we have to compare only vector estimates, thus, x, $y = \tilde{x}^{\{r_1,\ldots,r_n\}\{t_1,\ldots,t_l\}}$;

$$x_{r_1} = \cdots = x_{r_n}; \quad x_{t_1} = \cdots = x_{t_l}, \tag{43.4}$$

here $\tilde{x}^{\{r_1,\ldots,r_n\}\{t_1,\ldots,t_l\}}$ is the vector obtained from x by replacing each of n coordinates x_{r_i} by x_{t_1} and each of l coordinates x_{t_j} by x_{r_1}.

Definition 43.3: Homogeneous sets of criteria of the form (43.3) are equivalent to $\omega_3 = \{r_1, \ldots, r_n\}S\{t_1, \ldots, t_l\}$ when any two vector estimates (43.4) are the same by preference $(xI^{\omega_3}y)$.

Definition 43.4: From two homogeneous sets of criteria (43.3) the first is more important than the second $\omega_4 = \{r_1, \ldots, r_n\}B\{t_1, \ldots, t_l\}$ when from any two vector estimates (43.4) with $x_{r_1} > x_{t_1}$ the first is preferred over the second ($xP^{\omega_4}y$).

Practical problems make good use of transformations of the initial inhomogeneous criteria k_i into the homogeneous $\hat{k}_i$ criteria in the following way:

$$\hat{k}_i = \frac{k_i}{b_i}; \qquad \hat{k}_i = \frac{k_i - a_i}{b_i - a_i},$$

here b_i are some standard values of k_i; a_i are minimum feasible (possible) values of k_i.

By no means do we always manage to transform inhomogeneous criteria into homogeneous and then order them by importance. It is desirable to determine the relative importance of inhomogeneous criteria for which we need to correlate the influence on a common quality of the objects or phenomena under study, where changes in values of these criteria exists in the scales corresponding to the criteria on hand. It is called to compare by utility the differences of criteria values. Such comparisons take place only when the differences of criteria values are meaningful. The criteria of that kind are assumed to be identified as those possessing solid scales. The criteria with solid scales may be exemplified by the criteria with an absolute scale, a scale of relations (the unit of measurement can be fixed), a scale of intervals (a scale can be specified), the criterion with a rating scale where rates are prescribed according to established standards or assigned to initial non-numerical (clearly stated) estimates.

V. V. Podinovsky[11] investigated the problems of estimating relative importance of criteria or their combinations (ω-tuples). Furthermore, he studied not only superiority by importance but determines how much the superiority is.

§44. OPTIMIZATION OF HIERARCHICAL SEQUENCE OF QUALITY CRITERIA

Vector optimization problems were initially encountered in the works by Pareto.[24] The works[10,19,20,22] touched upon the issues of multicriterion problem, though the problem did not obtain further development.[15] A deeper insight into the vector optimization problem was gained by L. Zadeh,[26] who was the first to call the experts' attention to the fact that in the majority of practical cases, when developing complex systems, they tried to obtain the result possessing numerous optimal characteristics. Zadeh showed that exact optimization of the vector functional is generally unattainable. This means that if, by selecting the control, we manage to optimize some scalar functional, then, almost without exception, no possibility is left, with the same solution, for optimization of another scalar functional. The problem of vector optimization both in static and in dynamic systems was studied by A. Klinger, V. Nelson, M. Chang, Da Cuhn and E. Pollack, Don Hak Chuang, F. Waltz, and others.[16]

In the last few years the problem of vector optimization has been extensively developed since numerous problems of decision making in the national economy, engineering, military, art and social relations amount to problems of that kind.[4] At the same time, the up-to-date apparatus of vector optimization is still far from being perfect and continues to develop. Some theoretical results may be employed here as an instrument for argumentation on the decisions made.

We isolate the following three main groups of methods for the solution of vector optimization problems[15,16]:

- optimization of hierarchical sequence of quality criteria;
- definition of the set of unimprovable points;
- definition of the solution based on a particular trade-off.

We consider in greater detail the first group of methods, discussing them from the standpoint of methods for handling the criterion priority.

The priority of criteria may be specified by different parameters, from various points of view and with different accuracy. In this case the principal characteristics are as follows: a priority series $\vec{\mathcal{J}}$, a priority vector $\lambda = (\lambda_1, \ldots, \lambda_m)$ and a weight coefficient vector (a weight vector) $\alpha = (\alpha_1, \ldots, \alpha_m)$.

The priority series $\vec{\mathcal{J}}$ represents an ordered set of local criterion indices

$$\vec{\mathcal{J}} = \{1, 2, \ldots, m\}. \tag{44.1}$$

We postulate purely qualitative relations of dominance (preference relations, including also the binary ones) of criteria, thus the criterion y_1 is more important than y_2, y_2 is more important than y_3, etc. and we write them as:

$$\{\vec{y}_j\} = \{y_1 \succ y_2 \succ \cdots \succ y_m\}. \tag{44.2}$$

The quantitative side of dominance is not included here. In the criterion set, we may find groups of equivalent criteria, denoted by $\{y_j \sim y_{j+1} \sim$

$y_{j+2}\}$. If such groups are met in the priority series we isolate them with internal brackets, thus we have

$$\vec{\mathcal{J}} = \{1, 2, [3, 4], 5, \ldots, m\}\,. \tag{44.3}$$

In specifying the priority series, the following three main cases are possible:

- one priority series is specified without equivalent groups (44.1);
- one priority series is specified with equivalent groups (44.3);
- several variants of the priority series $\{\vec{\mathcal{J}}_\nu\}$ are specified.

The latter is associated with the presence of several points of view as to the importance of criteria. It should be understood that different interpretations of orderliness generally lead to quite different notions of optimality.

The priority vector $\lambda = (\lambda_1, \lambda_2, \ldots, \lambda_m)$ represents an m-dimensional vector with the components λ_j — binary relations of the priority ($\lambda_j \in [1, \infty]$). Here is reflected the priority quantitative characteristic, the binary relation of priority λ_j determining the degree of superiority, by importance, of two adjacent criteria from the priority series. If there is $y_j > y_{j+1}$, then λ_j determines the degree of superiority by importance of the jth criterion over the $(j+1)$ criterion. λ_j has the following meaning: let the criteria y_j and y_{j+1} be normalized, i.e. their scales be reduced to a unified scale (commonly dimensionless), then the solution x is better than $x'/x > x'/$ if the following condition is met

$$\lambda_j \big(y_j(x) - y_j(x')\big) \geq \big(y_{j+1}(x') - y_{j+1}(x)\big)\,.$$

In the case of equivalence of the criteria $y_j \sim y_{j+1}$, $\lambda = 1$, we set $\lambda_m = 1$ to carry out calculations conveniently.

In specifying the priority vector, the following cases are possible:

- λ is accurately specified;
- λ is specified approximately as confidence intervals $b_j < \lambda_j < a_j$;
- λ is specified approximately as several qualitative groups: $\{\lambda_j > 1\}$; $\{\lambda_j > a\}$; $\{\lambda_j > b\}$; $\{\lambda_j > c\}$, and the like, here $a \gg 1$; $b \gg a$; $c \gg b$ are the corresponding gradations of superiority;
- λ is specified as dependence of λ_j on the number in the series of preference $\lambda_j = \psi(j)$;

– for λ we prescribe the range $\lambda \in \Lambda$; the pairwise comparison of criteria allowing λ_j to be determined with adequate accuracy.

The weight vector $\alpha = (\alpha_1, \alpha_2, \ldots, \alpha_m)$ represents an m-dimensional vector determined in the region $\alpha \in A = \{\alpha | \alpha_j \in [0,1], \sum \alpha_j = 1, j \in \mathcal{J}\}$. Each component α_j has the sense of the weight coefficient characterizing relative superiority of the jth criterion over the others.

The priority vector is related to the weight vector by the relations:

$$\alpha_j = \sum_{j=1}^{m} \lambda_j \Big/ \sum_{\nu=1}^{m} \prod_{j=0}^{m} \lambda_j\,; \qquad \lambda_j = \frac{\alpha_j}{\alpha_j + 1}\,.$$

The vector α can be determined accurately or approximately, e.g. in the form of specification of a particular region in A or as a hypothesis of the character of changes of α_j from j, the procedure of specifying the weight vector being more complex than that for the priority vector. If we need to specify simultaneously m numbers from A, then in determining λ, we determine successively each component without any relation to others, that is why it is advisable to determine λ rather than α.

The vector optimization problems may have the following forms of specification of the priority characteristics: (1) $\vec{\mathcal{J}}$; (2) $\vec{\mathcal{J}}$, Λ; (3) $\vec{\mathcal{J}}$, $\lambda_j = \psi(j)$; (4) Λ; (5) λ or α. The procedures of priority incorporation may be quite different, though in transition from one of them to another, the principle of optimality is always corrected. Here are two possible approaches to handling the criterion priority — the principle of rigid priority and that of flexible priority. We consider the methods of handling rigid priority.

The principle of rigid priority relies on the criteria being arranged by importance in a series of the priority $\vec{\mathcal{J}}$; $y_1 > y_2 > \cdots > y_m$, on the basis of which we carry out successive optimization of criteria. The principle of sequential optimization (optimization of hierarchical sequence of quality criteria) based on rigid priority consists in preventing the level of less important criteria to be raised if it brings about even insignificant decrease in the level of a more important criterion from the priority series.[8] The idea of such an approach to the solution of the vector optimization problems was advanced in Ref. 23. According to this notion, we first search for a local optimum for the most important criterion, which is fixed as an auxiliary condition of the problem. We next determine the local criterion of the criterion second in importance, though for a new set of feasible programs; the procedure is repeated. In this way a feasible set is successively restricted to a unique optimal solution or optimal subset.* The described principle of ordering vector sets is assumed to be known as lexicographical.[8]

Lexicographical optimization problems are identified as multicriterion problems with criteria ordered strictly by importance. The lexicographical optimization problems have been carefully studied in Ref. 13.

Let U be a set of all strategies of behavior (solutions), the efficiency of which will be characterized by specific criteria $k_1, k_2, \ldots, k_s$ forming in totality a vector criterion.

Definition 1: Strategy u is not worse than strategy v in the sense of vector criterion ($u \gtrsim v$) if the following inequalities are satisfied:

$$k_r(u) \geq k_r(v), \quad r = 1, 2, \ldots, s. \tag{44.4}$$

Definition 2: If $k(u) = k(v)$, i.e. s in equalities of the form $k_r(u) = k_r(v)$ hold, then the strategies u and v are identified as equivalent to ($u \sim v$). Equivalent strategies should be regarded as such in the sense of the criterion k.

Definition 3: The strategy u is explicitly better and preferable to the strategy v ($u > v$) if there are inequalities of form (44.4), even one of them being strict.

Definition 4: The strategy v is termed efficient if there is no strategy u such that $u > v$, i.e. for which the inequalities of form (44.4) might be satisfied, if only one of them is strict.

In the determinate case, i.e. in the absence of any random and indeterminate factors, the chosen strategy unambiguously determines the result of operation. Therefore, in the determinate lexicographical problem, each strategy u is characterized by s numbers — values of specific criteria:

$$k_1(u), k_2(u), \ldots, k_s(u).$$

*Here a quasioptimal approach is employed. At each stage we carry out quasioptimization, i.e. the search for a domain of decisions close to optimal — a quasioptimal set rather than a unique exact optimum.[8]

All of the specific criteria forming the vector criterion $k_1 = (k_1, \ldots, k_s)$ are strictly ordered by importance, as a result, in comparison to the pair of strategies we primarily employ k_1, the first of them, and the strategy, for which the value of this criterion is larger, is assumed to be better. If the value of the first criterion proves to be equal for both strategies, then the second criterion is employed and the strategy, for which its value is larger, is preferred; if the second criterion does not allow the better strategy to be isolated, then the third (specific) criterion is brought into use and this process proceeds up to the criterion k_s. If the values of each specific criterion for the strategies under discussion turn out to be equal, then these strategies are referred to as equivalent in the sense of the vector criterion k.

Definition 5: The strategy u^*, which is not worse than any other strategy v in the sense of relation $\overset{\text{lex}}{\gtrsim}$ (lexicographically not less preferable), is called lexicographically optimal if $u^* \overset{\text{lex}}{\gtrsim} v$ is satisfied.

Let v^* be a set of all lexicographically optimal strategies. Furthermore, let all these strategies be equivalent among themselves so that the optimal value of k^* of the vector criterion k is given as $k^* = (k_1(u^*), k_2(u^*), \ldots, k_s(u^*))$, where u^* is any strategy from V^*.

The lexicographical optimization problem consists in finding optimal strategies: we can restrict ourselves, almost always, to finding merely the optimal, unique strategy rather than the entire set V^* just as in conventional optimization problems since all the above strategies are equivalent in the present case. The definition of optimal strategy implies that the set V^* may be specified by the following recurrent relations:

$$
\begin{gathered}
V_r^* = \{u : u \in V_{r-1}^*; k_r(u) = \sup_{v \in V_{r-1}^*} k_r(v) = k_r^*\}\,, \\
r = 1, 2, \ldots, s; \quad V_r^* = V\,; \quad V_s^* = V^*\,.
\end{gathered} \tag{44.5}
$$

The latter relations imply $V \supseteq V_1^* \supseteq V_2^* \supseteq \cdots \supseteq V^*$, i.e. each subsequent specific criterion restricts a set of strategies obtained through the use of the preceding specific criteria. Thus, for example, if for the original optimization problem with one scalar criterion there are several solutions and for further selection the auxiliary criteria are successively employed, then the strategies resulting from them will be optimal for the corresponding lexicographical problem with the vector criterion composed of all these alternately employed criteria. In the general case the set V^* may prove to be empty. It is common knowledge that optimal strategies may not exist also in the ordinary extremal problems with one scalar criterion k_1, and its solution then is taken as the maximizing sequence $\{u^k\} \subseteq V$ determined by the condition

$$\lim_{k\to\infty} k_1(u^k) = \sup_{u\in V} k_1(u)\,.$$

The notion of the maximizing sequence turns out to be extremely useful also where optimal strategies exist since for the analysis and numerical solution of extremal problems, extensive use is made of iterative methods. The notion of the maximizing sequence also covers the lexicographical optimization problem. In the absence of the optimal strategy, the least upper (lexicographical) bound of the criterion k, for which the designation k^* will be preserved, may also be determined by relations (44.5) since for the empty set we have $\sup\limits_{u\in V_r^*} k_r(u) = -\infty$. In the general case $-\infty$ and $+\infty$ may represent the components of the vector k^* (here $+\infty$ can be met but not more than once). We assume that the sets V_r^*, beginning with a particular $r = p \geq 1$, are empty.

Definition 6: The sequence of strategies $\{u^k\}$ is termed lexicographically maximizing if, for sufficiently large numbers k, the strategies u^k enter into the set V_{p-1}^* (so that with $p > 1$, $k_r(u^k) = k_r^*$ for $r = 1, 2, \ldots, p^{-1}$) and the equality $\lim\limits_{k\to\infty} k_p(u^k) = k_p^*$ is satisfied.

Definition 7: If the set of optimal strategies V^* is non-empty, then the sequence $\{u^k\}$ is identified as lexicographically maximizing and all of its terms, perhaps with the exception of V_{s-1}^* their finite number, are the elements of the set.

$$\lim_{k\to\infty} k_s(u^k) = k_s^*$$

is satisfied for this sequence.

Definition 8: The sequence of strategies $\{u^k\}$ is called lexicographically maximizing if for any vector $g \in \bar{R}^s$, which is lexicographically less than k^*, we obtain the number $N(g)$ such that with any $k > N(g)$ the vector $k(u^k)$ is lexicographically larger than g.

Thus, the lexicographical optimization problem consists in finding the lexicographically optimal strategies or the lexicographically maximizing sequences of strategies.

The stated problem is also termed the lexicographical optimization problem and, by employing the notation $k^* = \operatorname*{lex\,sup}_{u \in V} k(u)$, it is sometimes briefly written as follows:

$$\text{find} \qquad \operatorname*{lex\,sup}_{u \in V} k(u)\,. \tag{44.6}$$

In the case of existence of lexicographically optimal strategies, problem (44.6) may be written as follows:

$$\text{find} \qquad \operatorname*{lex\,max}_{u \in V} k(u)\,.$$

Sometimes the lexicographical problem is more conveniently stated so that all specific criteria should be alternately minimized. In this form, the problem represents a lexicographic minimization, and it can be presented as follows:

$$\text{find} \qquad \operatorname*{lex\,inf}_{u \in V} k(u)\,. \tag{44.7}$$

It is to be noted that there is the relation:

$$\operatorname*{lex\,sup}_{u \in V} k'(u) = -\operatorname*{lex\,inf}_{u \in V} k(u)\,.$$

The lexicographical problem will have a solution and a unique one at that (the set V^* consists only of one strategy), if some criterion k_r changes to the maximum on V^*_{r-1} merely at one point. For example, the problem cannot have more than one solution if V is a convex set, all $k_1, k_2, \ldots, k_{r-1}$ are quasi-concave (the sets $V_1^*, V_2^*, \ldots, V^*_{r-1}$ being convex), and k_r is strictly quasi-concave.

The dependencc of solutions of the lexicographical optimality problem on parameters (including the initial data) is complex and nonregular (nonsmooth), and this should always be kept in mind when considering specific lexicographic problems, especially non-discrete ones. Of practical interest is the possibility of transforming the vector criterion of the lexicographical problem. In this regard, we should refer to some very simple, but quite useful results.

We assume that the function f has been defined on the set $k_l(v)$ of values of the criterion k_l and is strictly increasing (i.e. for any x and y from $k_l(v)$ such that $x > y$, there is a strict inequality $f(x) > f(y)$). Then the vector criteria $k = (k_1, k_2, \dots, k_s)$ and $(k_1, \dots, k_{l-1}, f(k_l), k_{l+1}, \dots, k_s)$ are lexicographically equivalent, i.e. generate on V the same lexicographical relation of preference $\overset{\text{lex}}{\gtrsim}$, specifically, the sets of optimal strategies isolated by these criteria coincide. The latter enables one, say to construct, for the problem under consideration, new vector criteria lexicographically equivalent to the original, though more convenient in calculation.

We consider the problem of finding lexicographically optimal strategies. From the computational point of view, it is rather difficult to solve the lexicographic problem immediately by recurrent relations (44.5) since at each stage, perhaps with the exception of the latter, we need to completely construct the set V_r^*. It is more convenient to solve the following sequence of problems:

$$\begin{array}{lll}
(1) & \text{Find} & \sup\limits_{u\in V} k_1(u) = k_1^* \\
(2) & \text{Find} & \sup\limits_{u\in V} k_2(u) = k_2^* \\
\multicolumn{3}{c}{\text{-------------------------}} \\
(s) & \text{Find} & \sup\limits_{\substack{u\in V \\ r=1,2,\dots,s-1}} k_s(u) = k_s^*
\end{array} \tag{44.8}$$

The solutions of problems (s) are only all optimal strategies, but for construction of constraints in problem (s) we need to know merely $s-1$ number k_r^*, for which it suffices to successively find one solution at a time for problem (1), (s–1). If the (l) problem is the first in which the upper bound is not attained (this is possible if and only if V_l^* is empty, and V_{l-1}^* nonempty), then any sequence composed of strategies satisfying all constraints in the (l) problem and maximizing k_l, is lexicographically maximizing. In this case the sets of strategies satisfying the constraints in problems $((l+1), \dots, s)$ are empty, thus $k_r^* = -\infty$ for all $r = l+1, \dots, s$. The above multistage approach basically allows the lexicographically optimal strategies to be found, though it is rather complex since to solve s problems of optimization is rather burdensome. This procedure has one more appreciable disadvantage: addition of constraints $k_r(u) = k_r^*$ to problem (1) in relations (44.8) may result in the loss of its features and, as a consequence, in the impossibility to employ efficient computational algorithms. As far as the foregoing is concerned, the problem arises with respect to the possibility of a single-stage solution of lexicographical problems and its reduction to one extremal problem with the scalar criterion.[13]

The study[18] shows that the lexicographic ordering of criteria determines one of the possible procedures for convoluting the vector criterion. It is established that a unified criterion can be formed as the weighted sum of specific criteria with a definite relationship of weight coefficients. These results imply the inverse inference: if the unified criterion represents the weighted sum of specific criteria, then, under some conditions, the behavior of the operating side will be just the same as in the case of the lexicographically ordered criteria.

The literature provides many examples of analysis of multicriterion problems as well the choice of their adequate procedure for convolution of criteria.[1,5,11,13,14] Reference 3 contains the theorems of the unitility theory, from which it follows that many criteria cannot be always reduced to one criterion in principle.

Following the work, we will consider the lexicographical problems of optimization under risk. In this case we know the distribution of random parameters, which enables us to determine the probabilities of appearance of possible results of an event or operation. It is possible to distinguish two types of specific criteria in the problems of optimization under risk.

As in deterministic problems, the criteria formally are the ordinary functionals determined on the set of strategies, so that each strategy is characterized, through the use of such criterion, by a fairly definite number — value of this criterion. Such criteria, however, have a probabilistic sense: they evaluate the probabilities of emergence (or, on the contrary, non-emergence) of some events or operations, e.g. the probability of no-failure operation of an apparatus or system as a whole.

For the fixed strategy, the criteria of another type are the random values with distributions described by the known functions or distribution densities. The specific criterion of that kind characterizes the strategy by a set of numbers rather than by one number. In general, it is inconvenient to compare strategies with the aid of such criteria, therefore they should be transformed firstly by the procedures extensively applied for the scalar problems of optimization under risk.

The lexicographic problems under uncertainty amount to finding maximin strategies or a sequence of strategies being lexicographically maximizing for the criterion $\underline{k}(u)$. The basic principles of optimality suggested for decision making under uncertainty, are carefully analyzed in Refs. 7, 12, 13, 17, and others.

We will consider the methods of handling the flexible priority, whose principle presupposes specification of quantitative characteristics of the priority λ or α, which enables us to partly prefer more important criteria when choosing a solution. In practice, this leads to estimation of the quality of solution by the weighted vector criterion — vector pair (y, α) and to correction of the optimality principle. Basically, this results in transformation of the criterion space as well as in a corresponding change in the scale of criteria. Then comes the choice of the optimal decision based on one of the possible principles of optimality, though in a transformed space of criteria. The principle of flexible priority enables us to give preference, within reasonable limits, to more important criteria with regard to the degree of their relative importance, which should be placed among the merits of the given principle. However, definition of the numerical characteristics of the priority is associated with considerable difficulties. In the approximate specification of the priority characteristics, we have to make recourse to solving parametric problems and suggest in some cases several variants of optimal decisions.[8]

To obtain the data necessary for the justified decision making, Ref. 6 offers to carry out zoning–partitioning of the set of the nature state vectors into subsets, for each of which a definite alternative solution is optimal. It has been noted that stability of solutions is inherent in the problems with a finite number of operations and an uncountable set of state vectors: with changes in the state of nature or in distribution of the state probabilities, which may lead to changes in the result, a definite decision remains unaltered until a corresponding vector of the nature state moves inside the same subset. Upon further movement, the nature state vector corresponding to actual situation may pass into another subset and cause the necessity of changes in the decision.

Stability of the decision shows that the requirement for complete knowledge of the nature state vector is excessive: to make accurate decisions requires merely the data allowing the determination of which of the subsets the actual situation corresponds to. If the initial data are so approximate that we can indicate merely a group of subsets to which they correspond, then an accurate choice of decision is not ensured, but in many instances it is possible to carry out an approximate selection of decision, where maximal losses do not exceed a specified admissible limit.

The principle of rigid priority can also be implemented in the form of an equivalent scheme of the flexible one. To do this, we introduce the priority vector with the components that ensure the choice of decision based on the rigid priority of criteria.[8] References 2, 4, 13 offer the method of successive concessions for the case where all objectives are ordered by their importance.

Following the study,[13] we assume that the criterion k_1 is important for definiteness, k_2 is less important, and then come other specific criteria $k_3, k_4, \ldots, k_s$ according to the degree of their importance. Let the criterion k_1 and the first in importance at that be maximized, its largest value Q_1 being determined. We next assign the value of the "admissible" decrease (concession) $\Delta_1 \geq 0$ of the criterion k_1 and the largest value Q_2 of the second criterion k_2 provided that the value of the first criterion must not be less than $Q_1 - \Delta_1$.

We again assign the value of the concession Δ_2, by the second criterion, which, together with the first, is utilized in finding the relative maximum of the third criterion, and so on. Finally, the criterion k_s, which is the last in importance, is maximized provided that the value of each criterion from the preceding $s-1$ should be not less than the corresponding magnitude $Q_r - \Delta_r$; the resultant strategies (solutions) are assumed to be optimal.

Thus, we take as optimal any strategy being the solution of the last problem from the next sequence of problems (assuming that k_r are bounded from above). The latter ensures finiteness of all magnitudes Q_r, and consistency of constraints in all S problems with any $\Delta_r \geq 0$.

We have:

$$\begin{aligned}
&(1)\quad \text{Find}\quad Q_1 = \sup_{u\in V} k_1(u)\\
&(2)\quad \text{Find}\quad Q_2 = \sup_{\substack{u\in V\\ k_1(u)\geq Q_1-\Delta_1}} k_2(u)\\
&\text{-------------------------------------}\\
&(s)\quad \text{Find}\quad \sup_{u\in V} k_s(u) = Q_s\\
&\qquad k_r(u) \geq Q_r - \Delta_r, \quad r = 1, 2, \ldots, s-1\,.
\end{aligned} \tag{44.9}$$

If the criterion k_s on the set of strategies obeying the constraints of problem (s) does not achieve its largest value Q_s, then the solution of the multicriterion problem is thought to be the maximizing sequence of strategies $\{u^k\}$ from the above set ($\lim_{k\to\infty} k_s(u^k) = Q_s$).

It is advisable to consider similar maximizing sequences also for the case wherein the upper bound in problem (s) is achieved since extensive use is made of iterative methods for the solution of extremal problems.

Comparison of problems (44.9) with problems (44.8) shows that in the case of all Δ_r being zeros, the method of successive concessions isolates only lexicographical optimal strategies, these strategies delivering the value largest on V to the criterion k_1 which is the first in importance. In the other extreme case, where the values of concessions prove to be very large (e.g. such that $Q_r - \Delta_r \leq \inf_{u\in V} k_2(u)$, $r = 1, 2, \ldots, s-1$), the strategies obtained with the aid of this method, delivering the value largest on V to the specific criterion k_s, which is the last in importance. The values of concessions assigned for multicriterion problems, may be regarded as a peculiar measure of deviation of the priority (a degree of relative importance) of specific criteria from the rigid, lexicographical priority.

The approach to the process optimization problem depends basically on the possibility of estimating the measure of unequivalence of different specific criteria and specification methods for this unequivalence. The process optimization problem may be most easily solved if all specific criteria can be arranged by importance. We now develop the idea of the hierarchical scalar criterion ranking in the optimization problem. Let the mathematical model of the controlled system be specified, i.e. we specify:

(1) the equation of movement in the vector form

$$x = F(x, u, t); \quad x \in R^n, u \in R^m; \tag{44.10}$$

(2) the domain of its distribution

$$N(x, u, t) \geq 0; \tag{44.11}$$

(3) the time interval $\tau = (0, T)$ or $\tau = [0, \infty]$;

(4) the set V of piecewise continuous functions assuming the values from (44.11), which is identified further as the set of admissible controls;

(5) the function $F(x,u,t)$ smoothened by (x,u) in the domain (44.11).
Suppose we have specified the boundary conditions for the problem, and we denote them symbolically by the equality

$$(i,f)=0\,. \tag{44.12}$$

Furthermore, let the vector function be specified as

$$\mathcal{J}(u)=\phi(x,u,t) \tag{44.13}$$

with the components

$$\mathcal{J}_\alpha(u)=\phi_\alpha(x,u,t)\,,\quad (\alpha=1,\dots,k)\,. \tag{44.14}$$

On any feasible trajectory $x(t)$, $u(t)$ each function $\mathcal{J}_\alpha(u)$ assumes a definite numerical value and, consequently, $\mathcal{J}(u)$ represents a vector functional. We will consider the first, scalar component $\phi_1(x,u,t)$ of the vector criterion for optimization (44.13). Let the control $u^{(1)}$ transfer the minimum to this functional and assign the value $\phi_1^{(1)}=\phi_1(x,u^{(1)},t)$ to it. With $u^{(1)}$, we calculate the values of the remaining criteria $\phi_2^{(1)},\phi_3^{(1)},\dots,\phi_k^{(1)}$. Let the control $u^{(2)}$ transfer the minimum $\phi_r^{(2)}$ to the criterion ϕ_2, thus $\phi_2^{(2)}=\phi_2(x,u^{(2)},t)$. Calculate with $u^{(2)}$ the values of other criteria $\phi_2^{(2)},\phi_3^{(2)},\dots,\phi_k^{(2)}$. Iterate this operation till all scalar criteria are exhausted. After completing the operation, we obtain the matrix

$$\begin{pmatrix} \phi_1^{(1)} & \phi_2^{(1)} & \cdots & \phi_k^{(1)} \\ \phi_1^{(2)} & \phi_2^{(2)} & \cdots & \phi_k^{(2)} \\ \hdashline \phi_1^{(k)} & \phi_2^{(k)} & \cdots & \phi_k^{(k)} \end{pmatrix} \tag{44.15}$$

which will be identified as characteristic. As the matrix is finite, the analysis of numerical values of its elements and estimation of the gain (or loss) in values of different system quality indices enable one to make a definitive choice of the basic criterion. In this case the remaining criteria will obey hierarchically.

The hierarchical ranking idea suggested here, is alluring because of its simplicity for application in the simplest cases to solutions of the optimal trajectory programming problems. It can be useful wherever the characteristic matrix points to weak sensitivity of criteria to controls and reveals an explicit advantage of a particular criterion over the others. Such an advantage enables one to draw a conclusion about hierarchical subordination of criteria. The difficulty in choosing a preferable criterion is preserved in all other cases where the optimizing criteria are weakly sensitive to the choice of a control, and the matrix elements entering into the same column will not undergo substantial changes associated with those in their upper index, i.e. in control function (44.13). Investigations in this direction have been carried out in Refs. 21, 23, 25, etc. In Ref. 25 the basic criterion is obtained from the engineering analysis, and other criteria are stated as hierarchically subordinated to this criterion. Reference 13 studies some approaches to the solution of the process optimization problems by several successively utilized criteria as well as the available methods of process optimization by scalar criteria. Specifically, the problem of utilization of the Pontryagin maximum principle has also been discussed. In so doing, the lexicographical optimization problems for continuous and discrete processes were first studied, and then the method of successive concessions was analyzed.

Optimization of solutions in hierarchical systems or in their relation (the synthesis problem) gives rise to considerable difficulties, which is determined by special features of hierarchical systems: distribution of subsystems by different hierarchical levels, subordination of one level to another, multiplicity of objectives and other conditions for functioning. As a consequence of diversity of the nature of the vector optimization problems, it seems possible to largely overcome the above difficulties by utilizing formalism of the vector optimization.[8] The above paper discusses the method described below.

The behavior of the hierarchical system features a series of parameters. As first we consider the behavior of the system without conditions determining the hierarchy. We then consider the implemented problem with conditions determining the hierarchy. This enables us to carry out natural decomposition of the general problem into the vector optimization subproblems of two classes — the problems of optimization at each hierarchical level and the interlevel optimization problem. Next, as a result of convolution of vector criteria, each hierarchical level obtains a generalized scalar criterion, which enables one to solve the second class of subproblems — interlevel optimization. The latter problem applies to the class of vector optimization problems with priority. At first it must be solved conceptually. Convolution of the vector criterion is carried out with regard to the priority vector.

With considerable differentiation of priority relations, algorithmic solution of the lexicographical or quasilexicographical ordering type is also possible. Thus, the formalism of vector optimization enables one to construct a well-founded procedure of selecting the optimal solution in hierarchical systems. In this case the complex, original optimization problem reduces to the scalar optimization problem or a sequence of such problems.

§45. FINDING THE SET OF UNIMPROVABLE POINTS

In implementing the vector optimization problems, it is common to firstly isolate the tradeoff domain (the set of unimprovable points, Pareto-optimal solutions) since in the majority of vector optimization problems there is a certain range of solutions, where the criteria are consistent. The tradeoff domain obtained is presented in a number of cases as an optimal solution to the problem, e.g. "negotiation set" in the theory of cooperative games,[3] "efficient programs" in the Koopmans economic theory,[10] and the like.

We will consider the definition of the unimprovable system which has been first given by L. Zadeh.[16] Let C denote a subclass in Σ determined by a set of values of the system S. Introduce the following notation:

$\underset{>}{\Sigma}(S)$ – a set of all systems, each of which is better than the system S.

$\underset{\leq}{\Sigma}(S)$ – a set of all systems, each of which is worse than or equivalent to the system S.

$\underset{\sim}{\Sigma}(S)$ – a set of all systems, which cannot be compared with the system ...

Each system from Σ belongs to one of these sets given above, and the linkage of these sets determines Σ.

Definition 1: The system $S_0 \in C$ is unimprovable in C if an intersection of the sets C and $\underset{>}{\Sigma}(S_0)$ is empty. In other words, this means that in C there is no other system which might be better than system S_0.

The present definition is also correct for the case where Σ is determined by the vector index of quality. Thus, for example, let S be determined by the vector values $x\{x_1, \ldots, x_n\}$, and $C \in R^n$.

Next it is assumed that the quality indices of the system S are measured by components of the vector $\varphi(x)\,\{\varphi_1(x), \ldots, \varphi_m(x)\}$ wherein each $\varphi_i(x)$ is the real-valued function from x.

The system S_0 determined by the vector value $x_0 \in R^n$, is unimprovable in R^n if for any $x \in R^n$ there is $\varphi_i(x_0) \leq \varphi_i(x)$, $(i = 1, \ldots, m)$ (the minimization case of all quality indices $\varphi_i(x)$ being considered).

The control problem is stated as follows.[6] The controlled object is described by the differential equations:

$$\dot{x}_i = f(x, u, t)\,, \quad i = 1, \ldots, m\,,$$

where $x(x_1, \ldots, x_n)$ is the system state vector, $u(u_1, \ldots, u_m)$ the control vector. The functions $f_1(x, u, t), \ldots, f_n(x, u, t)$ are continuous by x, u, and continuously differentiable by x. We specify $x(t_1)$, and the time interval $T = \{t : t_1 \leq t \leq t_2\}$, which completely determines $x(t)$ with the specified $u(t)$. It is required that the control function $u(t)$ should satisfy the following three conditions:

(a) $u(t)$ belongs to the class of operative convex functions C at the interval T;
(b) the trajectory $x(t)$ corresponding to the control $u(t)$ must stay in the feasible domain X, i.e. $x(t) \in X$ with any $t \in T$;
(c) the trajectory $x(t)$ must take on the given value $x(t_2)$, where t_2 may be fixed or arbitrary, satisfying $t_2 \leq T$.

The controls satisfying only conditions (a) and (b) are referred to as the admissible controls, whereas those satisfying all three conditions are known as acceptable controls. The quality indices of the system are defined as

$$y_i(t) = \int_{t_1}^{t_2} g_i(x, u, t)dt\,, \quad i = 1, \ldots, N\,.$$

The acceptable control A is thought to be worse than the acceptable control B if there is the relation

$$y_i(t_2)_{|\,A} \geq y_i(t_2)_{|\,B}$$

in which the sign of inequality is preserved at least for one value of i.

The acceptable control is identified as unimprovable if it is not worse than the remaining acceptable controls. The unimprovable acceptable control determines the unimprovable system.

Theorem 1: Specify the fixed points $x(t_1)$ and $x(t_2)$, and let X be unbounded by the X-space. For the acceptable control $\hat{u}(t)$ and its corresponding trajectory $\hat{x}(t)$ to be unimprovable, it requires the existence of the vector of the function $\hat{\psi}(\hat{\psi}_1, \ldots, \hat{\psi}_n)$ and the function $H(\hat{\psi}, \hat{x}, \hat{u}, t)$ satisfying the following expressions:

$$H(\hat{\psi}, \hat{x}, \hat{u}, t) \equiv \sum_{i=1}^{n} \hat{\psi}_i(t) f_i(\hat{x}, \hat{u}, t) - \sum_{k=1}^{N} c_k g_k(\hat{x}, \hat{u}, t),$$
$$\partial_u \int_{t_1}^{t_2} H(\hat{\psi}, \hat{x}, t)dt \le 0, \tag{45.1}$$
$$\frac{d\hat{\psi}_i(t)}{dt} = -\frac{\partial H}{\partial \hat{x}_i} \quad (i = \overline{1,n}), \quad c_k \ge 0 \ (k = \overline{1,N}).$$

In the latter expression (45.1), the inequality sign cannot be preserved for all k values; we denote in terms of δ_u the first variation of the next integral corresponding to the infinitesimal variation $u(t)$ with the fixed $\hat{\psi}$, $\hat{x}$, t. The theorem stated above is titled "the general theory of optimal control". Its proof has been constructed in Ref. 6, which also gives the mathematical results stated as necessary conditions for unimprovable controls. However, the above general theorem and the subsequent ones from Ref. 6 cannot be successfully applied to solve the particular control problems with many criteria of quality since the problem of determining the coefficients c_k, $k = \overline{1,N}$ in (45.1) remains to be solved.[4]

The general vector optimization problem can be stated as follows.[2] Let X be the solution defined on the feasible solution set X. The quality of solutions is estimated by the local criteria $y_1, y_2, \ldots$ forming the efficiency vector $y = (y_1, y_2, \ldots, y_m)$. The vector y is associated with the solution mapping $x \to y = f(x)$ specified analytically, statistically and heuristically. The relative importance of local criteria is given by the priority vector $\lambda = (\lambda_1, \lambda_2, \ldots, \lambda_m)$, where $\lambda_j \in [1, \infty)$ is the binary relation of the jth criterion priority over the $(j+1)$ criterion. We need to find the optimal solution x^0 determined by two conditions:

- the solution must be feasible, i.e. belong to the feasible solution set X;
- the solution must be the best, i.e. optimize the efficiency vector y with regard to the criterion priority λ.

The present formulation corresponds to the general vector optimization model.

Model 1.

$$x^0 = f^{-1}\Big[\operatorname*{opt}_{x \in X} \big(y(x), \lambda\big)\Big] \tag{45.2}$$

or in the non-uniqueness case of the optimal solution, wherein the optimal solution subset X^0 is isolated as optimal, there is also the model of a more general type:

$$X^0 = \{x^0\} = \bigcup_{x \in X} f^{-1}\left[\operatorname*{opt}_{x \in X}(y(x), \lambda)\right], \tag{45.3}$$

where opt is the optimization operator determining the principle of optimality and having the sense of the order relation; f^{-1} is the inverse mapping $y \to x = f^{-1}(y)$.

Model 1 is the vector optimization model since it has the criterion of quality (or efficiency) being the vector, and its implementation results in finding the optimal solution x^0 or the optimal subset X^0.

In vector optimization problems there is no contradiction between local criteria. This contradiction is commonly non-strict. There is a certain domain of agreement X^S which has no contradiction, and the solution quality may be improved simultaneously by all local criteria. At the same time, we have the tradeoff domain X^C where there is contradiction even with one of the criteria. Here improvement of the solution quality by some criteria brings about deterioration of others, and the choice of any solution is based on tradeoff. Thus, the range of feasible solutions consists of two nonintersecting parts

$$X = X^C \cup X^S, \quad X^C \cap X^S = \varnothing. \tag{45.4}$$

The optimal solution should not belong to the agreement domain X^S since any solution from this domain may be improved by all criteria. Therefore it will be, without fail, in the tradeoff domain $x^0 \in X$, and the field of search for an optimal solution should be confined only to this domain. The problem thus arises with respect to determining the tradeoff domain, and its isolation from the range of possible solutions.

In models (45.2)–(45.4), this involves restriction of the range of feasible solutions and transition to the vector optimization problems with a strict criterion contradiction and new ranges of feasible solutions X^C and Y^C.

Definition 2: The agreement domain X^S is identified as a subset of the set of feasible solutions X featuring the possibility for each solution $x \in X^S$ to be improved without lowering the level of any one of the local criteria. In other words, if $x \in X^S$, then we always find in X a better solution x', for which $y(x') \geq y(x)$. The relation $\geq$ means that for all j, $y_j(x') \geq y(x)$ and even for one j, $y(x') > y(x)$.

Definition 3: The tradeoff domain X^C is identified as a subset of the set of feasible solutions X featuring that solution $x \in X^C$ cannot be improved without lowering the level even of one of the local criteria. Thus, if x, $x' \in X^C$, then in the case of two criteria $y_1(x) > y_1(x')$, $y_2(x) < y_2(x')$ or vice versa. In other words, for any two solutions from X^C we have without fail a contradiction of criteria, and the choice can be made only on the basis of tradeoff.

The points $x' \in X$ are also referred to as effective points, unimprovable solutions or Pareto-optimal solutions. The trade-off domain can be defined in two basically different ways:

- by defining the agreement domain X^S and eliminating it from the range of feasible solutions

$$X^C = X \backslash X^S \tag{45.5}$$

- by isolating the trade-off domain proceeding from its properties (Definitions 3, 4)

$$(x, y) \to X^C \tag{45.6}$$

The second way is preferable. It is associated only with one operation. Furthermore, it is common that $X^S \geq X^C$ and isolation of elements is more complex than X^C.

The model of choosing the trade-off domain, which is common to all classes of vector optimization problems, corresponds to Definition 3.

Model 2:

$$X^C = \{x | x \in X, \{x' | y(x') \geq y(x)\} \cap X = \varnothing\} \tag{45.7}$$

or in the space of criteria

$$Y^C = \{y | y \in Y, \{y' \geq y\} \cap Y = \varnothing\}\,. \tag{45.8}$$

At the basis of this model is the principle of optimality opt $y \equiv \max y$, i.e. that of dominance: the solution x is better than the solution x' if $y(x) \geq y(x')$.

For purposes of calculation, it is more convenient to employ the parametric model of optimality equivalent to (45.7).

Model 3:

$$X^C = \bigcup_{\alpha \in A} f^{-1}\left[\max_{x \in X} \sum_{j=1}^{m} \alpha_j y_j(x) \right]. \tag{45.9}$$

Here $\alpha = (\alpha_1, \alpha_2, \ldots, \alpha_m)$ is the vector parameter determined on the set $A = \{\alpha | \alpha > 0\,, \sum_{j=1}^{m} \alpha_j = 1\}$. For convex problems (where A is convex), implementation of Model 3 requires to find the local maximum of the linear form $L(x) = \sum_j \alpha_j y_j(x)$ in modifying $\alpha \in A$. For non-convex problems, we need to find all local maxima of the linear form $L(x)$ and test them for dominance conditions.

We consider some properties of the set of unimprovable points as well as the relation between the optimization procedure for hierarchical sequence of quality criteria and the procedures of defining unimprovable points following Ref. 1.

We assume that the system behavior is characterized by the n-dimensional vector $x = \{x_1, \ldots, x_n\}$ from a certain given set X of the Euclidean n-dimensional space E^n, and it is estimated by the m-dimensional vector function $\varphi(x) = \{\varphi_1(x), \ldots, \varphi_m(x)\}$, where $\varphi_i(x)$, $i = \overline{1,m}$ are the given real functions of the variable x. It is required to choose such point $x^0 \in X$ that the function value $\varphi(x)$ in a sense would be optimal.

The author isolates three approaches to the solution of the stated problem:

(1) Weighted criterion optimization. We choose the numbers of $\lambda_i > 0$, $i = \overline{1,m}$; $\sum_{i=1}^{m} \lambda_i = 1$ having the sense of relative utility of the ith component of the criterion (general utility is taken as equal to a unity), and we choose, as the problem solution, the vector $x^0 \in X$ minimizing on X the function $\varphi_\lambda(x) = \sum_{i=1}^{m} \lambda_i \varphi_i(x)$.

(2) Optimization of the hierarchical system of functionals. Among the vector components $\varphi(x)$, we specify the order of preference. To be specific, we take the following order:

$$\varphi_{i_1}(x), \varphi_{i_2}(x), \ldots, \varphi_{i_m}(x)\,. \tag{45.10}$$

As a solution, we assume the vector $x^0 \in X$ minimizing the hierarchical system of functions (45.10), i.e. x^0 satisfying the relations:

$$\varphi_{i_1}(x^0) = \min_{x \in X^0} \varphi_{i_1}(x) ,$$

$$\varphi_{i_2}(x^0) = \min_{x \in X^1} \varphi_{i_2}(x), \dots ,$$

$$\varphi_{i_m}(x^0) = \min_{x \in X^{m-1}} \varphi_{i_m}(x) ,$$

where $X^0 = X$, $X^S = \left\{x : \varphi_{i_s}(x) = \min_{x \in X^{s-1}} \varphi_{i_s}(x),\, x \in X^{s-1}\right\}$.

(3) Definition of unimprovable points. We introduce the notion of the set of points unimprovable in X with respect to $\varphi(x)$. The choice of unimprovable points is thought to be judicious with the vector criterion of quality. The set of unimprovable points was studied in Refs. 7, 9, 16, and others.

The work in Ref. 11 discussing some properties of the set of unimprovable points in the general dynamic problem with k criteria of quality has shown that the problem amounts to minimization of the linear form of the vector components $\varphi(x)$ with constant weight coefficients, i.e. minimization of the expression

$$\varphi_\lambda(x) = \sum_{i=1}^{m} \lambda_i \varphi_i(x) , \quad \lambda_i > 0, \quad i = \overline{1,m}; \quad \sum_i \lambda_i = 1 .$$

Concurrent with unimprovability, L. Zadeh has introduced the notion of optimization with a vector criterion, i.e. the point $x^0 \in X$ is called optimal in X with respect to $\varphi(x)$ if $\varphi_i(x^0) \leq \varphi_i(x)$ for all $x \in X$. Evidently, if x^0 is optimal, then x^0 is unimprovable, too. In general, the opposite is incorrect. However, if $\varphi_i(x)$ is a scalar function, then these notions coincide. Besides, the following property holds.

Property 1: If there is the point x^0 optimal in X with respect to $\varphi(x)$, then any unimprovable point is optimal, too. The proof of the property may be constructed by contradiction.

The following problems will be considered proceeding from the first two approaches.

Problem 1: Find $x^0 \in X$ such that $\varphi_\lambda(x^0) \leq \varphi_\lambda(x)$ for all $x \in X$, where λ is a given m-vector belonging to the set $\Lambda^+ = \{\lambda : \lambda \in E^m,\ \lambda_i > 0,\ i = \overline{1,m},\ \sum_{i=1}^{m} \lambda_i = 1\}$.

Problem 2: Let $\Pi(\mathcal{J})$ be a permutation specified on the set $\mathcal{J} = \{1, 2, \ldots, m\}$. Find $x^0 \in X$ being the solution of the minimization problem with the hierarchical system of functions, where the order of preference is specified according to $\Pi(\mathcal{J})$.

Problem 3: Let $\mathcal{J}_1, \ldots, \mathcal{J}_k$, $k \leq m$ be a specified, ordered partition of the index set $\mathcal{J}$. Specify the numbers $\lambda_i > 0$, $i = \overline{1,m}$ such that $\sum_{i \in \mathcal{J}_s} \lambda_i = 1$, $s = \overline{1,k}$. Find the point $x^0 \in X$ such that

$$\phi_1(x^0) = \min_{x \in X^0} \phi_1(x)\,,$$

$$\phi_2(x^0) = \min_{x \in X^1} \phi_2(x), \ldots, \phi_k(x^0) = \min_{x \in X^{k-1}} \phi_k(x)\,,$$

where $\phi_s(x) = \sum_{i \in \mathcal{J}_s} \lambda_i \varphi_i(x)$, $s = \overline{1,k}$, $X^0 = X$,

$$X^s = \{x : x \in X^{s-1}, \varphi_i(x) = \varphi_i(x^0), i = \mathcal{J}_s\}, \quad s = \overline{1,k}\,.$$

Problem 3 is more general than Problems 1, 2. Indeed, if partition of the index set consists of one set $\mathcal{J}_1 = \mathcal{J}$, then we get the problem $\mathcal{J}$. If each set $\mathcal{J}_s$ entering into the partition $\mathcal{J}$ is the single-element one, then Problem 3 becomes a variant of Problem 2. The following statement holds true.

Statement 1: If $x^0 \in X$ is the solution of Problem 3 for a certain partition $\mathcal{J}_1, \ldots, \mathcal{J}_k$ and some $\lambda_i > 0$, $i \in \mathcal{J}$, $\sum_{i \in \mathcal{J}_s} \lambda_i = 1$, $s = \overline{1,k}$ then x^0 is also the unimprovable point in X with respect to $\varphi(x)$. From this statement we have obtained the following corollaries.

Corollary 1: If $x^0 \in X$ is the solution of Problem 1 with a given λ, $\lambda \in \Lambda^+$ then x^0 is also unimprovable in X with respect to $\varphi(x)$.

Corollary 2: If $x^0 \in X$ is the solution of Problem 2 with a specified order of preference, then x^0 is also unimprovable in X with respect to $\varphi(x)$.

Corollary 3: For any $s \in \mathcal{J}$ there is such an unimprovable point x^0 in X with respect to $\varphi(x)$ that

$$\varphi_s(x^0) = \min_{x \in X} \varphi_s(x)\,. \tag{45.11}$$

Corollaries 1 and 3 stem from Problems 1 and 2 being special cases of Problem 3. To understand Corollary 3, we will consider point x^0 which constitutes the solution of Problem 2 with the order of preference corresponding to the permutation $\Pi(\mathcal{J}) = \{i_1, \ldots, i_m\}$, where $i_1 = S$. By Corollary 2, x^0 is then unimprovable and, besides, equality (45.11) is completed.

The solution of Problem 2 does not depend on the order of preference if the value assumed by $\varphi(x)$ on the solutions of Problem 2, is constant with all possible orders of preference. Denote in terms of N the set of all points unimprovable in X with respect to $\varphi(x)$. The following statements prove to be valid.

Statement 2: The solution of Problem 2 does not depend on the order of preference if and only if $\varphi(x) = \text{const}$ for all $x \in N$.

Statement 3: In order for the point x^0 optimal in X with respect to $\varphi(x)$ to exist, it is necessary and sufficient that $\varphi(x) = \text{const}$ for all $x \in N$.

Statement 4: In order for the point x^0 optimal in X with respect to $\varphi(x)$ to exist, it is necessary and sufficient that the solution of Problem 2 would not depend on the order of preference. Henceforth the function $\varphi_i(x)$, $i = \overline{1,m}$ and set X will be assumed to be convex. In this case we have the following statement.

Statement 5: In order for $x^0 \in X$ to be unimprovable in X with respect to $\varphi(x)$, it is necessary and sufficient that there would be such partition $\mathcal{J}_1, \ldots, \mathcal{J}_k$ of the set $\mathcal{J}$ and such $\lambda_i > 0$, $i \in \mathcal{J}$, $\sum\limits_{i \in \mathcal{J}_s} \lambda_i = 1$, $s = \overline{1,k}$ with which x^0 is the solution of Problem 3.

By employing the theorem of detachability of convex sets, we will study the sets of unimprovable points. To do this, we need to utilize the following auxiliary statements. Let Ω_i, $i = 0, 1, \ldots, m$ be the bounded, closed convex sets from E^n.

Lemma 1: In order that $\bigcap_{i=0}^{m} \Omega_i \neq \varnothing$, it is necessary and sufficient that

$$\max_{\|g\|_{E^{nm}}=1} \left\{ \min_{x\in\Omega_0} \left(\sum_{i=1}^{m} g_i \right)' x - \sum_{i=1}^{m} \max_{x\in\Omega_i} g_i'(x) \right\} \leq 0 . \tag{45.12}$$

Here $g \in E^{nm}$, $g = \{g_1, \ldots, g_m\}$, where $g_i \in E^n$, $i = \overline{1,m}$, E^{nm} is the $(n \times m)$ space.

Lemma 2: In order that $\Omega_0 \subset \bigcap_{i=1}^{m} \Omega_i$, it is necessary and sufficient that

$$\max_{\|g\|_{E^{nm}}=1} \left\{ \max_{x\in\Omega_0} \left(\sum_{i=1}^{m} g_i \right)' x - \sum_{i=1}^{m} \max_{x\in\Omega_i} g_i'(x) \right\} \leq 0 . \tag{45.13}$$

As before, it is assumed that the functions $\varphi_i(x)$, $i = \overline{1,m}$ and the set X are convex and, besides, the functions $\varphi_i(x)$, $i = \overline{1,m}$ are continuous. Introduce the sets $G_i(\delta_i) = \{x : x \in E^n,\ \varphi_i(x) \leq \delta_i\}$, $i = \overline{1,m}$, where δ_i is a certain real number, and the function $\Lambda(\delta)$ of the m-dimensional argument $\delta = \{\delta_1, \ldots, \delta_m\}$:

$$\Lambda(\delta) = \max_{\|g\|_{E^{nm}}=1} \left\{ \min_{x\in X} \left(\sum_{i=1}^{m} g_i \right)' x - \sum_{i=1}^{m} \max_{x\in G_i(\delta_i)} g_i' x \right\} .$$

The function $\Lambda(\delta)$ is continuous and nondecreasing in δ. We now consider the equation

$$\Lambda(\delta) = 0 . \tag{45.14}$$

Introduce the concept of unimprovable roots (45.14). We say that δ^0 is the unimprovable root if $\Lambda(\delta^0) = 0$ and there is no $\bar{\delta}$ such that $\Lambda(\bar{\delta}) = 0$ and $\bar{\delta}_i \leq \delta_i^0$, $i = \overline{1,m}$, even one inequality being strict.

Among unimprovable roots of this equation and unimprovable points there is a close relation revealed by the following statements.

Statement 6: In order for the point x^0 to be unimprovable in X with respect to $\varphi(x)$, it is necessary and sufficient that $\delta^0 = \varphi(x^0)$ would be the unimprovable root.

Therefore, the set of values assumed by the vector function $\varphi(x)$, when x runs through N, belongs to the set of unimprovable roots.

Statement 7: If δ^0 is the unimprovable root, then there is such a point x^0 unimprovable in X with respect to $\varphi(x)$ that $\varphi(x^0) = \delta^0$, x^0 satisfying the relation

$$\left(\sum_{i=1}^{m} g_i^0\right)' x^0 = \min_{x \in X} \left(\sum_{i=1}^{m} g_i^0\right)' x\,, \tag{45.15}$$

where $g^0 = \{g_1^0, \ldots, g_m^0\}$ is the vector transferring the maximum in the left-hand side (45.14) with $\delta = \delta^0$. Therefore, the set of values assumed by $\varphi(x)$ on N, coincides with that of unimprovable roots.

Let us consider the optimal control problem of the linear system described by the system of differential equations with the vector criterion of quality. Specify the following system:

$$(\dot{x}) = A(t)x + b(t,u)\,, \quad x(t_0) = x_0\,, \tag{45.16}$$

where x is the n-dimensional vector, and $u(t)$ the r-dimensional measurable vector function of control adhering to restrictions:

$$u(t) \in V \qquad \text{for all } t \in [t_0, T]\,. \tag{45.17}$$

V is the compact set from E^r, and the function in (45.17) is the n-dimensional continuous vector function. The quality of the process is estimated with the vector functional

$$\mathcal{J}(u) = \varphi\bigl(x(T)\bigr)\,, \tag{45.18}$$

where $\varphi(x)$ is the m-dimensional continuous convex vector function. The feasible control $u^0(t)$ is unimprovable for system (45.16) with restrictions (45.17) by vector criterion (45.18) provided there is no other feasible control $u(t)$ such that $\mathcal{J}_i(\bar{u}) \leq \mathcal{J}_i(u^0)$, $i = \overline{1,m}$, with even one inequality being strict.

Let $M = \{\delta : \delta \in E^m, \delta = \mathcal{J}(u), u \in \bar{N}\}$. The following statement provides constructive description of the set M.

Statement 8: The set M coincides with the set of unimprovable roots of the equation

$$\max_{\|g\|_{E^{nm}}=1} \left\{ \left(\sum_{i=1}^{m} g_i \right)' \phi(T, t_0) x_0 + \int_{t_0}^{T} \min_{u \in V} \left(\sum_{i=1}^{m} g_i \right)' \phi(T, t) b(t, u) dt - \sum_{i=1}^{m} \max_{x \in G_i(\delta_i)} g_i' x \right\} = 0 . \tag{45.19}$$

Here $\phi(t, \tau)$ is the matrix satisfying the conditions

$$d\phi(t, \tau)/dt = A(t)\phi(t, \tau), \quad \phi(\tau, \tau) = 0 .$$

Among the studies on finding the set of unimprovable points, it might be worthwhile to point out the work in Refs. 5, 8, 12–15. Reference 15 contains generalization for the nonlinear case and provides simpler proofs for some properties of unimprovable points obtained by other researchers for the linear case.

Suppose we specify a certain subset X of the Euclidean finite-dimensional space and p functions: $f_k : X \to R$, $k = \overline{1, p}$.

By employing the auxiliary problem $p(\bar{x})$: $\varphi(\bar{x}) = \inf \sum_{k=1}^{p} f_k(x)$, $f(x) \leq f(\bar{x})$, x, $\bar{x} \in X$ the authors show that x^0 is efficient if and only if x_i^0 is the optimal solution $p(x^0)$. This result enables one, by utilizing, say the Kuhn–Tucker conditions, easily to find whether the given point is efficient. On the contrary, if x^* is the optimal solution $p(\bar{x})$, then x^* is efficient and $f(x^*) \leq f(x)$. The problem of interrelation between efficient points proper is also discussed.

To find efficient points in the multicriterion control problem[12] suggests the method of internal penalty functions. The problem is solved for objective functions:

$$f_i(u) = \langle x(b) - \xi_i, W_i[x(b) - \xi_i] \rangle + \int_a^b \langle \tilde{x}_i(t) - c_i(t)x(t)Q_i(t) \times [\tilde{x}_i(t) - c_i(t)x(t)] \rangle dt + \int_a^b b_i(t)\|u(t)\|^2 dt , \quad (i = \overline{1, n})$$

with constraints

$$\dot{x} = Ax + Bu$$
$$x(a) = x_0 , \quad a \leq t \leq b , \quad x \in R^m .$$

Here the scalar function $b_i(t) > 0$, vectors $\tilde{x}_i(t)$ and ξ_i and matrices W_i, $c_i(t)$, $Q_i(t)$ are specified; $u(t)$ is the vector function measurable by Lehbeg on $[a, b]$. The efficient controls u^* are such that, for any feasible controls u the system of inequalities $f_i(u) \leq f_i(u^*)$, $(i = \overline{1, n}\,)$, from which at least one inequality is strict, is inconsistent. In the absence of constraints on u the problem is solved by common minimization of the scalar criterion $f^\alpha(u) = \sum_{i=1}^{n} \alpha_i f_i(u)$, where $\alpha_i \geq 0$ and $\sum_{i=1}^{n} \alpha_i = 1$. The explicit form $u^\alpha = u^*$ is extracted. Under restrictions of form $\|u(t)\| < 1$ with the aid of the penalty function the problem is reduced to minimization $f_k^\alpha(u) = f^\alpha(u) + \int_a^b \|u(t)\|^{2k} dt$. We extract the explicit form of controls u_k^α minimizing f_k^α and state that $f_i(u_k^\alpha) \to f_i(u^\alpha)$, $k \to \infty$. References 5, 8, 13, 14 and other studies deal with finding unimprovable points for multicriterion linear programming problems.

Reference 8 discusses the problem of finding the optimum of the Pareto objective function cx (where $x \in R^n$, c is the $k \times n$ matrix) on the polyhedron F determined by the system $Ax \leq b$, $x \geq 0$. Let the point $x^0 \in F$ be specified. Denote in terms of P_{x^0} the problem of maximization $l^T S$ under conditions $cx = \mathcal{J}_s + cx^0$, $Ax \leq b$, $x \geq 0$, $s \geq 0$.

Theorem 1: If the problem P_{x_0} has no finite maximum, then the set of efficient points is empty.

Reference 5 systematically studies the properties of the set of efficient solutions in the multicriterion linear problem. Consider the following cases:

(1) The feasible set is arbitrary.

(2) The feasible set is convex.

(3) The feasible set is a convex polyhedron.

In conclusion it may be said that a search for the set of unimprovable points is not always needed and well founded in solving the problems of vector optimization. It is not needed in the problems with strict contradiction of criteria. In these problems there is no agreement domain $X^s = \varnothing$ and the range of feasible solutions constitutes the tradeoff domain (the set of unimprovable points) $X = X^c$. Therefore, before defining the tradeoff domain, we have to make sure whether the above case exists here. Furthermore, definition of X^c is not necessary if the problem presupposes a search for the optimal tradeoff solution, especially if the algorithm is low-sensitive to the size of the range of search. Exact definition of the tradeoff domain by models 2 and 3 as well as by the Hermeyer models encounters serious computational difficulties. Hence, in a number of cases we may utilize approximation methods. Approximate definition of the tradeoff domain can be accomplished by following the procedures: scanning, directed search, probing, approximation, and the like. All of them are related to common properties of the tradeoff domain and directed towards ordering the choice of the parameter α in model 3, or towards simplification of the procedure of testing the dominance conditions.

In a number of cases, where the number of solution variants and local criteria is countable, the vector problem is formulated in the matrix form — as "matrix efficiency," the procedure of defining the tradeoff domain becomes more simplified. Here matrix algorithms are utilized to estimate dominance, which results in coming from the initial set of solutions to the tradeoff set.[2]

§46. DETERMINATION OF THE SOLUTION BASED ON A PARTICULAR TRADEOFF

It is commonly insufficient to be confined to isolating the tradeoff domain to solve specific vector optimization problems. It is generally required to determine a unique optimal solution or at least several solutions. This involves intrusion into the tradeoff domain as well as selection of the optimal solution based on a particular tradeoff scheme and its corresponding optimality principle. The number of various schemes, even in the case of two local criteria, is very large, which aggravates selection of the requisite tradeoff scheme corresponding to the real situation of decision making. Hence, in every specific case, we have to choose the tradeoff domain — optimality principle without fail.

Following the study,[7] we will discuss the basic tradeoff principles and their corresponding schemes. For simplicity, we assume that there is a normalized problem without priority.

I. Uniformity Principles

If criteria are normalized and the same in importance, it is natural to seek a uniform and harmonic improvement of the quality of all local criteria.

In this case the uniformity principle can be implemented in a variety of ways.

1. Equality principle

$$\mathrm{opt}_2\, y \equiv \{y_1 = y_2 = \cdots = y_m\} \in Y^c\,. \tag{46.1}$$

Maximization is carried out here provided the level of all criteria is equal. This principle is unduly "rigid." It may lead to a solution outside the tradeoff domain and may even have no solution at all, especially in discrete problems. "In the pure form," the principle of equality is primarily applied in optimization of solutions in cooperative and coalitional systems.

2. Principle of uniformity (maximin)

$$\underset{y\in Y^c}{\mathrm{opt}_3}\, y \equiv \max_{y\in Y^0} \min_j y_j\,. \tag{46.2}$$

Uniformity of raising the level of all criteria is realized by "pulling up" the worst of the criteria, i.e. those with the lowest level. Apart from uniformity, this principle also has another important meaning — the guaranteed level — $\min y_j$.

3. Principle of the best uniformity (sequential maximin)

$$\underset{y\in Y^c}{\mathrm{opt}_4}\, y \equiv \max_{y\in Y^c} \min_j{}_1 \max_{y\in Y^c} \min_j y_j \ldots\,. \tag{46.3}$$

The notion of uniformity is somewhat augmented as against (46.2). In the non-uniqueness case of solution by the maximin criterion, we define the second minimum and perform its maximization, etc. Thus the principle of sequential maximin is implemented and better matching and fuller ordering of Y^c is achieved, i.e. the unique optimal solution is defined. The optimality principles opt_3 and opt_4 can be implemented in the equivalent integral form.

$$\mathrm{opt}_5 \equiv \min_{y\in Y^c} \sum_{j=1}^{m} \bar{y}_j^s\,. \tag{46.4}$$

4. Principle of quasi-equality

$$\mathrm{opt}_6 \equiv \{y \| y_j - y_\nu\| \le \delta, j, \nu \in \mathcal{J}\} \cap Y^c\,. \tag{46.5}$$

All local criteria are maximized providing their level will not differ from one another, by not more than the magnitude δ.

5. Quality matching principle

Two theorems on average values of the highest steps underlie this principle.

S of average degree from the set $\{y_j\}$, $j \in \{1, 2, \ldots, m\}$ is identified as the expression

$$\Delta_S\{y_j\} = \left(\sum_{j=1}^{m} y_j^S \Big/ m\right)^{1/S} . \tag{46.6}$$

The following asymptotic transitions hold true for them

$$\lim_{s\to\infty} \Delta_S\{y_j\} = \max_j\{y_j\}\,; \qquad \lim_{s\to-\infty} \Delta_S\{y_j\} = \min_j\{y_j\}\,. \tag{46.7}$$

For the finite m and y_j, equalities (46.7) are satisfied with the finite $s > s^*$ (46.4). This property underlies a series of equivalent transformations of vector models and, specifically, construction of equivalent model (46.4). The quality criteria matching model may also be constructed proceeding from this feature

$$\text{opt}_7 \equiv \min_{y \in Y^c} \sum_{j=1}^{m} Y_j^{-s} , \tag{46.8}$$

where $s \in (1, s^*)$, $s^* = \log m / \log(1+\varepsilon)$. As the parameter s increases from $s = 1$ the matching of criteria levels is performed, and with $s > s^*$ we obtain the ideal matching equivalent to the sequential maximin model (46.3). However, by choosing s somewhere in the interval $(1, s^*)$, we get only partial matching and, therefore, take into account the uniformity principle rather approximately, which offers a means of achieving a high level of general integral efficiency.

II. Valid Concession Principles

All uniformity principles (46.1)–(46.8) pertain to the category of the "valid" ones. In these principles, the "validity" manifests itself in the tendency to match the level of all local criteria. We consider the approach based on estimating and comparing a decrease and increase in the level of local criteria, which are bound to occur in the tradeoff domain. This principle is identified as the concession estimation principle. It has two modifications: the principle of absolute and that of relative concession.

1. Absolute concession principle

$$\underset{y \in Y^c}{\mathrm{opt}_8}\, y \equiv \left\{ y \,\middle|\, \sum_{j \in {}^{+}\mathcal{J}} \Delta y_j \geq \sum_{j \in {}^{-}\mathcal{J}} \Delta y_j \right\} \cap Y^c , \tag{46.9}$$

where ${}^{+}\mathcal{J}$ is a subset of the majorized local criteria ($\Delta y_j > 0$); ${}^{-}\mathcal{J}$ — a subset of the minorized local criteria ($\Delta y_j < 0$); Δy_j — the absolute magnitude of an increase.

The model is based on the following tradeoff principle: the tradeoff, where the total absolute level of lowering one or several criteria does not exceed that of raising other criteria, is assumed to be valid. This principle corresponds to the criteria sum optimization model (that of integral efficiency).

$$\mathrm{opt}_9\, y \equiv \max_{y \in Y^c} \sum_{j = \mathcal{J}} y_j \,. \tag{46.10}$$

A serious disadvantage of the absolute concession principle is that it may admit pronounced differentiation in the levels of some criteria, i.e. a high value of the generalized integral criterion can be achieved at the cost of one criterion or a group of criteria with a rather low level of other criteria. However, it is suitable for implementation.

2. Relative concession principle

$$\mathrm{opt}_{10}\, y \equiv \left\{ y \,\middle|\, \sum_{j \in {}^{+}\mathcal{J}} \kappa_j \geq \sum_{j \in {}^{-}\mathcal{J}} \kappa_j \right\} \cap Y^c , \tag{46.11}$$

where κ_j is the modulus of relative changes — "the concession price." Suppose that all of the local criteria forming the efficiency vector have the same significance. Then we consider as valid the tradeoff where the total relative level of quality deterioration by one or several criteria does not exceed (less than or equal to) the total relative level of quality improvement by the remaining criteria. The model corresponding to the given validity principle, will be identified further as the valid relative concession model.

The absolute concession principle is indifferent to the real value of criteria and, hence, can admit greater uniformity of the criteria levels. Therefore it can be utilized only together with one of the uniformity principles.

The relative concession principle is rather sensitive to the value of criteria, the value of concession for large value criteria automatically decreasing due to concession relativity, and vice versa. This results in the considerable smoothing of criteria levels. An important advantage of the relative concession principle also is that it is invariant to the scale of criteria measurement. If the importance of criteria is not the same, the idea of the valid tradeoff based on estimation of concessions loses its lucidity, and argumentation of the choice of the weighting vector of importance distribution proves to be rather difficult, especially, where the number of the criteria involved is great.

III. Other Principles of Optimality

1. Principle of isolating the basic criterion

$$\mathrm{opt}_{12}\, y \equiv \max_{y \in Y^1} y_1 \,, \tag{46.12}$$

where $Y^1 = \{y | y_j \geq y_j^{\partial}, j \in \{1, 2, \dots, m\mathcal{J}\} \cap Y^c$. Here we isolate one local criterion as the basic and carry out its scalar maximization providing the level of other criteria is not less than the admissible one. Any criterion may be chosen as the basic, but it is more convenient to take the criterion for which an admissible level is difficult to define. By (46.12), we can implement, in principle, any tradeoff scheme and obtain any optimal solution in the tradeoff domain. However, in the overwhelming majority of cases it is impossible to give reasons for the choice of an admissible level of criteria y_j^{∂}. Moreover, it is also impossible to advance arguments for the choice of a particular notion of tradeoff or optimality by varying the magnitudes y_j^{∂}. In some cases, when implementing the given principle and advancing arguments for the choice of y_j^{∂}, it is wise to utilize equivalent reciprocal models, wherein the basic and additional criteria interchange their positions.

2. Principle of maximizing the weighted sum of criteria

$$\mathrm{opt}_{13}\, y \equiv \max_{y \in Y^c} \sum_{j \in \mathcal{J}} \alpha_j y_j \,, \tag{46.13}$$

where $\alpha_j \in [0, 1]$, $\sum_j \alpha_j = 1$, $j \in \mathcal{J} = \{1, 2, \dots, m\}$. The principle represents a modified absolute concession model for the priority case, where the weight coefficients α_j are introduced into the model to offer a relative advantage for more important criteria in the sense of concession. However this principle has somewhat universal meaning, too. This pertains to the fact that by employing the model, we define the tradeoff domain for convex problems via variation of weight coefficients. Thus, through the use of model (46.13) it is possible to obtain any tradeoff solution and, therefore, any tradeoff principle is implemented. In practice, it is impossible to advance arguments for the choice of weight coefficients to implement a certain tradeoff principle, therefore the present model plays, in most cases, a specific, local part of the tradeoff model based on estimation of absolute concession in the priority case.

3. Probability maximization principle of achieving an ideal quality

$$\mathrm{opt}_{14}\, y \equiv \max_{y \in Y^c} p\,, \quad (y \geq y^u)\,. \tag{46.14}$$

In the stochastic vector problem, we can often define the desired ideal value of all local criteria y_j^u and, therefore, that of the ideal quality y^u. In this case, (46.14) with a criterion is the probability of achieving a compound event $p(y \geq y^u)$. Practical methods of calculating the probability of events, even in the case of two or three events, are rather complex. Therefore the present method can be utilized only in some specific cases where $m \leq 3$ and calculation of $p(y \geq y^u)$ is carried out simply enough.

4. Optimality principle for multivector problems

Multivector problems in the class of vector problems are the most complex. In these problems, there are several reasons for transition to vector formalism, and, as a result, recourse is made to multivector criteria with components, which are vectors, too.

$$\begin{gathered} \mathop{\mathrm{opt}}_{y \in Y} y \\ y = (y_1, y_2, \ldots, y_m)\,, y_j = (y_{j_1}, y_{j_2}, \ldots, y_{j_l})\,, j \in \mathcal{J}\,, \end{gathered} \tag{46.15}$$

i.e. $y \in E^m \times E^l$.

Such problems are solved by employing a decomposition principle. The general multivector problem is broken down into a series of single vector problems (according to the vector order). Next we solve each subproblem in the ordinary, but only conceptual way, i.e. we define for each νth subproblem the principle of optimality $\mathrm{opt}^{(\nu)} y$. The general optimality principle results from aggregating the subproblem optimality principles.

For each subproblem, the situation has specific features and, in the general case, there is its own principle of optimality.

Modern literature outlines two basic approaches to the problem of seeking tradeoff solutions, which are the analytical and analytical-heuristic approaches. Both approaches are based on the central concept of tradeoff axiomatics. By the tradeoff axiomatics we mean a set of axioms strictly determining the tradeoff scheme principle, i.e. the principle on the basis of which the space of criteria is ordered and the optimal solution y^0 is chosen.

In the analytical approach, axiomatics is chosen and the procedure of seeking the optimal solution is implemented following the scheme

$$S \to A \to \text{opt} \to y^0 \to x^0 ,$$

where S is a certain decision making situation, A the tradeoff axiomatics. Thus, the given situation S determines the choice of the corresponding tradeoff axiomatics A on the basis of which we choose the tradeoff scheme opt y and then carry out the search for the optimal solution y^0 and x^0, i.e. the whole process of decision-making is carried out in a strictly formalized way.

In the analytical-heuristic approach, transition from S to A is carried out manually, analyzing the situation and choosing the most suitable from the suggested set of axiomatics.

We will consider both approaches in greater detail. To do this, we define, following Ref. 5, the multiobjective optimization problem as the set: $\langle N, x\{\underset{i}{\succ}\}_{i\in N}\rangle$, where $N = \{1, 2, \ldots, m\}$, $\underset{i}{\succ}$ is the ordering of alternatives by the criterion i, $\underset{N}{\succ} \supseteq \underset{i}{\cap} \succ i$. In the game terms, the problem may be characterized as follows: the feasible coalitions are the coalitions $\{i\}$, $i \in N$, and the large coalition N; only the latter can efficiently choose an alternative, whereas the choices of coalitions $\{i\}$ do not affect the result of chance. A planner, designer, governing body, and the like act as the coalition N in practical problems.

We consider the problem of finding Pareto-optimal alternatives for functions $f_i(x)$, $i \in N$ on the set of Euclidean space X. We set forth two results obtained by S. Karlin[9] and J.B. Germeyer.[6]

Lemma 46.1: If X is convex, $f_i(x)$ are concave, and x^0 is Pareto-optimal, then there are such λ_i, $\lambda_i \geq 0$, $\sum\limits_i \lambda_i = 1$ that

$$\max_{x\in X}\sum_i \lambda_i f_i(x) = \sum_i \lambda_i f_i(x^0)\,. \tag{46.16}$$

Lemma 46.2: If for some $\lambda_i > 0$, $i \in N$

$$\sum_i \lambda_i f_i(x^0) = \max_{x\in X}\sum \lambda_i f_i(x)\,, \tag{46.17}$$

then x^0 is Pareto optimal.

For the case where X is a convex polyhedron, and the functions f_i are linear, in Lemma 1 it may be stated[25] that all $\lambda_i > 0$ and for Pareto-optimality of the alternative x^0 it is necessary and sufficient that there would be such λ_i with which relation (46.17) takes place.

The above lemmas imply that the choice of a particular point in a Pareto-optimal set is equivalent, in a sense, to the indication of weights for each criterion. This statement underlies the majority of computational methods. It is also useful for understanding many heuristic methods of defining the optimum with the aid of weights.

Lemma 46.3: If x^0 is Pareto-optimal, here $f_i(x^0) > 0$ for all $i \in N$, then there are such λ_i, $\lambda_i > 0$, $i \in N$, $\sum_i \lambda_i = 1$, that

$$\max_{x\in X}\min_i \lambda_i f_i(x) = \min_i \lambda_i f_i(x^0)\,. \tag{46.18}$$

We consider the case, where, apart from the set of alternatives X and objective functions $f_1(x), \ldots, f_n(x)$ we specify some requirements for optimality further called axioms (of optimality). For application, we may need quite different axioms. However, literature still has few axiomatically defined solutions. Notwithstanding considerable differences in interpretation of the models, for which the systems of axioms were constructed, the axioms themselves are still similar. In all systems we find the variants of axiom symmetry, Pareto-optimality, covariance. Consider the system of axioms of G. Nash and K. Arrow, L. Gurwitz.[18,28]

G. Nash initially constructed his solution for the problem in transactions concluded by two individuals. Two-individual restriction is non-essential since the Nash axioms are readily generalized for any number of participants. But, in order to apply them to the above model, it must be supplemented with one parameter — status quo point (s.q.).

In the transaction problem, the s.q. point, which is denoted in terms of $f^* = (f_1^*, \ldots, f_n^*)$, characterizes the state occurring before the beginning of negotiations, and thereby in the case where no agreement is achieved. It is common knowledge that the unambiguous choice of the s.q. point is very hard to obtain. It is also difficult to get a sound answer as to its choice in the present case. It seems natural to choose $f_i^* = \min_{x \in X} f_i(x)$, though other variants more suitable for a specific problem are also possible.

The decision problem is described by the collection $\langle X, \{f_i\}, f^* \rangle$. Let $f^0 = (f_1^0, \ldots, f_n^0)$ be the optimal values of objective functions of this model. Consider the formulations of axioms to be satisfied.

Axiom 1: (Linear transformation). Let $L = (L_1, \ldots, L_n)$ be any linear positive transformation of the function f, then the optimal solution for the model $\langle X, \{L_i f_i\}, Lf^* \rangle$ is Lf^0. In other words, if we change the scale and reference unit of each criterion with respect to the s.q. point, then the optimal values of objective functions change in the same way too.

The axiom does not affect behavioral rationality, but merely postulates, in a sense, independence of the optimal solution from choosing the numerical values of utility. Therefore, it may be concluded that it is accomplished in all multiobjective optimization problems.

Axiom 2: (Pareto optimality). The optimal value of f^0 should satisfy the inequality $f^0 \geqq f^*$, i.e. $f_i^0 \geq f_i^*$ for all $i \in N$, f^0 must be attainable, i.e. $f^0 = f(x)$ for a particular $x \in X$, and, finally, there should not be such $x \in X$ that $f^0 \leqq f(x)$, i.e. $f_i^0 \leq f_i(x)$, for all $i \in N$ and $f_i^0 < f_i(x)$ for some of them.

Axiom 3: (Insignificant alternative independence). Let $\langle X, \{f_i\}, f^* \rangle$ and $\langle X', \{f_i\}, f^* \rangle$ be two models, and $X \supset X'$. If f^0 is optimal in the first model and $f^0 = f(x)$ for a certain $x \in X'$, then f^0 is optimal for the second model, too.

Accomplishment of this axiom should be carefully examined in each specific problem since it is possible to imagine economic situations, wherein restriction of choice opportunities results in drastic revision of values, optimal "in good life" though achievable in new circumstances, and it is not as required by the axiom. However, if the alternatives are actually not associated with the remaining ones and are insignificant with respect to both models, then the axiom seems to be satisfied.

Axiom 4: (Symmetry). If the model is symmetric, i.e. $f_i^* = f_j^*$ for all $i, j \in N$ and for any permutations π of the set N and $x \in X$ there is such $y \in X$ that $(f_i(x)) = (f_{\pi_i}(y))$ then $f_i^0 = f_j^0$ for all $i, j \in N$.

By interpreting the model under consideration as the transaction problem, we can reformulate the axiom as follows: if the status and opportunities of all negotiators are completely equal, so must be the optimal values of utility obtained by them.

Theorem 46.1: The solution f^0 satisfies axioms 1–4 if and only if $f^0 = f(x)$ and

$$\prod_{i=1}^{n}\left(f_i(x^0) - f_i^*\right) = \max_{x\in X}\prod_{i=1}^{n}\left(f_i(x) - f_i^*\right). \qquad (46.19)$$

To observe mathematical strictness, it should be noted that the Nash solution is given not for an individual model, but for a class of models. The Nash solution definition domain represents such set of models $\langle X, \{f_i\}, f^*\rangle$ that it may have feasible operations indicated in the axioms: linear positive transformations of the function f and point f^* as well as incorporation of the set of values of the function f on X into a symmetric set.

Following the Arrow–Gurwitz axiom system we will consider an arbitrary set Ω of objective functions $f_\omega(x)$ on the set of alternatives X. Since the set of objective functions is variable, Ω becomes the explicit component model of decision making. Two models $\langle X, \Omega_1, \{f_\omega(x)\}\rangle$ and $\langle X, \Omega_2, \{f_\omega(x)\}\rangle$ are referred to as isomorphic if the set constitutes permutation of the set Ω_1. If $f_\omega(x)$ is taken as the matrix with columns $\omega \in \Omega$ and rows $x \in X$, then the matrices of isomorphic models are derived from one another by rearranging columns. It is assumed that model 2 is derived from model 1 by removing recurrent objective functions if, after removing some $\omega \in \Omega_1$, model 1 is isomorphic with respect to model 2 and for each removed $\omega \in \Omega_1\backslash\Omega_2$ there is such $\omega' \in \Omega_2$ that $f_\omega(x) = f_{\omega'}(x)$ for all $x \in X$. The class of models P, on which the following axioms and, therefore, the solution are specified, is defined by the following three conditions:

(1) If X is finite and $\min_\omega f_\omega(x)$, $\max_\omega f_\omega(x)$ exist for all $x \in X$, then $\langle X, \Omega, f\rangle \in P$.
(2) If one model is derived from the other by withdrawing recurrent objective functions and the latter is contained in P, then the model derived must also be contained in P.

(3) If the model belongs to P, then for any $x \in X$ there must be $\min_{\omega} f_\omega(x)$ and $\max_{\omega} f_\omega(x)$.

The class P appears to be nonempty. In any event, the problems with a finite number of objective functions fall within this class because the minima and maxima indicated in the definition of P exist for them.

Denoting a set of optimal solutions by x^0, we state axioms.

Axiom 1: If $x_1 \subset x_2$ and $x_1 \cap x_2^0 \neq \varnothing$, then $x_1^0 = x_1 \cap x_2^0$. This axiom is similar to the Nash axiom of independence from insignificant alternatives differing only in that the Nash solution consists of a unique point whereas this is not needed here.

Axiom 2: The sets of optimal alternatives for isomorphic models are the same. In other words, the set of optimal alternatives does not vary with rearrangement of the columns ω of the matrix $\|f_\omega(x)\|$.

Axiom 3: If $x \in x^0$, $x' \in X$ and $f_\omega(x) \leq f_\omega(x')$ for all $\omega \in \Omega$, then $x' \in x^0$. Similarly, if $x \in x^0$, $x' \in X$ and $f_\omega(x) \geq f_\omega(x')$ for all $\omega \in \Omega$, then $x' \notin x^0$.

Conceptually, if a particular alternative is not worse by all criteria than an optimal one, then it is optimal, too. If it is not better by all criteria than a certain nonoptimal alternative, then it is also nonoptimal. This condition is utterly indisputable and very feeble as regards application. The result given below is an improvement of the Arrow–Gurwitz theorem.

Theorem 46.2: In order that the optimal solution satisfy axioms 1–4 in P, it is necessary and sufficient that

$$x^0 = \{x : (\min_{\omega} f_\omega(x), \max_{\omega} f_\omega(x) \gtrsim (\min_{\omega} f_\omega(x'), \max_{\omega} f_\omega(x')\} \tag{46.20}$$

for all $x' \in X$ where $\gtrsim$ is any ordering of the number pairs (m, M), $m \leq M$ satisfying the relationship

$$m_1 \geq m_2, \quad M_1 \geq M_2 \Longrightarrow (m_1, M_1) \gtrsim (m_2, M_2)\,. \tag{46.21}$$

In general, the optimal solution defined by axioms 1–3 is constituted by a set rather than an alternative, i.e. it is ambiguous. The solution is ambiguous in that there are several such sets. Each ordering $\gtrsim$ will correspond to its set x^0, here the more indefinite is the ordering (more equivalent pairs (m, M)) the larger is the set x^0. The gist of this result is that definition of optimality in the sense of axioms 1–3 for any multicriteria problem amounts to definition of optimality in the two-criteria problem with criteria $\min_\omega f_\omega(x)$ and $\max_\omega f_\omega(x)$.

We assume that in place of the objective functions $f_i(x)$, the preferences $\gtrsim$ are specified on X. The group decision is called the rule, following which every collection of preferences $(\gtrsim_i)_{i\in N}$ on X corresponds to the (transitive) preference $\gtrsim N$ on X. The group decision problem differs from the above problem primarily in that we consider, as a solution, not a set of alternatives, but their ordering. There are also other differences imposed by specific characteristics of application of group decisions to politics. K. Arrow enumerated requirements for the group decision rules:

(1) (Universality). The group decision rule is defined for various collections of individual preferences $\gtrsim_i$, considering that the number of individuals is not less than two, and that of alternatives not less than three.
(2) (Positive relation). If for a given set of individual preferences $x \gtrsim_N y$, then this relation is left also when the individual preferences in pair comparisons including x either do not change or change in favour of x, and other comparisons remain unchanged.
(3) (Independence from insignificant alternatives). If all comparisons between alternatives $x' \subset X$ coincide in two sets of individual preferences, then they have to coincide in group decisions, too.
(4) (Sovereignty of citizens). For each pair of alternatives x and y there is such a collection of individual preferences that $x \gtrsim_N y$.
(5) (Dictator absence). There must not be such individual i_0 that $x >_i y$ would imply $x >_N y$ irrespective of preferences of other individuals.

The well-known Arrow paradox resides in the fact that the above conditions are incompatible, i.e. there are no group decision rules satisfying all five conditions.

In practice, common use is made of the (simple) majority rule. Simplicity and efficiency of this rule does not need to be proved despite the fact that it sometimes leads to nontransitive social preferences. At present, we know the efficient conditions under which the solution is transitive.[1] Moreover, the problem of transitivity arises when there are only two alternatives, i.e. practically in all applications.

The basic features of the simple majority rule are as follows:

(1) Definiteness. In any collection of individual orderings, the rule indicates a unique decision for each pair of alternatives.
(2) Anonymity. The decision is independent of designations for individuals.
(3) Neutrality. The decision is independent of names for alternatives.
(4) Positive response. If for a particular set of individual preferences $x \gtrsim_N y$, and, with the other set, only one individual changes his preference in favour of X while the rest remains unchanged, then with a new set there should be $x >_N y$.

The meaning of the above conditions for the decision rules is quite evident. It should only be noted that anonymity and neutrality coincide with the corresponding two Arrow–Gurwitz axioms.

The following theorem[27] characterizes the simple majority rule in full.

Theorem 46.3: The simple majority rule represents a unique rule possessing the above features.

Let us consider the aggregative criteria problem. Suppose that instead of the alternative orderings $>_i$ by the i criteria, we specify the orderings $>_k$ by aggregative criteria $k \subset N$. In the general case, we cannot know, of course, the ordering for any $k \subset N$, therefore we denote by k the set of those $k \subset N$ for which the orderings $>_k$ are specified. The problem then is formally described by the collection $\langle X, K, \{>_k\}_{k \in K}\rangle$. The problem is to find a set of alternatives optimal by a set of criteria $\{>_k\}$, or, in a more general case, find the ordering of the set X by this set of criteria. The desired ordering is denoted as $>_N$. Such problem was first discussed in Ref. 4. The suggested approach can be described conceptually as follows.

The orderings are obtained by direct measurements or from individual orderings $>_i$, $i \in N$ by some aggregation rules. We need to extend these rules to cover the whole society N and thus obtain the ordering $>_N$. However, we know rather the results of aggregating $>_k$ than the rules themselves. Hence, we make it our first business to reveal these rules and find the comparative significance of orderings confined in the aggregative orderings $>_k$. Next, we need to construct such orderings $>_N$ (or their set) in which the revealed relative significance would be preserved and $>_N$ fit the most significant aggregative orderings well.

Considering that decision making must proceed from a set of alternatives, i.e. by choosing a point from the set of unimprovable points, in many practical situations it has been found advantageous to employ man-machine procedures. The idea to involve man in the process of decision making was expressed in Ref. 20.

The literature describes many procedures of decision making. We will describe some of the most familiar procedures and offer the scheme by which these procedures are constructed.

Most of the computational methods for finding the (Pareto) optimal points rely on maximization of the sum on X

$$\sum_{i \in N} \lambda_i f_i(x), \quad \lambda_i \geq 0 .$$

Various heuristic procedures share the same idea, too. For example, a researcher is informed about approximate values of the parameters λ_i. He finds the Pareto-optimal point x^0 corresponding to the particular weights and the rate of change of the function f_i depending on the point x^0 changing in different directions. Proceeding from this information, a decision-maker amends the weights λ_i, and the problem is solved all over again.

B. Roy[29] suggested the following general approach to construct the decision-making procedures.

The binary relation $\gtrsim$ on X is called the outranking relation providing that:

(1) It is reflexive ($x \gtrsim x$ for all $x \in X$).
(2) For any x, y, $z \in X$ the following statements hold true: $x \gtrsim y$, $f_i(y) \gtrsim f_i(z)$ for all $i \in N \Longrightarrow x \gtrsim z$, $f_i(x) \geq f_i(y)$ for all $i \in N$, $y \gtrsim z \Longrightarrow x \gtrsim z$. The outranking relation has the same sense as the above ordering: if $x \gtrsim y$, then x is not less preferable than y. Condition (2) implies (with $x = y$) that $f_i(y) \geq f_i(z)$ for all $i \in N \Longrightarrow y \gtrsim z$, i.e. the outranking relation satisfies the Pareto–optimality axiom if the condition for definition $f_i(x) \geq f_i(y)$, $i \in N$ is supplemented with the strictness condition even for one inequality and then the consequence is substituted by $x \gtrsim z$ and $z \not\gtrsim x$, which is quite reasonable.

B. Roy presents several specific outranking relations and procedures for their utilization. The general description is as follows. We construct a certain set $C \subset X \times X$ which, by analogy with statistics, may be identified as critical. If $(x, y) \in C$, then $x \succsim y$. The procedures rely on construction of the sequence $C_1 \subset C_2 \subset \cdots C_n$, and thereby on construction of the sequence of outranking relations $\succsim_1 \subset \succsim_2 \subset \cdots \subset \succsim_r$. The set C may be constructed according to available data in a variety of ways. Consider an example. Let the suggested weights p_i of criteria f_i be known. Assume that

$$N^+(x,y) = \{i \in N : f_i(x) > f_i(y)\},$$
$$N^=(x,y) = \{i \in N : f_i(x) = f_i(y)\},$$
$$N^-(x,y) = \{i \in N : f_i(x) < f_i(y)\}.$$

Consider the functions

$$P^+(x,y) = \sum_{i \in N^+(x,y)} p_i,$$
$$P^-(x,y) = \sum_{i \in N^-(x,y)} p_i,$$
$$P^=(x,y) = \sum_{i \in N^=(x,y)} p_i.$$

Define the critical set as the set of pairs (x, y) satisfying the inequalities

$$P^+(x,y) + P^=(x,y) \geq C, \quad P^+(x,y) \geq P^-(x,y).$$

Evidently, the larger is C, the less is the risk (due to the weights p_i being not well established) when eliminating alternatives from further consideration, because the advantages of a particular alternative over eliminated ones are more significant. Suppose that, with a certain C_1 and $\succsim_1$, alternatives of the set X_1 remain uneliminated. If the choice from X_1 is hard to make, we have to run the risk of choosing $C_2 < C_1$. Repeat calculations and reasonings with C_2 in place of C_1 and X_1 in place of X etc.

The method ELECTRA (Elimination and Choice Reflecting the Reality)[15] involves the procedure of multicriteria selection of the "best" from the set of objects under consideration. For each of the criteria, we introduce a discrete scale of possible values of this criterion (discernable gradations — estimations by this criterion). Thus, each object from the given set E corresponds to the "n-times" Cartesian product of criteria scales. For each of the criteria, we construct a graph, the vertices of which are separate objects of the set, and the edges point upon a dominance relation among the objects according to the present criterion. Next, we introduce weight coefficients (indices of relative importance) of the adopted criteria. Proceeding from the estimates of objects of the set treated by all criteria with respect to their importance, we calculate the matrices of special coefficient values identified as the indices of agreement or disagreement (to introduce the disagreement index we need first to place in agreement the employed "lengths" of criteria scales.)

Further, for each pair of objects (l, l') the outranking relation of, say l over l', is assumed to be established if the value of the corresponding agreement index is larger than a certain threshold value, and the disagreement index is less than the corresponding threshold value. Proceeding from the thus constructed binary relation of outranking, we construct a generalized graph (of outranking), the structure of which essentially depends on the chosen threshold values. The graph kernel contains the best of the available objects as well as those of the objects which are "incomparable" with the chosen thresholds. With the threshold values changing, the number of elements remaining in the kernel is changing, too.

The basic advantage of the given approach is likely to reside in the fact that a decision-maker is faced only with the objects, the comparison of which requires a more careful and deeper analysis. Furthermore, variation of threshold values enables one to isolate different collections of objects containing the "best" object and those most unlike it, though featuring the available estimates by the entire totality of criteria does not provide a foundation for these objects to be classed with the deliberately "bad" ones. The possibility of revealing such collections to be considered by the decision-making person seems to be quite reasonable from the standpoint of a more substantiated choice of the "best" object.

The method of pairwise comparison of objects is based on the following assumptions:

- a decision-maker can always and without difficulty compare two values by any one criteria;
- there is the scalar function of utility with continuous first derivatives, by means of which it is possible to carry out complete ordering of all objects being compared. The method involves procedures of successive comparing of object pairs with each other. The most difficult and

interesting is the case whereby some portion of criteria object A outranks B, and by another — quite the reverse. The method is discussed at greater length in Ref. 8.

A group of methods[8,12,20] has been developed as applied to the problem of linear programming with many criteria. Here an object is taken as the solution belonging to a domain of feasible solutions, due to which the number of objects is not limited and objects are not predetermined. The suggested methods involve a collection of analytical parts, each of which implies presentation of partial results to a decision-maker and utilization of new information obtained from him.

The paper of R. Kiney[11] deals with the problem of choosing the best action out of the feasible set, where the results of these actions bear a probabilistic character, i.e. each action features its vector of probabilities for all feasible alternative results. One of the most elaborate theories of choosing the best action in these conditions, generally, is the probabilistic utility theory of von Newman and Morgenstern. By this theory, to evaluate relative preferability of each alternative result under risk we introduce a special utility function, the values of which for each alternative are established on the basis of the decision-maker's suggestions. Based on the functions of each alternative utility and probabilities of their implementation, we form the utility function for each action as a whole. Next we choose the action securing the maximum utility function thus obtained. The alternatives, however, may be distinguished by a number of factors or criteria. Such alternatives have received the name "multidimensional", and each multidimensional alternative, in turn, is evaluated by the multidimensional utility function.

Proceeding from the assumption of mutual independence of the factors by utility characterizing the alternatives, R. Kiney suggests the procedure of constructing the multidimensional utility function which demands from a decision-maker actually the same (in practice, minimal possible) body of information as in the additive case, though the field of its application is much wider.

The above methods of decision-making with many criteria employ, in one way or another, the additional information obtained from a decision-maker. In these methods, the stages of carrying out formal mathematical operations (calling for application of computers with large N and n)

alternate with those of dialogues with a decision-maker, which allows the procedures of that type to be identified as man-machine procedures.

A new approach to solve the multicriteria optimization problem has been advanced in Ref. 16. The approach is based on the notion of determining an ideal (utopic) point in the space of quality criteria and on introduction of a norm into this space. The tradeoff solution obtained here is the Pareto-optimal solution and secures maximal proximity of quality criteria to their best values. The approach given above is free from choosing weight coefficients, which is difficult especially in solving the large system control problems. A similar idea was almost simultaneously expressed in Ref. 30 for static problems of vector optimization. The same idea was applied in Ref. 17 to solve the optimal design problem for a universal system.

For the case of static systems, the above approach has been developed in Refs. 32 and 34. In these works, a suggestion is made to isolate the dominating solutions determined by minimizing the norm more general than the square one and choosing the final solution out of this dominant set proceeding from additional reasonings and information obtained from the person making the final decision.

The work in Ref. 32 discusses the properties of tradeoff solution in vector optimization problems. Let X be a solution space, and the space of solution results

$$Y = \{y = f(x) = \{f(x_1), \ldots, f_N(x)\} | x \in X\}$$

is assumed to be compact. Introduce a group regret (for the parameter P)

$$R_P(y) = \left[\sum_{j=1}^{N} (y_j^* - y_j)^P \right]^{1/P},$$

where $P \geq 1$, $y_j^* - \sup\{f_j(x)|x \in X\}$, $y \in Y$. The point y^P of the function minimum $R_P(y)$ on the set Y and the corresponding point $x^P = f^{-1}(y^P)$ is referred to as the tradeoff solution for the parameter P. The case $P = 1$ leads to maximization of the total utility $\max_y \sum y_j$; the case $P = \infty$ corresponds to $\min_\rho \max_j \{y_j^* - y_j | j = 1, \ldots, N\}$; the case $p = 2$ corresponds to the square criterion of M. Salukwadze. It has been proved that the tradeoff solution y^p has the following properties: feasibility, minimal group regret, absence of dictator, Pareto-optimality, independence from inconsistent alternatives, symmetry and, in some cases, uniqueness.

Further we introduce the concept of a dominance structure which includes the tradeoff solution. Suppose we have two results y^1 and y^2 in the space Y and $d = y^2 - y^1$. If for any $\lambda > 0$ the result $y^1 + \lambda d$ proves to be preferable than y^1, then the vector d is termed the "dominance factor" for the given result y. Let $D(y)$ be a set of all dominance factors for the given result y supplemented with the null vector in R^N ($D(y)$ is a cone). It is assumed that the structure of dominance is specified on Y if for each $y \in Y$ the cone of dominance $D(y)$ is given. Then y^1 is dominated by the result y^2, if $y^1 \in y^2 + D(y^2) = \{y^2 + d | d \in d(y^2)\}$. The result y^0 is the nondominated solution if there is no result $y^1 \in Y (y_1 \neq y^0)$ such that $y^0 \in y^1 + D(y^1)$. Evidently, the concept is transferred into the solution space too.

The studies in Refs. 33, 34, and the like, have also been carried out in that direction.

In conclusion we will discuss the solution optimization problem in hierarchical systems.[7]

Optimization of solutions in hierarchical systems or with respect to them (synthesis problem) encounters many difficulties, which is determined by specific character of hierarchical systems: distribution by different hierarchical levels, subordination of one level to another, multiplicity of objectives and other conditions for functioning. As a result of diversity of the nature of vector optimization problems, it is not improbable that the above difficulties can be largely obviated by employing the vector optimization formalism.[2,3]

Let X be a hierarchical system, the behavior of which is determined by the parameters:

(1) $\mathcal{J} = \{1, 2, \ldots, n\}$ — a set of subsystems;

(2) $\{x \in X_i\}_{i \in \mathcal{J}}$, $x \in X = \bigcap_{i \in \mathcal{J}} X_i$ feasible sets of subsystems and the whole system;

(3) $\{x \to y_i(x)\}_{i \in \mathcal{J}}$, $y = (y_1, \ldots, y_n)$ the scalar criteria of subsystems and the vector criterion of the system. Conditions 1–3 determine behavior of any system. Introduce additional conditions determining a hierarchical structure of the system;

(4) $\mathcal{J} = \{1, 2, \ldots, m\}$ is a set of hierarchical levels;

(5) $\{\mathcal{J}_j\}_{j\in\mathcal{J}}$ is distribution of subsystems by levels;
(6) $\{\lambda_j\}_{j\in\mathcal{J}}$, $\lambda = (\lambda_1, \lambda_2, \ldots, \lambda_m)$ is a set of binary relations of the priority of levels or the priority vector.

Behavior of the hierarchical system is determined by six parameters:

$$\langle \mathcal{J}, \{X_i\}_{i\in\mathcal{J}}, \{y_i(x)\}_{i\in\mathcal{J}}, \mathcal{J}, \{\mathcal{J}_j\}_{j\in\mathcal{J}}, \{\lambda_j\}_{j\in\mathcal{J}}\rangle \tag{46.22}$$

on the basis of which the search is carried out for the optimal system X^0. We first consider behavior of the system without Conditions 4–6. In this case we have the general problem of vector optimization of systems by the criterion $y(x) = (y_1, y_2, \ldots, y_n)$. Without detriment to Conditions 4–6 it is possible to define the tradeoff domain, owing to which the field of search for an optimal system is restricted. Furthermore, in a number of cases we have to restrict optimization of the system by this

$$X^c = \{x | x \in X, \{x' | y(x') \geq y(x)\} \cap X = \varnothing\}\,. \tag{46.23}$$

Model (46.23) can be implemented either accurately or approximately, say, by employing the notion of the directed search or the properties of local optima. In a number of cases, model (46.23) may be employed to evaluate satisfactoriness of the system structure from the condition:

$$X^s = X \backslash X^c \neq \varnothing\,, \tag{46.24}$$

which means that in the systems of a given structure there is no strict rivalry of subsystems, and the structure is feasible. Condition (46.24) may be rather simply tested by evaluating variations in local optima.

Further intrusion into the tradeoff domain to search for a unique optimal system should be carried out with compulsory utilization of Conditions 4–6. They allow a general problem (46.22) to be naturally decomposed into two classes of vector optimization subproblems — those for optimization at each hierarchical level and that of interlevel optimization.

We consider subproblems of the first class. At each jth hierarchical level there is the vector optimization problem without priority of the form:

$$\langle \mathcal{J}_j, \{X_i\}_{i\in\mathcal{J}}, y^{(j)}(x)\rangle\,. \tag{46.25}$$

Here $y^{(j)}$ is a vector criterion for the jth hierarchical level with components-local criteria of all subsystems of a given level.

Problem (46.25) is solved only conceptually, i.e. the search for an optimum on the feasible set $X_j = \bigcap_{i\in\mathcal{J}_j} X_i$ by the criterion $y^{(j)}$ is not carried out. We perform merely a convolution of the vector criterion into a generalized scalar criterion on a basis of tradeoff axiomatics and its corresponding computational scheme

$$y^{(j)} \to \varphi_j = \varphi_j(y^{(j)})\,, \tag{46.26}$$

e.g. starting from the "valid concession" axiomatics

$$\varphi_j(x) = \prod_{i\in\mathcal{J}_j} y_i(x) \tag{46.27}$$

or from the "uniformity" axiomatics

$$\varphi_j(x) = \min_j y_j(x)\,. \tag{46.28}$$

As noted above, the tradeoff axiomatics may be chosen either analytically or heuristically, commonly by employing additional information on the system.

As a result of vector criteria convolution, each hierarchical level obtains a generalized scalar criterion $\varphi_j(x)$, which enables one to go over to solution of the second class subproblems of interlevel optimization

$$\langle \mathcal{J}, \{x_j\}_{j\in\mathcal{J}}, \{\varphi_j(x)\}_{j\in\mathcal{J}}, \{\lambda_j\}_{j\in\mathcal{J}}\rangle \tag{46.29}$$

The problem refers to the class of vector optimization problems with priority. It should be first solved conceptually, too. Convolution of the vector criterion $\varphi = (\varphi_1, \varphi_2, \ldots, \varphi_m)$ is carried out with regard to the priority vector λ

$$(\varphi, \lambda) \to \omega = \omega(\varphi, \lambda)\,. \tag{46.30}$$

In considerable differentiation of priority relations, it is also possible to construct an algorithmic solution of the lexicographic or quasilexicographic ordering type.

Thus utilization of vector optimization allows a sound procedure of choosing the optimal solution in hierarchical systems to be constructed.

In this case the complex original optimization problem (46.22) reduces to the trivial scalar optimization problem (or a sequence of such problems).

$$X^0 = f^{-1}\Big[\max_{x\in X} \omega\big((\varphi, x), \lambda\big)\Big]. \tag{46.31}$$

To illustrate, in utilizing the axiomatics of the valid concession tradeoff for interlevel optimization and optimization at each hierarchical level there is the scalar problem:

$$X^0 = f^{-1}\left[\max_{x\in X} \prod_{j\in \mathcal{J}} \left(\prod_{i\in \mathcal{J}} y_i(x)\right)\alpha_j\right]. \tag{46.32}$$

Here

$$\alpha_j = \prod_{j=1}^{m} \lambda_j \Big/ \left(\sum_{\nu=1}^{m} \prod_{j=\nu}^{m} \lambda_j\right); \quad j \in \mathcal{J}.$$

In this problem, when constructing the procedure of choosing a solution, account was taken of two features of hierarchical systems — hierarchism and multiplicity of objectives. In this case, on the way to solution (46.31), the formalism of vector optimization was utilized three times: in evaluating the entire system as a whole regardless of the hierarchical structure; in defining the decision procedure at each hierarchical level (46.25); for interlevel optimization (46.29). Similarly, account may also be taken of other factors of the given class systems.

CONCLUSION

Starting in the middle of the 60s, the trend was integration of different specialized and sufficiently developed theories with the object of founding a more abstract theory for a particular class of systems. However, due to the fact that the formal apparatus of such theory requires a fairly rich mathematical structure for adequate description of large-scale or purposeful systems, we may presently discuss essential alterations in the available formal apparatus rather than development of the new one capable of rising up to the level of generalization needed here. In our opinion, proceeding from unified positions and relying basically on a unified mathematical structure employed in systems description, a new formal apparatus is to investigate, in the hope of obtaining conceptual results, such different problems as that of existence and minimality for a set of axioms providing an opportunity for representations in the state space, the necessary and sufficient conditions for controllability and optimization of multidimensional systems, the problem of minimal implementation of regularities relating input actions to output values, the necessary and sufficient conditions for stability in a wide range of understanding of this category. To be investigated is the Gödel consistency and completeness theorem, the multimeasure system non-interaction problem, the Krione–Rodes decomposition theorem and the system classification problem based on utilization of the apparatus of a new, formal systems theory.

The present monograph touched upon only the most fundamental concepts and aspects of the systems investigation and optimization. Many theories of specific type systems have been developed for a fairly extended time. We attempted to consider the problems of analysis and synthesis and those of the systems multiobjective optimization, taking them as basic and common to more specialized disciplines.

Of vital importance is the possibility of studying the behavior of a system, investigating the decision-making processes proceeding in this system or the mechanisms ensuring the purposefulness of its behavior.

Research in the problems of the systems optimization methodology may be beneficial in investigating the systems under uncertainty where information on the system and regularities of its functioning proves to be incomplete and insufficient for constructing its detailed mathematical model. In the case of investigating large-scale and complex systems, it is possible to raise the efficiency of analysis and synthesis of the system behavior or just to provide such possibility if the system optimization methodology be employed, admitting development of a less structured model which merely utilizes key factors. In general, structural considerations have a dominant role both in analysis and in synthesis of the different types of systems. The systems optimization methodology may serve as a language of interdisciplinary exchange since it is sufficiently general for not introducing its constraints and to be considered, at the same time, as a unified foundation for more narrow directions of systems investigation.

REFERENCES

References to Chapter 2

1. Bacon R., "Large Writing", *Anthology of World Philosophy*, Moscow, Misl, 1969, vol. I, part 2, p. 865.
2. *Anthology of World Philosophy*, Moscow, Misl, 1969–1972, vol. I–IV, vol. 3, pp. 313–314.
3. Vindelband V., *History and Natural History: Preludes: Philosophical Articles and Discourses*, St. Petersburg, 1904.
4. Rickert H., *Sciences of Nature and Sciences of Culture*, St. Petersburg, 1911.
5. *Kant Philosophy and Modern Times*, Moscow, Misl, 1974.
6. Von Neumann J. and O. Morgenstern, *Theory of Games and Economic Behavior*, Moscow, Nauka, 1970.
7. Suppes P. and G. Zines, *Fundamentals of Measurement Theory. Psychology of Measurements*, Moscow, Mir, 1970.
8. Tarski A., *Logic, Semantics and Metamathematics*, Oxford University Press, 1956, p. 418.
9. Taube M., *Computers and Common Sense*, Moscow, Progress, 1964, p. 110.

References to Chapters 3–7

1. Allen R., *Mathematical Economics*, Foreign Literature Publishing House, 1963.
2. Ben-Israel A. and P.D. Roberts, "A decomposition method for interval linear programming", *Manage. Sci.* **16**, 5 (1970).
3. Bulavsky V.A., R.A. Zviagina, and M.A. Yakovleva, *Numerical Methods for Linear Programming*, Moscow, Nauka, 1977.
4. Vasilyev F.P., *Numerical Methods for Solving Extremal Problems*, Moscow, Nauka, 1980.
5. Vatel I.A. and J.A. Fliorov, "A model for annual planning in an industry", in *Program Method of Control*, Moscow, C.C. in the USSR Academy of Sciences, part 3, 1976.

6. Bakhutinsky I.J., L.M. Dudkin, and B.A. Shchennikov, "Iterative aggregation in some optimal economic models", *Economics and Mathematical Methods* **IX**, 3 (1973).
7. Verina L.F. and V.S. Tanayev, "Decomposition approaches in solving mathematical programming problems", *Economics and Mathematical Methods* **XI**, 6 (1975).
8. Williams A.C., "A treatment of transportation problem by decomposition", *SIAM J. Appl. Math.* **10**, 1 (1962).
9. Goldstein J.G. and D.B. Yudin, *New Trends in Linear Programming*, Moscow, Sov. Radio, 1966.
10. Goldstein I.G., "On the possibility of expanding applicability of special methods for linear programming", in *Planning and Economic-Mathematical Methods*, Moscow, Nauka, 1971.
11. Dantzig G., *Linear Programming, its Generalizations and Application*, Moscow, Progress, 1966.
12. Yermolyev J.M. and L.G. Yermolyeva, "Parametric decomposition method", *Cybernetics* 2 (1973).
13. Yermolyev J.M., "Methods for solving non-linear extremal problems", *Cybernetics* 4 (1966).
14. Cornai I. and T. Liptack, "Two-level planning" in *Application of mathematics in economic studies*, vol. 3, Moscow, Misl, 1965.
15. Kutikov L.M., "Decomposition of linear programming block problems with weakly associated blocks", *Economics and Mathematical Methods* **IX**, 4 (1973).
16. Kutikov L.M. and S.A. Gesan, "Simplex method for solving linear dynamic programming problems", *Economics and Mathematical Methods* **XVI**, 6 (1980).
17. Lebedev V.J., "Decomposition method for solving the linear programming block problems with the coupling variables", *Journal Computational Mathematics and Mathematical Physics* **21**, 4 (1981).
18. Lebedev V.J., "Scheme for solving partially integer problems of mathematical programming", *Journal Computational Mathematics and Mathematical Physics* **17**, 5 (1977).
19. Lesdon L.S., Optimization of large systems, Moscow, Nauka, 1975.
20. Malinnikov V.V., "Method for expansion in solving large linear programming problems with a block structure", *Economics and Mathematical Methods* **VII**, 5 (1971).
21. Mauer I., "Penalty constant in block programming", *Proc. of the Estonian Acad. of Sciences, Physics, and Mathematics*, vol. 20, no. 4, 1971.
22. Mauer I., "On utilization of penalty constant in decomposition of mathematical programming problems", *Proc. of the Estonian Acad. of Sciences, Physics, and Mathematics*, vol. 20, no. 4, 1971.

23. Mednitsky V.G., "On optimality of aggregation in the linear programming block problem" in *Mathematical Methods for Solving Economic Problems*, part 3, 1972.
24. Meressoo T.H., "On one decomposition method for the resources allocation problem", *Proc. of the Estonian Acad. of Sciences, Physics, and Mathematics*, vol. 27, no. 2, 1978.
25. Mesarovich M., D. Mako, and I. Takahara, *Theory of Hierarchical Multilevel Systems*, Moscow, Mir, 1973.
26. Mikhalevich V.S., J.M. Yermolyev, V.V. Shkurba, and N.Z. Shor, "Complex systems and solutions of extremal problems", *Cybernetics* 5 (1967).
27. Moiseyev N.N., J.P. Ivanilov, and J.M. Stolyarova, *Optimization Methods*, Moscow, Nauka, 1978.
28. Pervozvanskaya T.N. and A.A. Pervozvansky, "The algorithm of search for the optimal centralized resources allocation", *Proc. of the USSR Academy of Sciences, Engineering Cybernetics*, no. 3, 1966.
29. Pervozvanskaya T.N., "Optimal planning in the linear dynamic model with regard to uncertainty", in *Problems of Applying Mathematics in the Socialist Economy*, no. 2, Leningrad University, 1965.
30. Pervozvanskaya A.A. and V.G. Gaytzgori, *Decomposition, Aggregation and Approximate Optimization*, Moscow, Nauka, 1979.
31. Pshenichny B.N. and J.M. Danilin, *Numerical Methods in Extremal Problems*, Moscow, Nauka, 1975.
32. Ritter K., "A decomposition method for linear programming problems with coupling constraints and variables", Mathematics Research Center, University of Wisconsin, Rep. 739, 1967.
33. Rosen J.B., "Primal partition programming for block diagonal matrices", *Numerische Mathematik* 6 (1964).
34. Sekine J., "Decentralized optimization of an interconnected system", *IEEE Trans. Circuit Theory*, June (1963).
35. Ulm S., "Componentwise descent and hierarchical optimization", *Proc. of the Estonian Academy of Sciences, Physics, Mathematics*, vol. 18. no. 1, 1969.
36. Ulm S., "Conditional gradient method and decomposition of optimal control problem", *Proc. of the Estonian Academy of Sciences, Physics, Mathematics*, vol. 18, no. 1, 1969.
37. Ulm S., "On decomposition of the Resources allocation problem based on the interactions prediction principle", in *Models and Methods for Analysis of Economic Purposeful Systems*, Novosibirsk, Nauka, 1977.
38. Ulm S. and T.H. Meressoo, "On decomposition of the resources allocation type problem", in *Methods for Interaction Analysis in Economic Systems*, Novosibirsk, Nauka, 1980.
39. Hadley G., *Linear Programming*, Addison-Wesley Publ. Co., Reading, Mass., 1962.

40. Tsurkov V.I., *Decomposition in Large Dimension Problems*, Moscow, Nauka, 1981.
41. Shor N.Z., "Application of the generalized gradient descent in block programming", *Cybernetics* 3 (1967).
42. Shor N.Z., "On convergence of the generalized gradient method", *Cybernetics* 3 (1968).
43. Shchennikov B.A., "Block method for solving a system of large dimension linear equations", *Economics and Mathematical Methods* **I**, 6 (1965).
44. Shchennikov B.A., "Application of the iterative aggregation method for solving systems of linear equations", *Economics and Mathematical Methods* **II**, 5 (1966).
45. Shchennikov B.A., "Aggregation method for solving systems of linear equations", *Latvian Acad. of Sci. of USSR* **173**, 4 (1967).

References to Chapter 8

1. Rohlich, Breianer Das Getreide, Verlag A.W. Hayni Arben, Berlin 5036.
2. Freidrich H.P., Der Endproduzent in der Getrlide–noivtschaft, Leitschrift Agroforum 6/7, 1968.
3. Kolbin V.V., *Macromodels of the National Economy of the USSR*, D. Reidel Publishing Company, 1985.
4. Statisticher Yahrbuch der VEAB und KFM, 1967.
5. Walter Ulbricht: Rede auf dem VII Parteitag der SED.
6. Yudin D.B. and J.G. Goldstein, *Linear Programming. Theory and Finite Methods*, M., 1963.

References to Chapter 9

1. Arrow K.I., "Rational choice functions and orderings", *Econometrica* **26** (1959) 121–127.
2. Arrow K.J., *Social choice and individual values*, 2nd edn., Yale Univ. Press, New Haven, 1970, p. 124.
3. Arrow K.J. and L. Hurwitz, "An optimality criterion for decision-making under ignorance", in *Uncertainty and Expectation in Economics*, ed. by C. F. Carter and J.L. Ford, Oxford: Blackwell and Mott, 1972, p. 299.
4. Khachiyan L.G., "Convexity and complexity of solving polynomial programming problems", *Proc. of the USSR Academy of Sciences, Engineering Cybernetics*, no. 6, 1982.
5. Khachiyan L.G., "Complexity of convex problems for integer polynomial programming", Auther's Abstract of Dissertation, Moscow, All-Union Central Information Agency, USSR, 1984.
6. Khomenyuk V.V., *Optimal Control Systems*, Moscow, Nauka, 1977, p. 152.
7. Khomenyuk V.V., *Elements of the Multiobjective Optimization Theory*, Moscow, Nauka, 1983, p. 123.

8. Kolbin V.V., *Methods for Multiobjective Optimization*, All-Union Institute of Scientific and Technical Information, 1987.
9. Milnor J., "Games against nature", in *Decision Processes*, ed. by R.M. Thrall, C.H. Coombs, and R.L. Davis, Wiley; L.: Chapman and Hall, 1954, p. 331.
10. Nash J.F., "The bargaining problem", *Econometrica* **18** (1950) 155–162.
11. Podinovsky V.V., "On the relative importance of criteria in multicriteria decision problems", in *Multicriteria Decision Problems*, Moscow, Mashinostroyeniye, 1978.
12. Sen A.K., *Collective Choice and Social Welfare*, San Francisco: Holden Day; Edinburg; London; Oliver and Boyd, 1970, p. 225.
13. Sen A.K., "Choice functions and revealed preference", *Rev. Econ. Stud.* **38** (1971) 307–317.

References to §44

1. Benayun R., O.I. Larichev, G. Mongolfye, and G. Terny, "Linear programming with multiple criteria", *Automatics and Telemety*, 8 (1971) 108–115.
2. Ventzel J.S., "Introduction to operations research", Moscow, Sov. Radio, 1964.
3. Vilkas E., "Theory of utility and decision-making", in *Mathematical Methods in Social Sciences*, no. 7, Vilnus, 1971, pp. 13–60.
4. Vilkas E., "Multiobjective optimization", in *Mathematical Methods in Social Sciences*, no. 7, Vilnus, 1976, pp. 16–67.
5. Germeyer J.B., *Introduction to Theory of Operations Research*, Moscow, Fizmatgiz, 1971.
6. Diner I.Y., "Zoning the nature state vector set and the decision choice problem–Problems of general and social prediction", no. 2. *Operations Res. Information Bulletin*, Moscow, 1969, no. 14.[29]
7. Dubrov J.A., L.A. Grechina, and V.T. Kulik, "On decision making under uncertainty (systems approach)", *Trans. 7th Symp. on Cybernetics*, Tbilisi, 1972.
8. Yemelyanov S.V., V.I. Borisov, A.A. Malevich, and A.M. Cherkashin, "Models and methods of vector optimization" – *Results of Science and Engineering, Engineering Cybernetics*, Vol. 5, Moscow, 1973.
9. Yemelyanov S.V., J.B. Dudin, O.I. Larichev, A.A. Malevich, E.L. Nappelbaum, and V.M. Ozernoy, "Preparation and decision making in the organizing control systems", *Results of Science and Engineering, Engineering Cybernetics*, Moscow, 1971.
10. Karlin S., *Mathematical Methods in the Game Theory, Programming and Economics*, Moscow, Mir, 1964.
11. Lebedev B.D., V.V. Podnovsky, and R.S. Styrikovich, "Optimization problem for ordering a set of criteria", *Economics and Mathematical Methods*, Vol. VII, no. 4, 1971.

12. Lewis R. and X. Raypha, *Games and Solutions*, Moscow, FL., 1961.
13. Podinovsky V.V. and V.M. Gavrilov, *Optimization by Successively Applied Criteria*, Moscow, Sov. Radio, 1975.
14. Podinovsky V.V., "Multicriteria problems with homogeneous criteria ordered by importance", *Automatics and Telemetry*, no. 1, 1976.
15. Salukvadze M.J., *Problems of Vector Optimization in Control Theory*, Tbilisi, Metsniereba, 1975.
16. Salukvadze M.J., "Optimal control problems with several quality criteria", Author's Abstract of the Dissertation, Moscow, 1974.
17. Trukhayev R.I., *Research Methods for Decision Making under Uncertainty*, Leningrad, 1972.
18. Federov V.V., *Numerical Maximin Methods*, Moscow, Nauka, 1979.
19. Von Nauman G. and N.O. Morgenstern, *Theory of Games and Economic Behavior*, Moscow, Nauka, 1970.
20. Debrez G., Theory of Value, John Wiley, New York, 1959.
21. Dong Hak Chyung, "Optimal systems with multiple cost functionals", *J. SIAM Control* **5**, 3 (1967).
22. Kuhn H.W. and A.W. Tucker, "Nonlinear programming", *Proc. Second Berkeley Symp. on Mathematic Statistics and Probability*, University of California Press, Berkeley, California, 1951.
23. Nelson W.L., "On the use of optimization theory for practical control system design", *IEEE Trans. Automatic Control* **AC-9**, 4 (1964).
24. Pareto V., Cour's d'Economie Politique, Lausanne. Rouge, 1896.
25. Waltz F.M., "An engineering approach: Hierarchical optimization criteria", *IEEE Trans. Automatic Control* **AC-12**, 2 (1967).
26. Zadeh L.A., "Optimality and non-scalar valued performance criteria", *IEEE Trans. Automatic Control* **AC-8**, 1 (1963).

References to §45

1. Gorokhovik V.V., "To the vector optimization problem", *Proc. of the USSR Acad. of Science, Engineering Cybernetics*, no. 6, 1972.
2. Yemelyanov S.V., V.I. Borisov, A.A. Malevich, and A.M. Cherkishin, "Models and methods for vector optimization", *Results of Science and Engineering, Engineering Cybernetics*, vol. 5, 1973.
3. Von Neuman G. and O. Morgenstern, *Theory of Games and Economic Behavior*, Moscow, Nauka, 1970.
4. Salukvadze M.J., "Problems of vector optimization in control theory", Metsniereba, Tbilisi, 1975.
5. Bragaard L. and J. Vangeldere, "Points efficaces en programmation a objectives multiples", *Bull. Soc. Roy. Sci. Liege* **46**, 1–2 (1977) 27–41.
6. Chang and Sheldon S.L., "General theory of optimal process", *J. SIAM Control* **4**, 1 (1966).

7. Da Cunha N.O. and E. Polak, "Constrained minimization under vector-valued criteria in linear topological spaces", *Mathematical Theory of Control*, Academic Press, 1967.
8. Ecker J.G. and J.A. Kanada, "Finding efficient points for linear multiple objective programs", *Math. Program* 8 (1975) 375–377.
9. Klinger A., "Vector valued performance criteria", *IEEE Trans. Automatic Control* **AC-8**, 1 (1963).
10. Koopmans T.C., Analysis of Production. An Efficient Combating of Activities. Activity Analysis of Production, New York, 1951.
11. Reid R. and S.J. Citron, "On noninferior performance index vectors", *J. Optimization Theory and Application* **7**, 1 (1971).
12. Stern Ronald J. and Adi. Ben-Israel, "An interior penalty function method for the construction of efficient points in a multicriteria control problem", *J. Math. Anal. Appl.* **46**, 3 (1974) 768–776.
13. Steuer Ralph E., "Multiple objective linear programming with interval criterion weights", *Manage. Sci.* **23**, 3 (1976) 305–316.
14. Steuer Ralph E., "A five phase procedure for implementing a vector-maximum algorithm for multiple objective linear programming problems", *Lect. Notes Econ. and Math. Syst.*, (1976).
15. Wendell R.E. and D.N. Lee, "Efficiency in multiple objective optimization problems", *Math. Program* **12**, 3 (1977) 406–414.
16. Zadeh L.A., "Optimality and non-scalar valued performance criteria", *IEEE Trans. Automatic Control.* **AC-8**, 1 (1963).

References to §46

1. Apinis A., "Transitivity of group decisions according to the simple majority rule", *Lithuanian Mathematical Problems*, vol. 14, no. 4, 1974.
2. Borisov V.I., "Vector optimization of systems", in *Systems Research*, Moscow, 1971.
3. Borisov V.I., "Vector optimization solutions in organizing systems." Heads of reports at the Int. Symp. on the Organization Control Problems and Hierarchical Systems, Baku, Moscow, 1971.
4. Vilkas E., "Co-ordination of coalition orderings", in *Achievements of the Game Theory*, Mintis, Vilnus, 1973, pp. 83–92.
5. Vilkas E., "Multiobjective optimization", in *Mathematical Methods in Social Sciences*, no. 7, Vilnus, 1976, pp. 16–67.
6. Germeyer J.B., *Introduction to Operations Research*, Nauka, Moscow, 1971.
7. Yemelyanov S.V., V.I. Borisov, A.A. Malevich, and A.M. Cherkishin, "Models and methods for vector optimization", *Summary of Science and Engineering, Engineering Cybernetics*, vol. 5, 1973.

8. Yemelyanov S.V., V.M. Ozernoy, and O.I. Larichev, "Problems and methods of decision making", International Centre of Scientific and Technical Information, Moscow, 1973.
9. Karlin S., *Mathematical Methods in the Game Theory, Programming and Economics*, Mir, 1964.
10. Kafarov V.V., G.B. Lazarev, and V.I. Avdeyev, "The method of solving the multicriteria control problems for a complex chemical-technological system", USSR Acad. of Sci., vol. 198, no. 1, 1971.
11. Kiny R., "Utility functions of multidimensional alternatives", in *Problems of Analysis and Procedures of Decision Making*, Mir, Moscow, 1976.
12. Larichev O.I., "Man-machine procedures of decision making", *Automatics and Telemetry*, no. 12, 1971.
13. Lews R. and X. Raifa, *Games and Decisions*, Moscow, Foreign Literature, 1961.
14. Von Neuman G., Morgenstern O., *Theory of Games and Economic Behavior*, Moscow, Nauka, 1970.
15. Roy B., "Classification and choice with several criteria (ELECTRA method)", in *Problems of Analysis and Procedures of Decision Making*, Mir, Moscow, 1976.
16. Salukvadze M.J., *Problems of Vector Optimization in Control Theory*, Metsniereba, Tbilisi, 1975.
17. Tokarev V.V., "Optimization of parameters of the dynamic system universal for a series of manoeuvres with different degrees of information", *Automatics and Telemetry*, no. 8, 1971.
18. Arrow K.J. and L. Hurwitz, "An optimality criterion for decision-making under ignorance", in *Uncertainty and Expectation in Economics*, Basil Blackwell and Mot, Oxford, 1972.
19. Arrow K.J., *Social Choice and Individual Values*, Cowels Comission Monograph, 12, Wiley, N.Y., 1951.
20. Benayoun R. and J. Tergny, "Critéres multiples et programmation mathematique: une solution dans le cas linear", R.I.R.O., no. V–2, 1969.
21. Bergstresser K., A. Charnes, and P.L. Yu, "Generalization of domination structures and nondominated solutions in multicriteria decision making", *J. Optim. Theory Appl.* **18**, 1 (1976) 3–13.
22. Black D., *The Theory of Committees and Elections*, Cambridge University Press, 1958.
23. Bogaard P.J.M. and J. van den Versluis, "The design of optimal committee decision", *Statistical Neerlandica*, v. 16, 1962.
24. Fishburn P.C., "Additive utilities with incomplete product sets: Application to priorities and assignments", *Oper. Res.* **15**, 3 (1967).
25. Focke J., "Vector maximum problem und parametrische optimierung", *Math. Operations forsch. und Statist.* **4**, (1973) Heft 5, 365–369.

26. Harsanyi J.C., "Cardinal welfare. Individualistic ethics and interpersonal comparisons of utility", *J. Political Economy* 63 (1955).
27. MAYKO, "A set of independent necessary and sufficient conditions for simple majority decision", *Econometrica* **20** (1952) 680–684.
28. Nash J., "The bargaining problem", *Econometrica* **18** (1950) 155–162.
29. Roy B., "How outranking relation helps multiple criteria decision making: METRA", *Rapport de Recherche*, no. 67, Sept., (1972).
30. Salama A.J. and V. Gurishankar, "Optimal control of systems with a single control and several cost functionals", *Int. J. Control* **14**, 4 (1971).
31. Shlaifer R.O., *Analysis of Decision under Uncertainty*, McGraw-Hill, 1967.
32. Yu P.L. and G. Leitmann, "Compromise solutions, domination structures and Sulakvadse's solution", *Multicriteria Decision Making and Differential Games*, 1976, pp. 85–101.
33. Yu P.L. and G. Leitmann, "Nondominated decision and cone convexity in dynamic multicriteria decision problems", *Multicriteria Decision Making and Differential Games*, 1967, pp. 61–72.
34. Yu P.L., "Cone convexity, cone extreme points and nondominated solutions in decision problems with multiobjectives", *Multicriteria Decision Making and Differential Games*, 1976, pp. 1–59.
35. Zadeh L.A., "Optimality and non-scalar valued performance criteria", *IEEE. Trans. Automatic Control* **AC-8**, 1 (1963).